Methods in Molecular Biology™

Series Editor
John M. Walker
School of Life Sciences
University of Hertfordshire
Hatfield, Hertfordshire, AL10 9AB, UK

For other titles published in this series, go to
www.springer.com/series/7651

Chemical Library Design

Edited by

Joe Zhongxiang Zhou

Department of Pharmacology, University of California, San Diego, CA, USA

Humana Press

Editor
Joe Zhongxiang Zhou
Department of Pharmacology
University of California
La Jolla, CA 92093, USA
zjoe.zhou@gmail.com

ISSN 1064-3745 e-ISSN 1940-6029
ISBN 978-1-4939-6154-2 ISBN 978-1-60761-931-4 (eBook)
DOI 10.1007/978-1-60761-931-4
Springer New York Dordrecht Heidelberg London

Printed on acid-free paper

Humana Press is part of Springer Science+Business Media (www.springer.com)

Preface

Over the last two decades we have seen a dramatic change in the drug discovery process brought about by chemical library technologies and high-throughput screening, along with other equally remarkable advances in biomedical research. Though still evolving, chemical library technologies have become an integral part of the core drug discovery technologies. This volume primarily focuses on the design aspects of the chemical library technologies. Library design is a process of selecting useful compounds from a potentially very large pool of synthesizable candidates. For drug discovery, the selected compounds have to be biologically relevant. Given the enormous number of compounds accessible to the contemporary synthesis and purification technologies, powerful tools are indispensible for uncovering those few useful ones. This book includes chapters on historical overviews, state-of-the-art methodologies, practical software tools, and successful applications of chemical library design written by the best expert practitioners.

The book is divided into five section. **Section I** covers general topics. **Chapter 1** highlights the key events in the history of high-throughput chemistry and offers a historical perspective on the design of screening, targeted, and optimization libraries. **Chapter 2** is a short introduction to the basics of chemoinformatics necessary for library design. **Chapter 3** describes a practical algorithm for multiobjective library design. **Chapter 4** discusses a scalable approach to designing lead generation libraries that emphasize both diversity and representativeness along with other objectives. **Chapter 5** explains how Free–Wilson selectivity analysis can be used to aid combinatorial library design. **Chapter 6** shows how predictive QSAR and shape pharmacophore models can be successfully applied to targeted library design. **Chapter 7** describes a combinatorial library design method based on reagent pharmacophore fingerprints to achieve optimal coverage of pharmacophoric features for a given scaffold.

Three chapters in **Section II** focus on the methods and applications of structure-based library design. **Chapter 8** reviews the docking methods for structure-based library design. **Chapters 9** and **10** contain two detailed protocols illustrating how to apply structure-based library design to the successful optimization of lead matters in the real drug discovery projects.

Section III consists of three chapters on fragment-based library design. **Chapter 11** describes the key factors that define a good fragment library for successful fragment-based drug discovery. It also provides a summary view of the fragment libraries published so far by various pharmaceutical companies. **Chapter 12** shows how a fragment library is used in fragment-based drug design. **Chapter 13** introduces a new chemical structure mining method that searches into a huge virtual library of combinatorial origin. The method uses fragmental (or partial) mappings between the query structure and the target molecules in its initial search algorithms.

Chapter 14 in **Section IV** describes a workflow for designing a kinase targeted library. It illustrates how to assemble a lead generation library for a target family using known ligand–target family interaction data from various sources.

Section V contains four chapters on library design tools. PGVL Hub described in **Chapter 15** is an integrated desktop tool for molecular design including library design. It streamlines the design workflow from product structure formation to property calculations, to filtering, to interfaces with other software tools, and to library production management. An application of PGVL Hub to the optimization of human CHK1 kinase inhibitors is presented in **Chapter 16**. **Chapter 17** is a detailed protocol on how to use library design tool GLARE to perform product-oriented design of combinatorial libraries. Finally, **Chapter 18** is a detailed protocol on how to use the library design tool CLEVER to perform library design and visualization.

Joe Zhongxiang Zhou

Contents

Contributors

CAROLYN BECK • *Department of Industrial and Enterprise Systems Engineering, University of Illinois at Urbana Champaign, Urbana, IL, USA*
PAUL H. BERNARDO • *Institute of Chemical and Engineering Sciences, Singapore, Singapore*
NIKLAS BLOMBERG • *DECS GCS Computational Chemistry, AstraZeneca R&D Mölndal, Mölndal, Sweden*
CLAUDIO N. CAVASOTTO • *School of Biomedical Informatics, The University of Texas Health Science Center at Houston, Houston, TX, USA*
CHRISTINA L.L. CHAI • *Institute of Chemical and Engineering Sciences, Singapore, Singapore*
BO CHAO • *PGRD-La Jolla, Pfizer Inc., San Diego, CA, USA*
HONGMING CHEN • *DECS GCS Computational Chemistry, AstraZeneca R&D Mölndal, Mölndal, Sweden*
JOHN CLARK • *PGRD-La Jolla, Pfizer Inc., San Diego, CA, USA*
BOB CONER • *PGRD-La Jolla, Pfizer Inc., San Diego, CA, USA*
JASON B. CROSS • *Cubist pharmaceuticals, Inc., Lexington, MA, USA*
ROLAND E. DOLLE • *Department of Chemistry, Adolor Corporation, Exton, PA, USA*
KLAUS DRESS • *Oncology Medicinal Chemistry, La Jolla Laboratories, Pfizer Inc., San Diego, CA, USA*
JERRY O. EBALUNODE • *Department of Pharmaceutical Sciences, BRITE Institute, North Carolina Center University, Durham, NC, USA*
JEFF ELLERAAS • *Oncology Medicinal Chemistry, La Jolla Laboratories, Pfizer Inc., San Diego, CA, USA*
OLA ENGKVIST • *DECS GCS Computational Chemistry, AstraZeneca R&D Mölndal, Mölndal, Sweden*
ERIC FEYFANT • *Pfizer Global R&D, Cambridge, MA, USA*
QIYUE HU • *Pfizer Global Research and Development, La Jolla Laboratories, San Diego, CA, USA*
BUWEN HUANG • *Oncology Medicinal Chemistry, La Jolla Laboratories, Pfizer Inc., San Diego, CA, USA*
SHOGO ITO • *PGRD-La Jolla, Pfizer Inc., San Diego, CA, USA*
THERESA L. JOHNSON • *Pfizer Research Technology Center, Cambridge, MA, USA*
CHRISTOS C. KANNAS • *Department of Computer Science, University Of Cyprus, Nicosia, Cyprus; Noesis Chemoinformatics, Nicosia, Cyprus*
DAVID KLATTE • *PGRD-La Jolla, Pfizer Inc., San Diego, CA, USA*
JAMES KONG • *PGRD-La Jolla, Pfizer Inc., San Diego, CA, USA*
JAROSLAV KOSTROWICKI • *Pfizer Global Research and Development, La Jolla Laboratories, San Diego, CA, USA*
ATSUO KUKI • *Pfizer Global Research and Development, La Jolla Laboratories, San Diego, CA, USA*
TZE HAU LAM • *Data Mining Department, Institute for Infocomm Research, Singapore, Singapore*

ELIZABETH A. LUNNEY • *PGRD-La Jolla, Pfizer Inc., San Diego, CA, USA*
SARATHY MATTAPARTI • *PGRD-La Jolla, Pfizer Inc., San Diego, CA, USA*
JAMES NA • *Pfizer Global Research and Development, La Jolla Laboratories, San Diego, CA, USA*
CHRISTOS A. NICOLAOU • *Noesis Chemoinformatics, Nicosia, Cyprus*
GENEVIEVE D. PADERES • *Cancer Crystallography & Computational Chemistry, La Jolla Laboratories, Pfizer Inc., San Diego, CA, USA*
KEVIN PARIS • *Pfizer Global R&D, Cambridge, MA, USA*
TOM PAULY • *Oncology Medicinal Chemistry, La Jolla Laboratories, Pfizer Inc., San Diego, CA, USA*
ZHENGWEI PENG • *Pfizer Global Research and Development, La Jolla Laboratories, San Diego, CA, USA*
SHARANGDHAR S. PHATAK • *School of Biomedical Informatics, The University of Texas Health Science Center at Houston, Houston, TX, USA*
PAUL A. REJTO • *Oncology, La Jolla Laboratories, Pfizer Inc., San Diego, CA, USA*
SRINIVASA SALAPAKA • *Department of Mechanical Science and Engineering, University of Illinois at Urbana Champaign, Urbana, IL, USA*
SIMONE SCIABOLA • *Pfizer Research Technology Center, Cambridge, MA, USA*
NUNZIO SCIAMMETTA • *PGRD-La Jolla, Pfizer Inc., San Diego, CA, USA*
ROBERT SELLIAH • *Drug Design Consulting, Irvine, CA, USA*
PUNEET SHARMA • *Integrated Data Systems Department, Siemens Corporate Research, Princeton, NJ, USA*
THOM SHULOK • *PGRD-La Jolla, Pfizer Inc., San Diego, CA, USA*
ROBERT V. STANTON • *Pfizer Research Technology Center, Cambridge, MA, USA*
THOMAS THACHER • *PGRD-La Jolla, Pfizer Inc., San Diego, CA, USA*
JOO CHUAN TONG • *Data Mining Department, Institute for Infocomm Research, Singapore, Singapore; Department of Biochemistry, Yong Loo School of Medicine, National University of Singapore, Singapore, Singapore*
ALEXANDER TROPSHA • *Laboratory for Molecular Modeling and Carolina Center for Exploratory Cheminformatics Research, School of Pharmacy, University of North Carolina at Chapel Hill, Chapel Hill, NC, USA*
JEAN-FRANÇOIS TRUCHON • *Chemical Modeling and Informatics, Merck Frosst Canada, Kirkland, QC, Canada*
DÉSIRÉE H.H. TSAO • *Pfizer Global R&D, Cambridge, MA, USA*
CHRIS WALLER • *PGRD-La Jolla, Pfizer Inc., San Diego, CA, USA*
HUALIN XI • *Pfizer Research Technology Center, Cambridge, MA, USA*
SHUNQI YAN • *Drug Design Consulting, Irvine, CA, USA*
BO YANG • *PGRD-La Jolla, Pfizer Inc., San Diego, CA, USA*
WEIFAN ZHENG • *Department of Pharmaceutical Sciences, BRITE Institute, North Carolina Center University, Durham, NC, USA*
JOE ZHONGXIANG ZHOU • *PGRD-La Jolla, Pfizer Inc., San Diego, CA, USA; Department of Pharmacology, University of California, San Diego, CA, USA*

Section I

General Topics

Chapter 1

Historical Overview of Chemical Library Design

Roland E. Dolle

Abstract

High-throughput chemistry (HTC) is approaching its 20-year anniversary. Since 1992, some 5,000 chemical libraries, prepared for the purpose of biological intestigation and drug discovery, have been published in the scientific literature. This review highlights the key events in the history of HTC with emphasis on library design. A historical perspective on the design of screening, targeted, and optimization libraries and their application is presented. Design strategies pioneered in the 1990s remain viable in the twenty-first century.

Key words: High-throughput chemistry, chemical library, random library, targeted library, optimization library, library design, biological activity, drug discovery.

1. Milestones in High-Throughput Chemistry

High-throughput chemistry (HTC) is a widely used technology for accelerating the synthesis of chemical compounds, in particular the synthesis of biologically active compounds. HTC originated in the early 1990s. Its development and application was largely driven by the pharmaceutical industry. In the years leading up to the introduction of HTC, the pharmaceutical industry had been transformed by advances in molecular biology. Routine cloning and expression of molecular targets enabled medicinal chemists to optimize the potency of chemical leads directly against an enzyme or receptor prior to in vivo testing. Brimming with molecular targets and nascent high-throughput screening technology, there was a demand to access large compound collections to discover new drug leads. Vintage industrial compound collections generated over many decades amounted to

J.Z. Zhou (ed.), *Chemical Library Design*, Methods in Molecular Biology 685,
DOI 10.1007/978-1-60761-931-4_1, © Springer Science+Business Media, LLC 2011

less than a few hundred thousand molecules and the perceived diversity of such collections was low. Accelerating the synthesis of new analogs during lead optimization was desired. The lack of medicinal chemistry resource was a frequent bottleneck in drug discovery programs. The benchmark at the time was that a chemist required on average 2 weeks to synthesize a single analog at an estimated cost of $5,000–$7,000 per compound. Hence, the prospect of HTC potentially creating "chemical libraries" of hundreds of thousands of structurally diverse compounds formatted for high-throughput screening and the potential to prepare analogs in half the time at half the cost had overwhelming appeal. As such, HTC promised to revolutionize medicinal chemistry just as molecular biology ushered in the era of molecular-based drug discovery. The amalgamation of these technologies was thought to dramatically reduce the cost and time to bring a drug to market, increasing the overall efficiency of the drug discovery process. For these reasons, the pharmaceutical industry invested heavily in HTC.

Figure 1.1 offers a perspective on selected major events in HTC. Most of the innovations in HTC were made during the 1990s. In 1992, Ellman published a report in the *Journal of the American Chemical Society* describing the solid-phase-assisted synthesis of benzodiazepinones (**Fig. 1.2**) (1). This was hailed as the first example of accelerated synthesis of small molecule, non-peptide drug-like compounds. Within a year, DeWitt and coworkers at Parke-Davis introduced Diversomer® (2). The paper, appearing in the *Proceedings of the National Academy of Sciences*, described the first apparatus specifically designed to carry out HTC (**Fig. 1.3**). It was a rather simple device consisting of eight gas dispersion tubes for loading solid-phase resin. It was used to prepare parallel arrays of hydantoins and benzodiazepines. In retrospect, these HTC milestones seem insignificant relative to the advances made in the field over the past 20 years. At the time they served to fuel the excitement of HTC. Today, they serve as an early example of what would become one of the recurring themes in library design: chemical libraries modeled after known biologically active scaffolds. Solid-phase and solution-phase synthesis techniques are used to prepare libraries (3). In solid-phase synthesis, building blocks are immobilized on resin through a cleavable linker. Reactants and reagents are used in excess to speed synthesis and then simply rinsed away from resin eliminating tedious purification of intermediates. Target compounds are detached from the linker and eluted from the resin and tested for biological activity. The utility of solid-phase synthesis was greatly enhanced when electrophoric tags were invented to index the reaction history on a single resin bead (4). This advance enabled binary encoded split-pool synthesis, i.e., the combining of building blocks in true combinatorial fashion to give tens

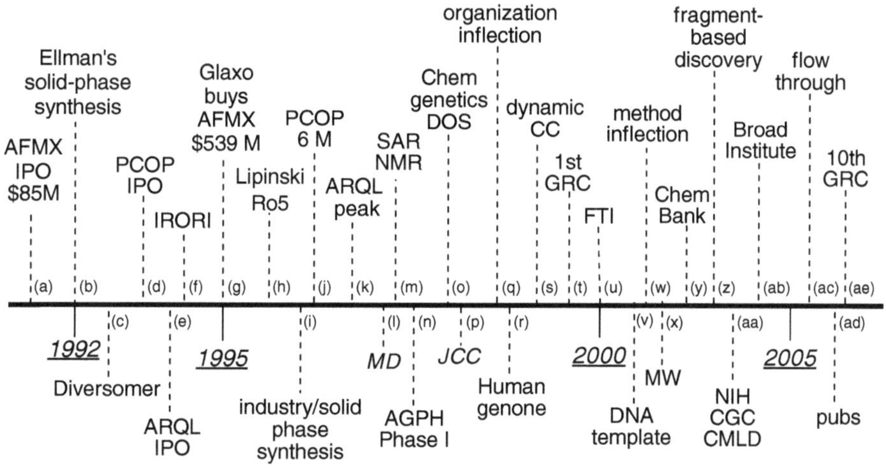

Fig. 1.1. Time chart showing selected events in the history of HTC. Key: (**a**) Affymax is the first combinatorial chemistry company to go public. (**b**) Ellman's solid-phase parallel synthesis of benzodiazepines fuels HTC. (**c**) Parke-Davis introduces Diversomer®, apparatus for solid-phase synthesis of small molecules. (**d**) Pharmacopeia licenses Columbia University's encoded split synthesis technology and company goes public a year later (NASDAQ symbol: PCOP). (**e**) ArQule goes public (NASDAQ symbol: ARQL) with its industrialized solution-phase synthesis of discrete purified compounds. (**f**) IRORI introduces radio frequency (Rf) encoding technology for solid-phase synthesis in "cans" containing reusable Rf chips. (**g**) Glaxo Wellcome buys Affymax for $539 M in cash. (**h**) Lipinski publishes landmark correlation of physiochemical properties of drugs – "Rule of 5" (Ro5) has profound impact on library design. (**i**) 1992–1996: 80% of published libraries are from industry; 75% using solid-phase synthesis. (**j**) Pharmacopeia generates 6 M encoded compounds. (**k**) ArQule has the largest number of collaborations (27) reported for a combichem company. (**l**) Inaugural issue of *Molecular Diversity*, the first journal dedicated to HTC. (**m**) SAR by NMR – compounds binding to proximal subsites of a protein are linked and optimized using HTC. (**n**) Agouron Pharm. moves human rhinovirus 3C protease inhibitor into the clinical trials; HTC played a key role in its discovery. (**o**) S. Schreiber introduces the concept of chemical genetics and diversity-oriented synthesis (DOS). (**p**) A. Czarnik editor of a new ACS journal: *Journal of Combinatorial Chemistry*. (**q**) Academia overtakes industry library synthesis publications for the first time. (**r**) Human genome sequence is published in *Science*. (**s**) Dynamic combinatorial chemistry. (**t**) First Gordon Research Conference entitled combinatorial chemistry: High Throughput Chemistry & Chemical Biology. (**u**) D. Curran develops fluorous reagents and tags and launches Fluorous Technology Inc. (FTI). (**v**) DNA-templated synthesis. (**w**) Solution-phase overtakes solid-phase in library synthesis. (**x**) Microwave-assisted synthesis gains momentum in HTC. (**y**) ChemBank public database established. (**z**) First reports of fragment-based drug discovery. (**aa**) NIH Roadmap defined. NIH funds the Chemical Genomics Center and Molecular Library Initiative, establishing 10 chemical methodology and library design centers throughout the US. (**ab**) Broad Institute established, furthers application of DOS in chemical biology. (**ac**) Flow through synthesis for HTC gains in popularity. (**ad**) Of the 497 library publications reported in 2008, 90% originated from academic labs; >80% were made by solution-phase chemistry. (**ae**) HTC Gordon Research Conference celebrates tenth anniversary and revises conference title: High Throughput Chemistry & Chemical Biology.

of thousands of compounds per library with a minimal number of synthetic steps. Encoding technology was honed at Pharmacopeia, Inc., one of the early HTC startups. Within just a few years the company had amassed over six million compounds. Simultaneous with these developments were advances in solution-phase synthesis. Resin-bound reagents were developed to assist in common reaction transformations. Spent resins are filtered from reaction mixtures aiding in product isolation. Similarly, scavenger resins were invented to clean up reaction mixtures also aiding in the isolation of target molecules. ArQule, Inc. embraced

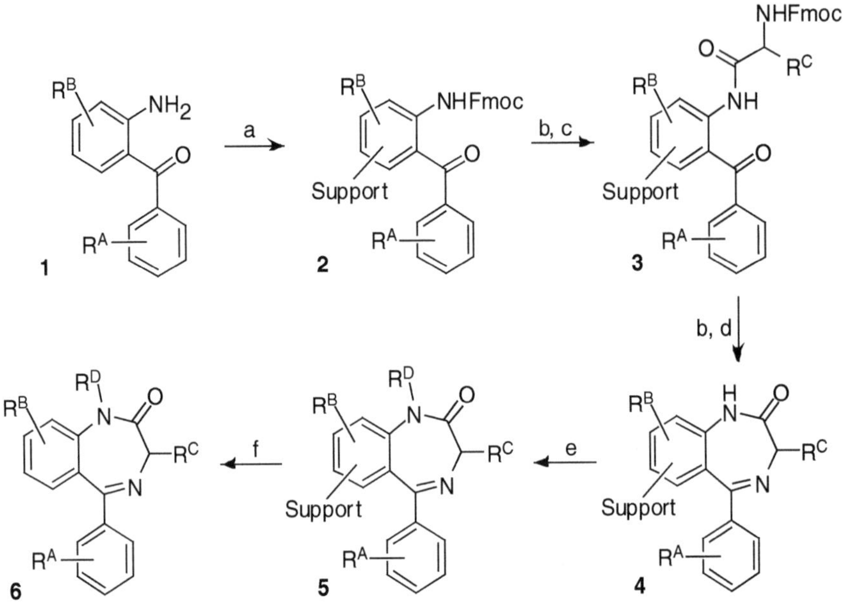

Fig. 1.2. One of the first nonpeptide library synthesis (reprinted ("adapted" or "in part") with permission from *Journal of the American Chemical Society*. Copyright 1992 American Chemical Society).

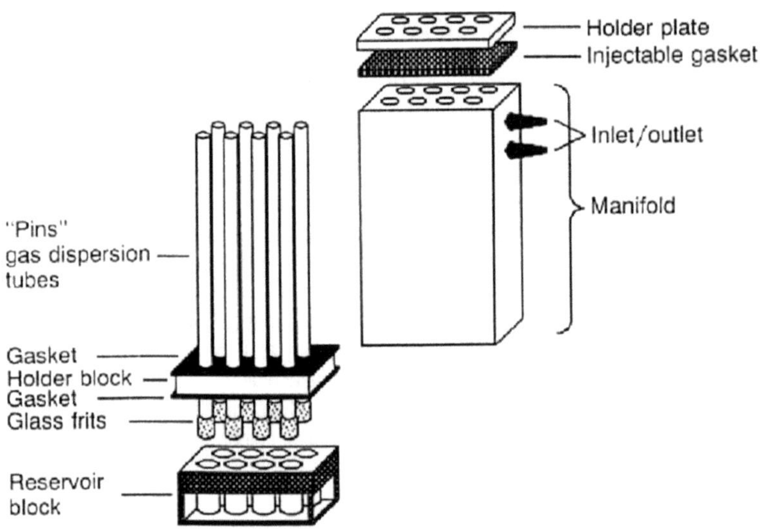

Fig. 1.3. One of the first devices for HTC (copyright (1993) National Academy of Sciences, USA).

solution-phase parallel synthesis on a massive scale. **Table 1.1** shows the number (27) of collaborations ArQule enjoyed in the mid-1990s as companies flocked to design and purchase parallel libraries (5). ArQule's solution-phase approach made available milligram quantities of discrete purified compounds for screening and immediate resupply.

Table 1.1
ArQule collaborations 1996–1997

Pharmaceutical companies

Abbott Laboratories	ACADIA Pharmaceuticals	Fibrogen
Monsanto Company	Aurora Biosciences	Genome Therapeutics
Pharmacia Biotech AB	Cadus Pharm. Corp.	GenQuest
Roche Biosciences	Cubist Pharm., Inc.	Genzyme
Solvay Duphar B.V.	DGI Biotechnologies	Immunex Corp.
Amersham Pharmacia Biotech	ICAgen, Inc.	Ontogeny
American Home Products	Scriptgen Pharm., Inc.	Ribogene
Sankyo Company	Signal Pharm., Inc.	Sepracor, Inc.
T Cell Sciences, Inc.	SUGEN, Inc.	ViroPharma

Library design was less important than library size and >3-point scaffold diversification was a common practice invariably producing physicochemically-challenged compound arrays. However, refocus on design occurred in 1996 when Lipinski linked certain physicochemical properties with orally active drugs (6). Lipinski's "Rule of 5" (Ro5; molecular weight (MW) <500, clogP <5, total number of hydrogen bond (H-bond) acceptors <5, total number of H-bond donors and rotatable bonds each <10) was rapidly adopted into library design. Ro5 put an abrupt end to the practice of numbers inflation. A similar analysis of the physicochemical properties yielding productive leads, i.e., those that led to marketed drugs, gave rise to the "Rule of 4" (7). This correlation underscored the concept that the preferred leads are those in which MW, H-bond donor/acceptor counts, and rotatable bonds can be *increased* during optimization as opposed to trimming these parameters from ligands. These and related correlations had a profound impact on chemical library design. During its first decade, while development of HTC was driven by the pharmaceutical industry, interest of academic researchers in HTC was bolstered in 2004 by the creation of the Chemical Genomics Center and allied high-throughput academic screening centers [Combinatorial Molecular Design Centers (CMLDs)]. Under the auspices of the National Institutes of Health (NIH), their mission is to identify small-molecule probes to establish the function of all proteins in the proteome. Also in 2004, the Broad Institute was established, strengthening the resolve to apply diversity-oriented synthesis (DOS) in chemical biology. Over 5,000 libraries have been reported in the literature from 1992 to 2008 (8).

2. Historical Library Designs

The objective of creating a chemical library for drug discovery, regardless of its size or method of synthesis, is to supply biologically active compounds. For the purpose of this text, chemical libraries can be classified into one of two categories: screening libraries and optimization libraries. The screening library category is further subdivided into (a) random libraries, collections with a unique design theme that has a distant, if any, relation to known biologically active agents, and (b) targeted libraries where the link with other biologically active structures is clearly evident. Targeted libraries generally contain a known pharmacophore, i.e., a set of structural features in a molecule that is recognized at the molecular target (enzyme, receptor, etc.) and is responsible for that molecule's biological activity (9). They may also contain structural scaffolds that interact with a variety of molecular targets, commonly referred to as privileged scaffolds (10). Optimization libraries, on the other hand, function primarily to enhance the biological activity of an existing lead. Potency, selectivity, and metabolic stability are examples of deficits in leads which can be addressed using optimization libraries. The term *lead* is defined here as a biologically active molecule that has emerged from a high-throughput screen or reported in the scientific or patent literature.

2.1. Historical Designs: Random Screening Libraries

Peptide libraries. The very first examples of random screening libraries were massive collections of peptides. Although amino acid monomers and peptides are endowed with biological activity and therefore may be thought of as privileged structures, it is the scale and extensive screening of these libraries that they are considered random libraries. Researchers at Affymax developed a process for generating and screening peptide libraries on microchips (11). Houghten conceived the technique of positional scanning to create synthetic peptide combinatorial libraries (SPCLs; **Fig. 1.4**) (12). In positional scanning, each amino acid in a given peptide sequence is sequentially held constant while the other amino acid positions are randomized. In this way peptide mixtures are formed and screened in solution for biological activity. Deconvolution and resynthesis of single peptides are necessary to confirm the activity of the screening results. Peptide coupling reactions were initially carried out in hand-labeled "tea bags." Libraries of several hundred thousand to millions of members are attainable. Naturally occurring L-amino acids and unnatural D-amino acids are employed in SPCLs. In the example

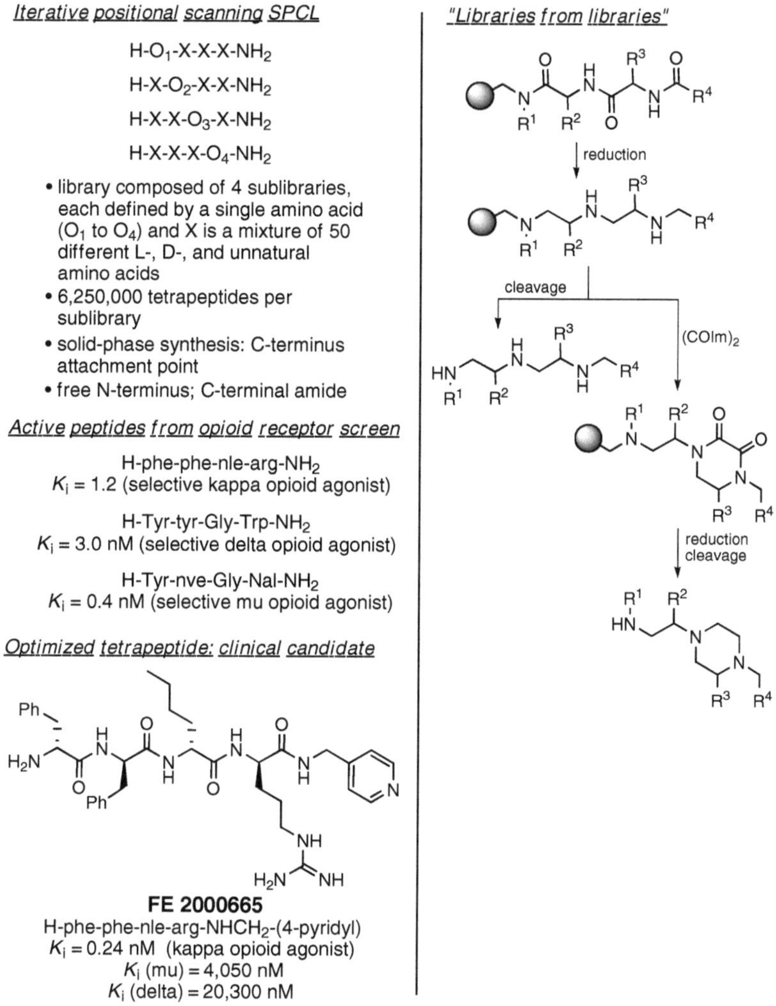

Fig. 1.4. Synthetic peptide combinatorial libraries (SPCLs).

of **Fig. 1.4**, a *ca.* 25 million member tetrapeptide SPCL was screened against the mu (μ), kappa (κ), and delta (δ) opioid receptors (13). Peptide sequences with high affinity and selectivity for each of the receptor types were found. One of the all D-amino acid-containing peptides, H–phe–phe–nle–arg–NH_2, identified as a selective κ receptor agonist, was further optimized to the C-terminal-modified analog, H–phe–phe–nle–arg–$NHCH_2$(4-pyridyl). This agent, also known as FE 2000665, is currently undergoing evaluation in human clinical trials as an analgesic. Chemical modification of SPCLs, for example, through a borane-mediated reduction reaction (amide bond → CH_2NH bioisostere) affords "libraries from libraries." These new random derivative libraries are useful in the discovery of biologically active compounds (14). The SPCLs and their derivative libraries have provided ligands for numerous molecular targets.

Peptoid libraries. A variant of peptide libraries, known as peptoids, was designed at Chiron (15) and then explored by many other research groups (**Fig. 1.5**). In peptoids, the amino acid side chains are relocated from the α-carbon to the nitrogen atom; hence, *N*-substituted glycines are monomeric building blocks. Peptoid sequences are synthesized on solid support from immobilized α-bromoacetic acid and primary amines thus giving rise to structural diversity. In contrast to amino acids, peptoids are not recognized as substrates for proteolytic-type metabolizing enzymes. Peptoids were thought to be superior to peptides as drug leads because of their perceived metabolic stability in vivo.

Nonoligomeric libraries. Peptide and peptoid libraries are examples of oligomeric (polymeric) libraries made up of repeating monomers (α-amino acids, *N*-substituted glycines). Random libraries composed of nonoligomeric compounds have been extensively explored. One illustration comes from the former laboratories at Organon (**Fig. 1.6**) (16). Thirteen different secondary amino-phenol inputs were attached to solid support by reaction with REM resin yielding resin-bound β-amino propionates. Two-site derivatization was then used to drive library diversity. The free phenolic OH was subjected to O-alkylation, O-acylation, O-sulfanylation, O-triflation/Suzuki coupling followed by N-quaternization (six inputs) and Hofmann elimination to release a 3,042-member library of tertiary amino aryls. One advantage of small-molecule nonoligomeric libraries

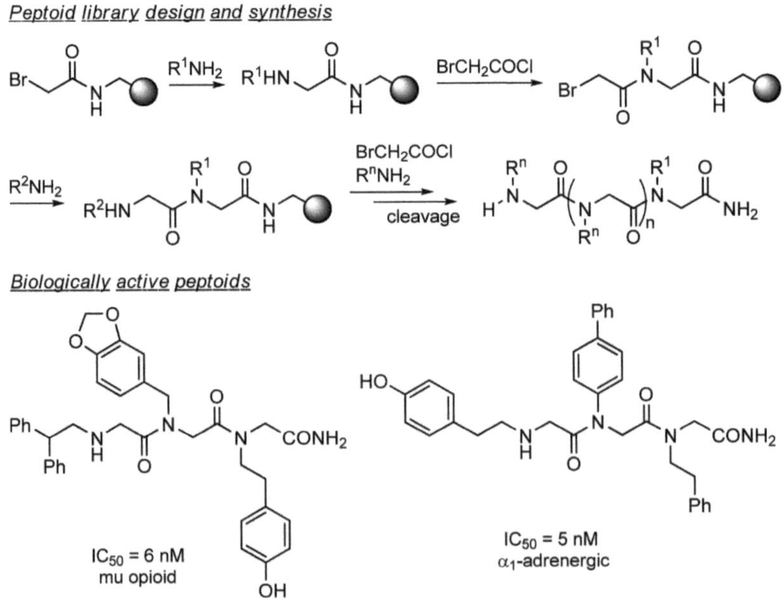

Fig. 1.5. Peptoid libraries.

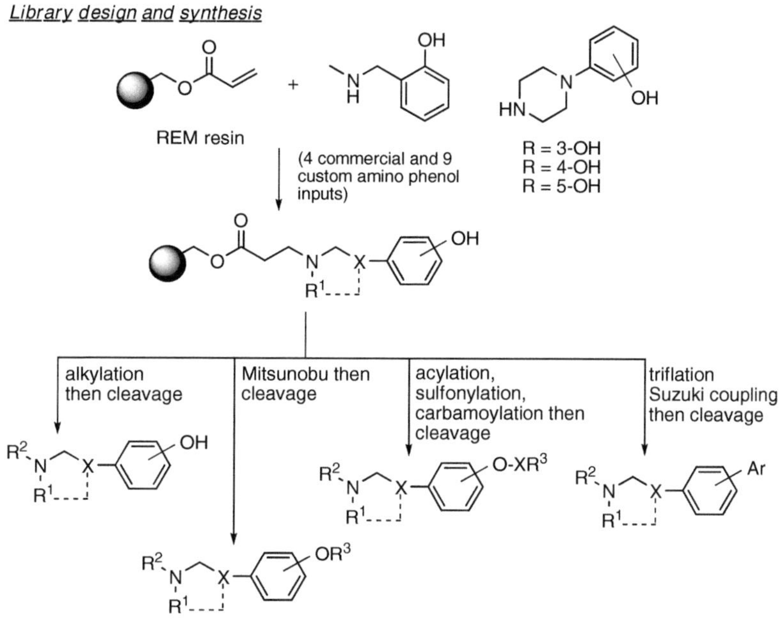

Physicochemical properties	Lipinski Ro5	>75% members
MW	≤ 500	≤ 450
ClogP	≤ 5	≤ 6
No. H-bond donors	≤ 10	≤ 1
No. H-bond acceptors	≤ 5	≤ 3
No. rotatable bonds	≤ 10	≤ 1

Fig. 1.6. Nonoligomeric library (reprinted ("adapted" or "in part") with permission from *Journal of the American Chemical Society*. Copyright 2002 American Chemical Society).

versus oligomeric libraries is the control over design and physicochemical properties. In the example of **Fig. 1.6**, >75% of the library members fell well within the Ro5 and successfully targeted central nervous system (CNS) property space. Screening the library against a variety of biological targets revealed a *ca.* 1 μM lead against the glycine-2 transporter.

Diversity-oriented synthesis (DOS) libraries. DOS libraries are a special class of nonpolymeric libraries distinguished by their synthetic design. Emphasis is placed on complexity-generating reactions to drive structural complexity in combination with branching pathways to drive structural diversity (**Fig. 1.7**) (17). A single library will contain multiple stereochemically rich molecular frameworks incorporating multiple building blocks and functional groups. Less emphasis is placed on physicochemical properties. They are intended for application in chemical biology. Originally prepared using encoded split-pool synthesis, DOS libraries are now prepared as discrete compounds on multimilligram scale. Build-couple-pair is the current paradigm for constructing DOS libraries (18).

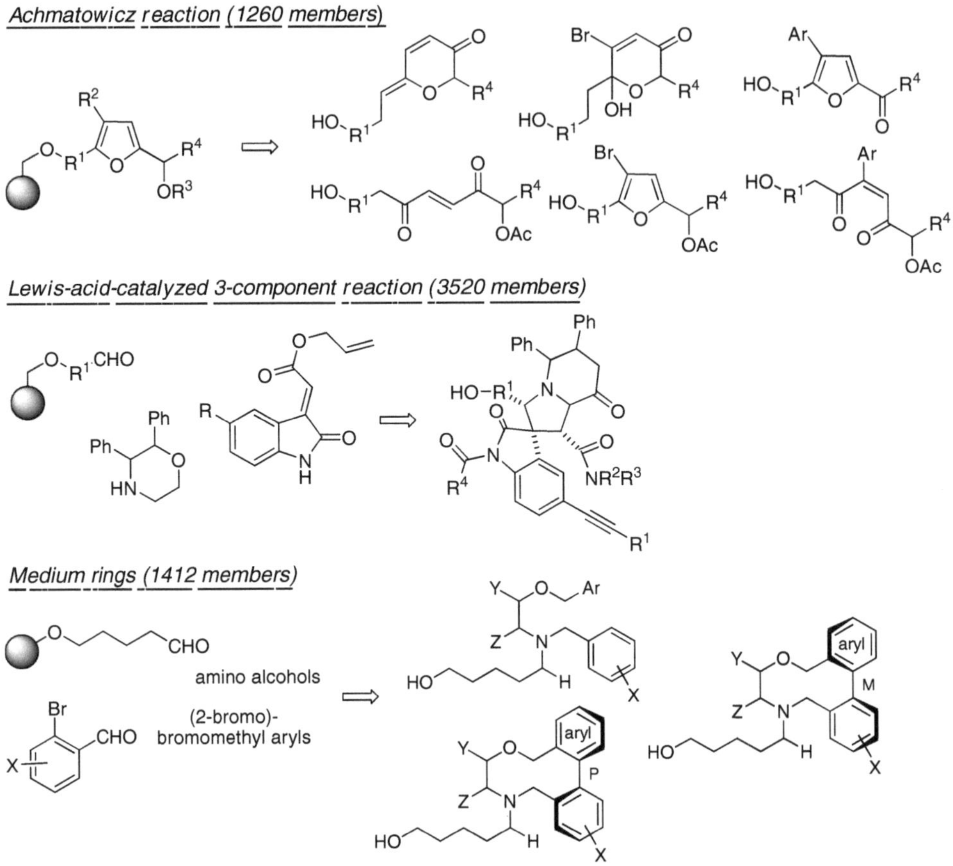

Fig. 1.7. Diversity-oriented synthesis (DOS) libraries (reprinted ("adapted" or "in part") with permission from *Journal of the American Chemical Society*. Copyright 2005 American Chemical Society).

2.2. Historical Designs: Targeted Screening Libraries

The design, synthesis, and evaluation of targeted libraries have been described much more frequently in the literature than random libraries. As mentioned above, the distinguishing feature of targeted libraries is the intentional inclusion of a known pharmacophore or privileged structure. Targeted library design revolving around these motifs can dramatically increase the odds of finding valuable leads. Examples of just a few such motifs used in library generation are shown in **Fig. 1.8**.

Mercaptoacyl pharmacophore library. Zinc metalloproteases are inhibited by small molecules that contain mercaptans (thiols; $-CH_2SH$), carboxylic acids ($-CO_2H$), and hydroxamic acids (–CONHOH). These functional groups chelate the active-site metal disrupting normal enzyme function. The angiotensin-converting enzyme (ACE) inhibitor Captopril® is an example of a thiol-based metalloprotease inhibitor. Thiols, carboxylic acids, and hydroxamic acids are consequently affirmed pharmacophores for this protease family. A historical example of a pharmacophore-

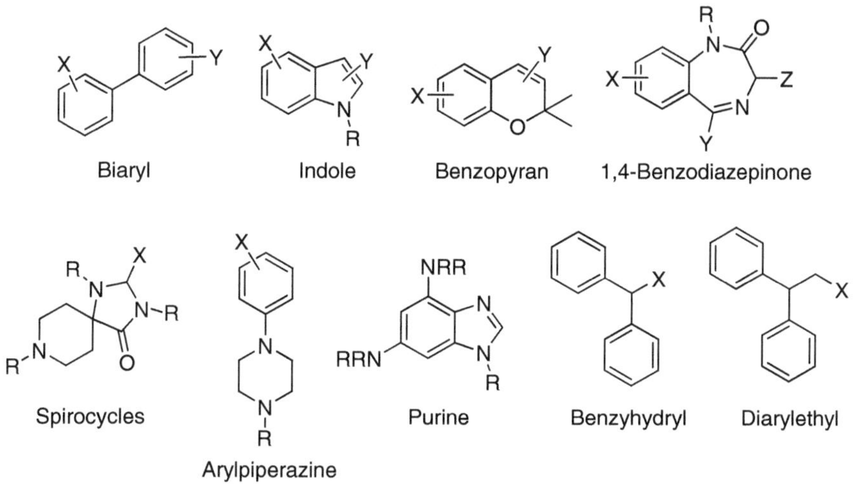

Fig. 1.8. Privileged scaffolds and pharmacophores found in libraries.

based library was described by Affymax (**Fig. 1.9**) (19). An encoded pool of highly substituted prolines was prepared utilizing the 1,3-dipolar cycloaddition reaction of resin-bound azomethine ylides and electron-deficient olefins. A mercapto pharmacophore was then introduced via N-acylation with a series of *S*-acetyl protected mercaptoacyl chlorides. S-deprotection and cleavage from resin afforded a *ca.* 500-member library of substituted prolines bearing a CO–Y–CH_2SH functional group. The library was assayed against ACE. Several inhibitors were found with one possessing extraordinary potency: K_i = 160 pM. A closely related diastereomer was >1,000-fold less active indicating a preferred stereochemical display of pyrrolidine ring substituents for high-affinity binding. The –CH_2SH pharmacophore was similarly introduced as a final step in a dipeptide amide library from which potent matrix metalloprotease-1 inhibitors were discovered (20).

Statine pharmacophore library. Aspartic acid protease-mediated peptide bond hydrolysis occurs via the addition of water to the amide carbonyl. The newly formed high-energy tetrahedral intermediate, tightly bound to enzyme, is stabilized through hydrogen bonding with aspartic acid residues in the active site. Collapse of the tetrahedral intermediate completes hydrolysis releasing the corresponding C-terminal acid and N-terminal amine peptide fragments. Statine, (3*S*,4*S*)-4-amino-3-hydroxy-6-methylheptanoic acid, may be considered a mimic of the putative high-energy tetrahedral intermediate, and when embellished with appropriate functionality, potently inhibits aspartic acid proteases. Therefore, 4-substituted-4-amino-3-hydroxybutanoic acids are pharmacophores for this class of protease. Researchers at Pharmacopeia designed a library using statine and an analog,

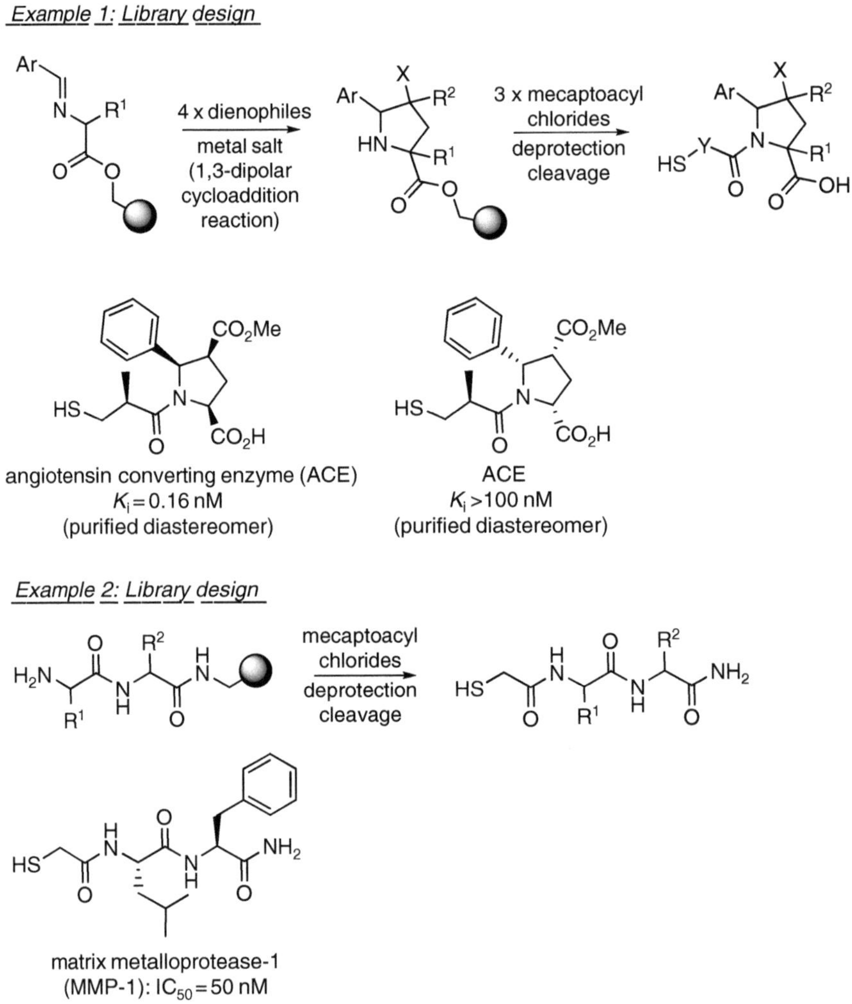

Fig. 1.9. Targeted library containing the mercaptoacyl pharmacophore.

4-amino-3-hydroxy-5-phenylpentanoic acid (**Fig. 1.10**) (21). Encoded split-pool synthesis was utilized in its construction, generating all possible combinations of compounds from the 2 × statines, 10 × N-terminal capping groups, 63 × C-terminal amino acids and 40 × C-terminal capping groups. The 25,200-member library was screened for aspartic acid protease inhibition, in particular for inhibitory action against plasmepsin II (plm II) and human cathepsin D (cat D). Plm II is a protease found in the malaria (*Plasmodium*) parasite and functions to degrade hemoglobin, an energy source for the maturing organism. It is a potential molecular target for malaria intervention. A large number of active compounds were found. Following bead decoding and compound resynthesis, agents with balanced inhibitory

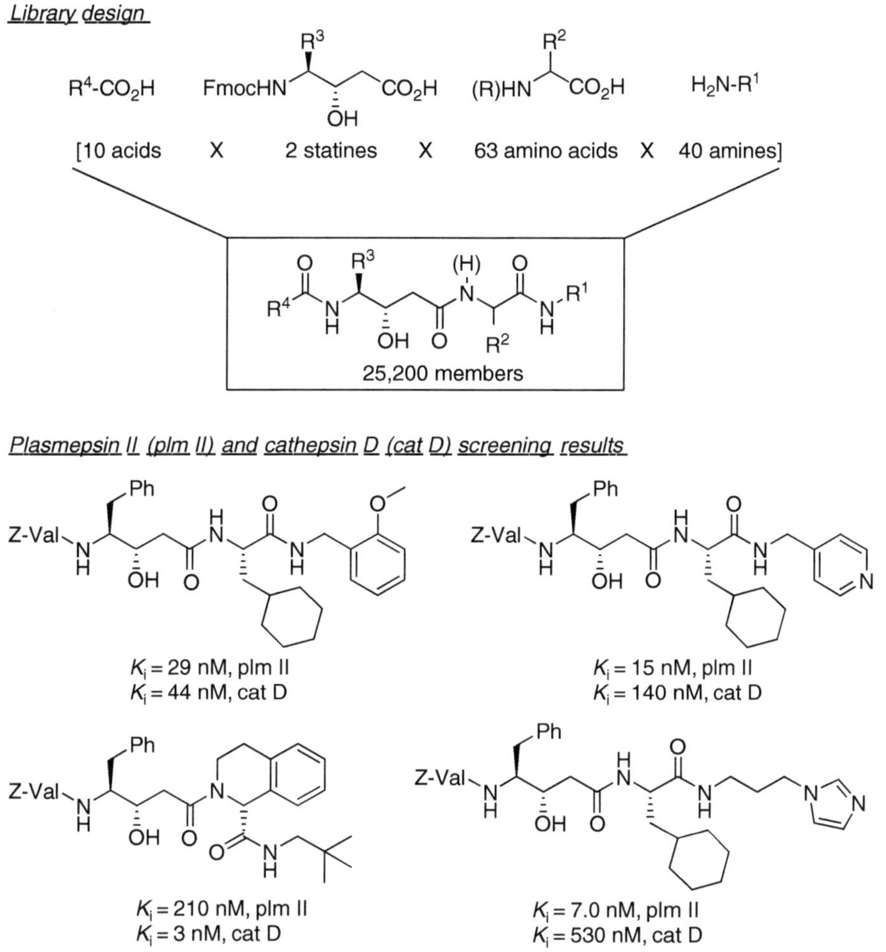

Fig. 1.10. Statine pharmacophore library targeting aspartic acid proteases (reprinted ("adapted" or "in part") with permission from *Journal of the American Chemical Society*. Copyright 2001 American Chemical Society).

activity at the two enzymes were identified as well as agents showing up to 75-fold selectivity for plm II versus cat D.

Affymax's thiolacyl library (**Fig. 1.9**) and Pharmacopeia's statine library (**Fig. 1.10**) are pharmacophore-based libraries; however, their design is different. In the former library, a pool of advanced library intermediates are derivatized with the pharmacophore (thiolacylation) as the final step in library construction, while in the latter library the pharmacophore (statine) is derivatized with synthons as part of library construction.

2-Aryl indole as a G-protein-coupled receptor (GPCR) privileged scaffold. The indole ring is a premier example of a privileged scaffold. The heterocycle is present in a profusion of medicinally important natural products and pharmaceutical substances, and it is associated with an extraordinary manifold of biological activity. Indoles have been extensively modified to exploit their inherent therapeutic properties. In many instances, these properties are manifested through interaction with GPCRs. The privileged nature of the heterocyclic system was adroitly demonstrated in a library of 2-aryl indoles (**Fig. 1.11**) (22). The library was prepared using combinatorial mixture and deconvolution techniques. Twenty arylalkyl keto acids were anchored to Kenner's safety catch resin. These resin pools were subjected to Fisher indole synthesis with a selection of 20 aryl hydrazines. Upon activation of the sulfonamide linker, the indoles were cleaved from resin with 80 amines yielding an indole amide library. Half of the library was further treated with a reducing reagent to furnish the corresponding amine indole library. In total, 128,000 compounds were generated. The judicious choice of synthons introduced substitutes at the indole 4-, 5-, 6-, and 7-positions as well as variation of aryl substitution at the indole 2-position. Evaluation of the compound collection was conducted across an array of GPCR binding assays. Remarkably, potent ligands were found for many of the receptors. The 0.8 nM human neurokinin-1 (hNK_1) ligand proved to be receptor subtype selective, devoid of affinity for hNK_2 or hNK_3. Several selective serotonin receptor ligands were uncovered representing potential leads for medicinal chemistry. Interestingly, with the exception of the NK_1 ligand emerging from the amide library, all of the other reported active compounds were 2-aryl indoles bearing a 3-alkylamine.

Purine as a privileged scaffold. The purine ring is another example of a ubiquitous, biologically-active heterocycle. It is readily identified as a substructure in adenine, one of the base units in DNA/RNA, and the nucleotide adenosine triphosphate (ATP). Among its many roles, ATP is the high-energy phosphate donor in phosphorylation reactions mediated by a large number of kinases. As a result, functionalized purines interact with a vast number of enzymes, receptors, and other biomolecules and satisfy the definition of a privileged structure. A purine derivative library was designed by Schlutz and coworkers (**Fig. 1.12**) (23). A series of *N*-9-substituted 2,4-dichloropurines were prepared by the direct alkylation (R^1-halogen) or Mitsunobu reaction (R^1-OH) of 2,6-dichloropurine. Reaction of these custom inputs with a selection of acid-labile amine resins resulted in selective displacement of the C-6 chlorine atom and simultaneous anchoring of the purine inputs to resin. This avails the C-2 position to a range of derivatization chemistries including nucleophilic displacement with amines, alcohols, phenols, thiols, and Pd-catalyzed Suzuki coupling with aryl boronic acids (carbon–carbon bond formation). Treatment of the penultimate resin intermediates with trifluoroacetic acid releases the final products from solid support for biological testing. This chemistry is sufficiently versatile to be

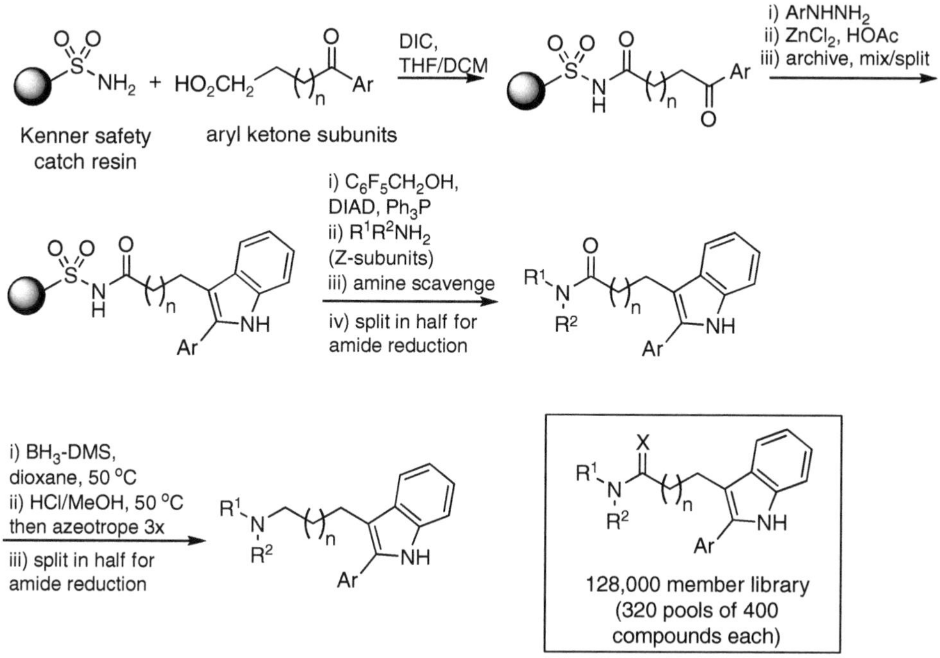

Selected (-NR[1] R[2]) pools and biological activity:
(Numbers in columns are % inhibition values at the given screening concentration)

Assay (concn, uM)	(R^{1}R^{2}N-)-subunit 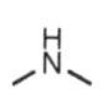	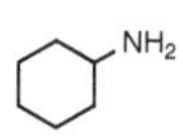	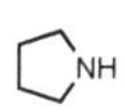	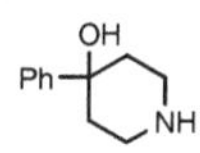	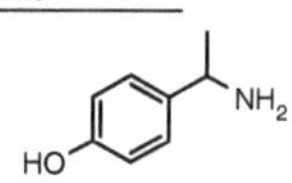	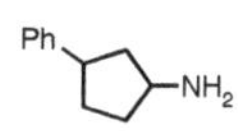	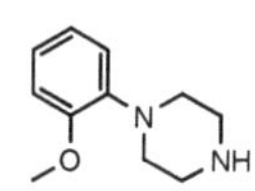
5-HT$_6$ (5)	**97**	76	95	44	87	68	42
MCR-4 (2)	10	**62**	17	--	5	23	--
5-HT$_{2a}$ (0.1)	81	14	**82**	54	45	63	0
GnRH (1)	4	7	4	**66**	6	6	--
NPY$_5$ (2)	82	89	85	--	**98**	96	23
CCR$_5$ (8)	1	21	4	10	0	**62**	--
NK$_1$ (1)	7	23	17	--	2	42	**92**

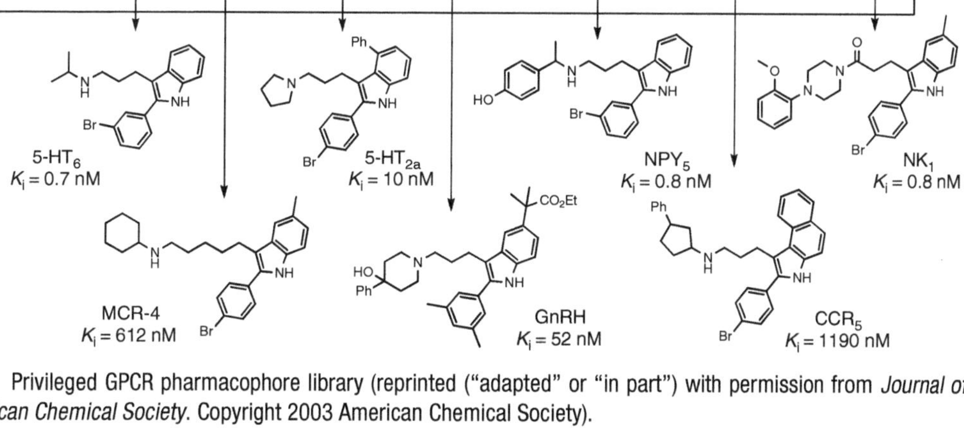

Fig. 1.11. Privileged GPCR pharmacophore library (reprinted ("adapted" or "in part") with permission from *Journal of the American Chemical Society*. Copyright 2003 American Chemical Society).

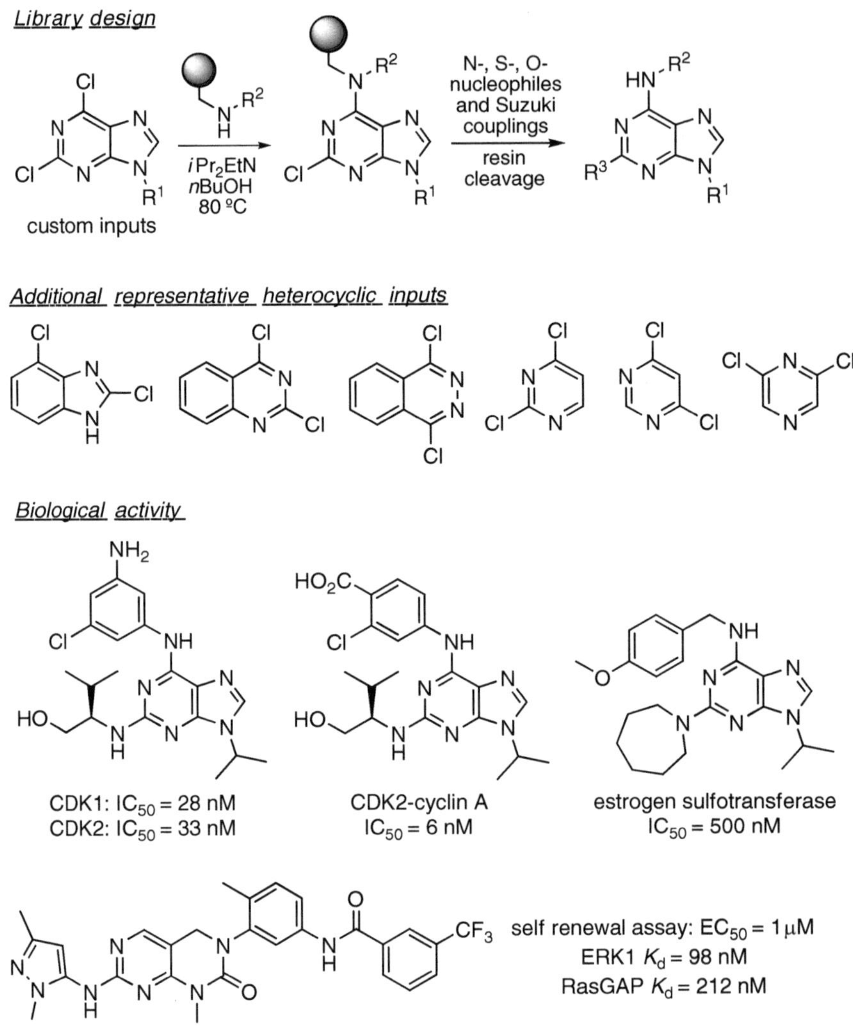

Fig. 1.12. Privileged purine library.

applied to structurally related halogenated heterocycles expanding chemotype diversity beyond the purine scaffold. Approximately 45,000 substituted purines and related derivatives were synthesized in total. Screening the library afforded potent cyclin-dependent kinase-1 (CDK1; $IC_{50} = 28$ nM) and CDK2 ($IC_{50} =$ 6 nM) inhibitors. These kinases utilize ATP to phosphorylate proteins on serine and threonine amino acid residues regulating cell division. Inhibitors of estrogen sulfotransferase ($IC_{50} = 500$ nM) (24) and enzymes involved in cell regeneration were found (25). Estrogen sulfotransferase catalyzes the transfer of a sulfuryl group from 3′-phosphoadenosine 5′-phosphosulfate (PAPS) to estrogen regulating hormone homeostasis.

2.3. Historical Designs: Optimization Libraries

In the previous examples, random and targeted libraries are used to discover leads. Optimization libraries are employed when lead structures have already been identified, serving to improve the potency, selectivity, or other characteristics of the molecule.

Human rhinovirus 3C protease inhibitor. The former Agouron Pharmaceuticals research laboratories identified a tripeptidyl Michael acceptor as a lead structure in a rhinovirus 3C protease inhibitor program (**Fig. 1.13**) (26). This agent was an irreversible inhibitor (second-order rate constant: $k_{obs}/I = 280{,}000\ M^{-1}s^{-1}$) of the enzyme which is essential for viral replication. One issue with the compound was the *N*-benzylthiocarbamate, an N-terminal capping group potentially undergoing metabolism leading to a short half-life of the agent in vivo. An optimization library was designed to find an N-terminal amide to replace the benzylthiocarbamate that would provide the necessary metabolic stability. The lead possessed a modified glutamate residue which proved to be a strategic asset for crafting an optimization library. A modified *N*-Fmoc glutamic acid bearing an α,β-unsaturated ethyl ester was attached to solid support through a Rink amide linker. A multistep elaboration completed the assembly of the penultimate tripeptide intermediate with a free N-terminus. This resin-bound intermediate was then acylated with *ca.* 500 carboxylic acids and acid chlorides. Cleavage of the analogs from the resin and evaluation in a high-throughput enzyme assay led to the discovery of 5-methylisoxazole-3-carboxamide as an ideal surrogate for the benzylthiocarbamate. The 5-methylisoxazole-3-carboxamide analog was essentially equipotent ($k_{obs}/I = 260{,}000\ M^{-1}s^{-1}$) with the original lead. This compound showed antiviral activity without cytotoxicity in cell culture and also exhibited broad-spectrum antiviral activity. The advanced lead was subjected to traditional optimization ultimately giving rise to AG7088 ($k_{obs}/I = 1{,}470{,}000\ M^{-1}s^{-1}$). AG7088 was nominated for development and subsequently advanced into human clinical trials (27). The *N*-(5-methylisoxazole-3-carboxamido) group was retained in the clinical candidate.

Kappa opioid receptor antagonist. There are three opioid receptors, mu (μ), kappa (κ), and delta (δ). 4-(3-Hydroxyphenyl)-*trans*-3,4-dimethylpiperidine is an opioid receptor antagonist pharmacophore originally discovered in the 1970s. All piperidine nitrogen analogs of this scaffold reported in the literature displayed no receptor selectivity with the exception of the μ selective agent shown in **Fig. 1.14**. This suggested to Carroll and coworkers that given the appropriate *N*-substituent, it may be possible to obtain selective antagonists for the κ and δ opioid receptors (28). In a program to investigate selective κ antagonists for the treatment of drug abuse,

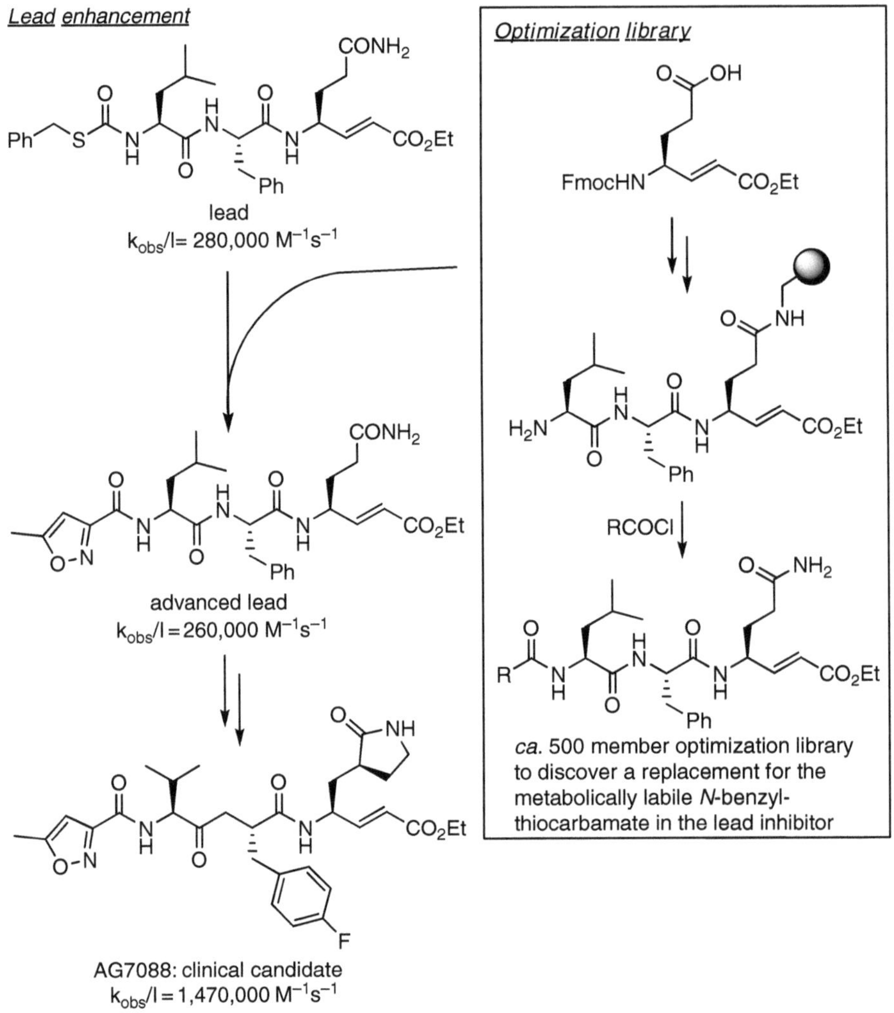

Fig. 1.13. Optimization library for human rhinovirus 3C protease (reprinted ("adapted" or "in part") with permission from *Journal of the American Chemical Society*. Copyright 2001 American Chemical Society).

an optimization library was conceived to search for such agents. (+)-4-(3-Hydroxyphenyl)-*trans*-3,4-dimethylpiperidine was coupled to 11 amino acids and reduced to give a series of piperidines with a newly appended primary or secondary amine. These inputs were subjected to solution-phase acylation reactions using substituted benzoic, phenylacetic, phenyl cinnamic, and (3-phenyl)propionic acids. The acylation reaction was sufficiently clean that following aqueous workup, the 288 library products were screened directly, without purification, against the three opioid receptors. A κ selective agent was discovered (K_i = 7 nM; 57-fold and >825-fold selective versus μ and δ, respectively) and its binding and antagonist activity confirmed upon retesting the purified compound. A remarkable range of potency and selectivity

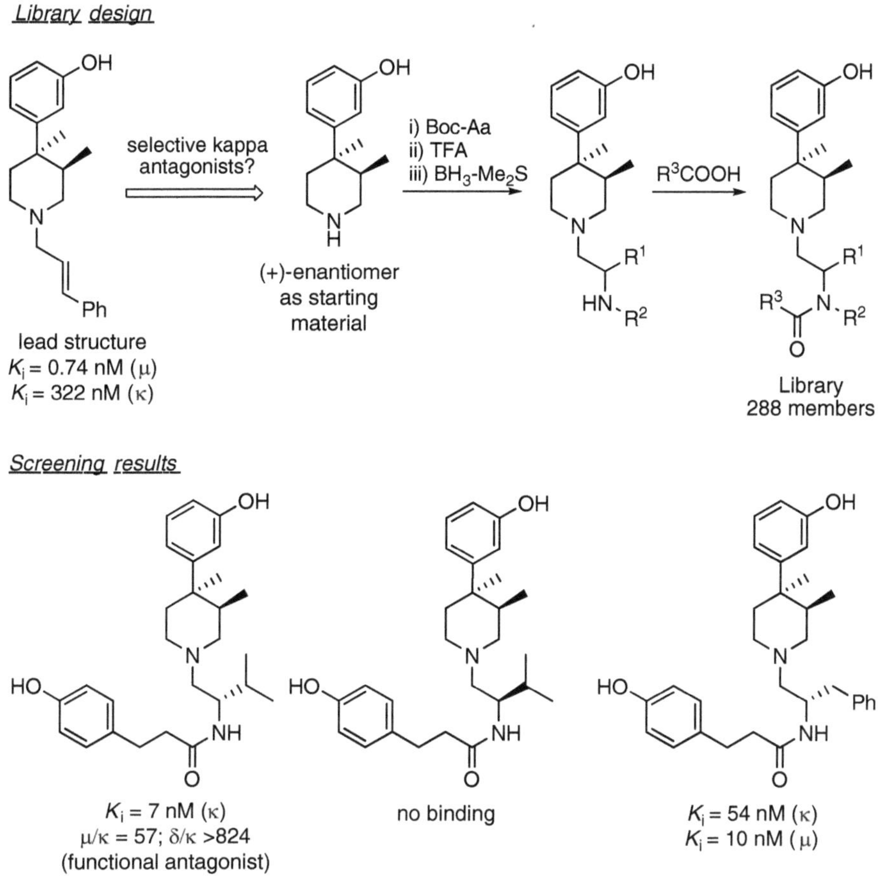

Fig. 1.14. Kappa (κ) opioid receptor antagonist optimization library (reprinted ("adapted" or "in part") with permission from *Journal of the American Chemical Society*. Copyright 1999 American Chemical Society).

was also observed. Structure–activity relationship (SAR) data obtained from the library accentuates the critical role of the isopropyl group. This is corroborated both in terms of stereochemistry, (*S*)-configuration necessary for affinity, and selectivity as the isopropyl → benzyl exchange resulted in μ selective antagonists. Using a molecule described in the literature as a starting point for library (analog) synthesis is an example of a knowledge-based approach to lead optimization.

Raf kinase inhibitor. High-throughput screening of a chemical collection at the former Bayer Research Center turned up 3-thienyl urea as a modestly potent inhibitor of p38 kinase (IC_{50} = 290 nM) possessing comparatively weak activity at raf kinase (IC_{50} = 17 μM; p38/raf = 0.017; **Fig. 1.15**) (29). Because of its low activity, many laboratories would have discounted this compound as a raf kinase inhibitor lead. Smith and coworkers applied HTC techniques in an attempt to improve

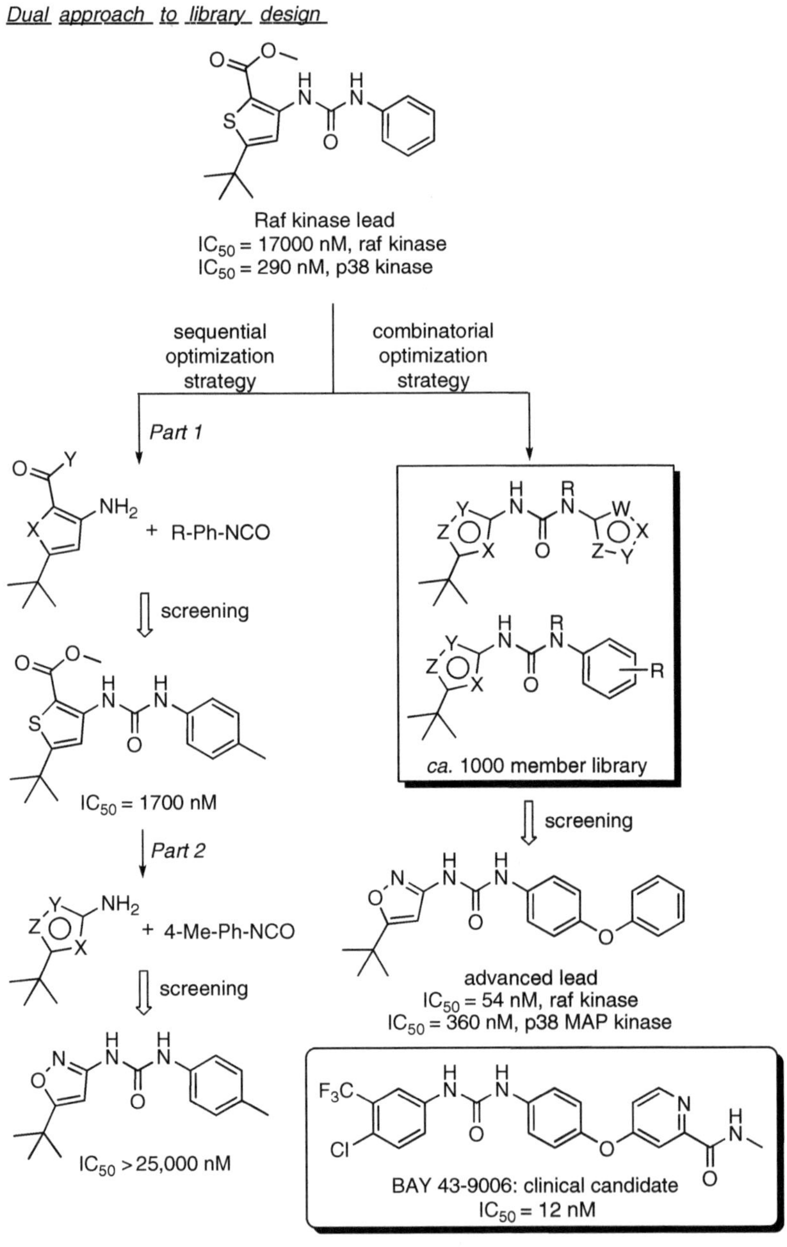

Fig. 1.15. Contrasting raf kinase inhibitor optimization strategies (reprinted ("adapted" or "in part") with permission from *Journal of the American Chemical Society*. Copyright 2002 American Chemical Society).

both the inhibitory potency and selectivity of the urea against raf kinase. A two-part sequential optimization strategy was devised. In part one, coupling conservatively altered 3-aminothienyls with phenyl-substituted isocyanates was carried out. A *ca.* 10-fold improvement in activity over the original lead was obtained with a 4-methyl group in the phenyl ring. In part 2, the "optimized"

4-methylphenyl portion of the molecule was held constant and a broad range of heterocycles was explored to optimize the 3-thienyl moiety. This resulted in no further improvement in activity. The sequential two-part optimization strategy *failed* to meet the objective. This was followed by a combinatorial strategy in which 300 anilines/heterocyclic amines were combined with 75 aryl/heteroaryl isocyanates to produce an array of a *ca.* 1,000 compounds. Evaluation of these compounds resulted in the identification of the advanced lead, 1-(5-*tert*-butylisoxazol-3-yl)-3-(4-phenoxyphenyl)urea: $IC_{50} = 54$ nM) possessing 7-fold selectivity over p38 kinase. This agent represented a significant 314-fold increase in raf kinase potency versus the original lead. The result was unanticipated. The 5-*tert*-butyl-3-aminoisoxazole present in the advanced lead was considered an *inactive* heterocycle based on the SAR data generated from the original sequential optimization strategy. Further optimization of the advanced lead was achieved, identifying a clinical candidate [IC_{50} (raf kinase) = 12 nM] displaying sufficient potency and favorable kinase enzyme selectivity. Key structural elements present in the advanced lead are retained in the clinical candidate. This library design example beautifully underscores the advantage of combinatorial versus the traditional step-wise approach to lead optimization.

3. Summary

HTC originated in the early 1990s in response to unprecedented access to molecular targets, advances in high-throughput screening technology, and the demand for new chemical compound collections. Approaching two decades of application, there are over 5,000 chemical libraries reported in the literature (8). Initial design strategies based on oligomeric and nonoligomeric libraries with multiple (>3) points of diversity have progressed toward more carefully crafted molecules with attention paid to physicochemical and toxiphoric properties. Today, library compounds are typically synthesized on a milligram scale (10–100 mg), purified, and evaluated not only against the primary target but also in selectivity assays including (a) in vitro drug metabolism pharmacokinetic (DMPK) assays which measure a compound's metabolic stability and interaction with cytochrome P450 metabolizing enzymes, and (b) ion channels associated with cardiac function. Libraries are being used to generate multiple SARs to efficiently identify and simultaneously address compound liabilities. Library designs incorporating pharmacophores (19, 21) and privileged structures (22, 23) have historically been successful in lead finding. New chemotypes are needed to investigate previously unexplored diversity space to discover fresh leads. Identifying a metabolically stable surrogate for the *N*-benzylthiocarbamate in the rhinovirus 3C protease inhibitor (25), generating a series of selective kappa opioid receptor antagonists starting with a non-selective opioid ligand (27), and enhancing the potency and selectivity of a marginally active raf kinase inhibitor by combinatorializing synthons when traditional medicinal chemistry failed (28) serve as historical references to the successful application of HTC in lead optimization. Such references are valuable lessons in library design that can still be considered in contemporary HTC.

References

1. Bunin, B. A., Ellman, J. A. (1992) A general and expedient method for the solid phase synthesis of 1,4-benzodiazepine derivatives. *J Am Chem Soc* **114**, 10997–10998.
2. DeWitt, S. H., Kiely, J. S., Stankovic, C. J., Schroeder, M. C., Cody, D. M. R., Pavia, M. R. (1993) "Diversomers": an approach to nonpeptide, nonoligomeric chemical diversity. *Proc Natl Acad Sci USA* **90**, 6909–6913.
3. Terrett, N. (1998) *Combinatorial Chemistry.* Oxford University Press, Oxford, UK.
4. Ohlmeyer, M. H. J., Swanson, R. N., Dillard, L., Reader, J. C., Asouline, G., Kobayashi, R., Wigler, M., Still, W. C. (1993) Complex synthetic chemical libraries indexed with molecular tags. *Proc Natl Acad Sci USA* **90**, 10922–10926.
5. Data taken from ArQule's 10 K annual reports for years ending 1996–1997. http://www.sec.gov/Archives/edgar/data.
6. Lipinski, C. A., Lombardo, F., Dominy, B. W., Feeney, P. J. (1997) Experimental and computational approaches to estimate solubility and permeability in drug discovery and development settings. *Adv Drug Delivery Rev* **23**, 3–25.
7. Teague, S. J., Davis, A. M., Leeson, P. D., Oprea, T. (1999) The design of leadlike combinatorial libraries. *Angew Chem, Int Ed* **38**, 3743–3748.
8. Dolle, R. E., Le Bourdonnec, B., Goodman, A. J., Morales, G. A., Thomas, C. J., Zhang, W. (2009) Comprehensive survey of chemical libraries for drug discovery and chemical biology: 2008. *J Comb Chem* **11**, 755–802.
9. Gund, P. (1977) Three-dimensional pharmacophoric pattern searching. *Prog Mol Subcell Biol* **5**, 117–143.
10. Hajduk, P. J., Bures, M., Praestgaard, J., Fesik, S. W. (2000) Privileged molecules for protein binding identified from NMR-based screening. *J Med Chem* **43**, 3443–3447.
11. Fodor, S. P. A., Read, J. L., Pirrung, M. C., Stryer, L., Lu, A. T., Solas, D. (1991) Light-directed, spatially addressable parallel chemical synthesis. *Science* **251**, 767–773.
12. Dooley, C., Houghten, R. (1993) The use of positional scanning synthetic peptide combinatorial libraries for the rapid determination of opioid receptor ligands. *Life Sci* **52**, 1509–1517.
13. Dooley, C. T., Ny, P., Bidlack, J. M., Houghten, R. A. (1998) Selective ligands for the μ, δ, and κ opioid receptors identified from a single mixture based tetrapeptide positional scanning combinatorial library. *J Biol Chem* **273**, 18848–18856.
14. Ostresh, J. M., Husar, G. M., Blondelle, S., Dorner, B., Weber, P. A., Houghten, R. A. (1994) "Libraries from libraries": chemical transformation of combinatorial libraries to extend the range and repertoire of chemical diversity. *Proc Natl Acad Sci USA* **91**, 11138–11142.
15. Zuckermann, R. N., Martin, E. J., Spellmeyer, D. C., Stauber, G. B., Shoemaker, K. R., Kerr, J. M., Figliozzi, G. M., Goff, D. A., Siani, M. A., Simon, R., Banville, S. C., Brown, E. G., Wang, L., Richter, L. S., Moos, W. H. (1994) Discovery of nanomolar ligands for 7-transmembrane G-protein-coupled receptors from a diverse *N*-(substituted)glycine peptoid library. *J Med Chem* **37**, 2678–2685.
16. Barn, D., Caulfield, W., Cowley, P., Dickins, R., Bakker, W. I., McGuire, R., Morphy, J. R., Rankovic, Z., Thorn, M. (2001) Design and synthesis of a maximally diverse and druglike screening library using REM resin methodology. *J Comb Chem* **3**, 534–541.
17. Burke, M. D., Berger, E. M., Schreiber, S. L. (2004) A synthesis strategy yielding skele-

tally diverse small molecules combinatorially. *J Am Chem Soc* **126**, 14095–14104.
18. Nielsen, T. E., Schreiber, S. L. (2008) Towards the optimal screening collection. A synthesis strategy. *Angew Chem, Int Ed* **47**, 48–56.
19. Murphy, M. M., Schullek, J. R., Gordon, E. M., Gallop, M. A. (1995) Combinatorial organic synthesis of highly functionalized pyrrolidines: identification of a potent angiotensin converting enzyme inhibitor from a mercaptoacyl proline library. *J Am Chem Soc* **117**, 7029–7030.
20. Lynas, J. F., Martin, S. L., Walker, B., Baxter, A. D., Bird, J., Bhogal, R., Montana, J. G., Owen, D. A. (2000) Solid-phase synthesis and biological screening of *N*-α-mercaptoamide template-based matrix metalloprotease inhibitors. *Comb Chem High Throughput Screening* **3**, 37–41.
21. Dolle, R. E., Guo, J., O'Brien, L., Jin, Y., Piznik, M., Bowman, K. J., Li, W., Egan, W. J., Cavallaro, C. L., Roughton, A. L., Zhao, W., Reader, J. C., Orlowski, M., Jacob-Samuel, B., DiIanni Carroll, C. (2000) A statistical-based approach to assessing the fidelity of combinatorial libraries encoded with electrophoric molecular tags. Development and application of tag decode-assisted single bead LC/MS analysis. *J Comb Chem* **2**, 716–731.
22. Willoughby, C. A., Hutchins, S. M., Rosauer, K. G., Dhar, M. J., Chapman, K. T., Chicchi, G. G., Sadowski, S., Weinberg, D. H., Patel, S., Malkowitz, L., Di Salvo, J., Pacholok, S. G., Cheng, K. (2001) Combinatorial synthesis of 3-(amidoalkyl) and 3-(aminoalkyl)-2-arylindole derivatives: discovery of potent ligands for a variety of G-protein-coupled receptors. *Bioorg Med Chem Lett* **12**, 93–96.
23. (a) Ding, S., Gray, N. S., Ding, Q., Wu, X., Schultz, P. G. (2002) Resin-capture and release strategy toward combinatorial libraries of 2,6,9-substituted purines. *J Comb Chem* **4**, 183–186. (b) Ding, S., Gray, N. S., Wu, X., Ding, Q., Schultz, P. G. (2002) A combinatorial scaffold approach toward kinase-directed heterocycle libraries. *J Am Chem Soc* **124**, 1594–1596.
24. Verdugo, D. E., Cancilla, M. T., Ge, X., Gray, N. S., Chang, Y. -T., Schultz, P. G., Negishi, M., Leary, J. A., Bertozzi, C. R. (2001) Discovery of estrogen sulfotransferase inhibitors from a purine library screen. *J Med Chem* **44**, 2683–2686.
25. Chen, S., Do, J. T., Zhang, Q., Yao, Q., Yao, S., Yan, F., Peters, E. C., Schoeler, H. R., Schultz, P. G., Ding, S. (2006) Self-renewal of embryonic stem cells by a small molecule. *Proc Natl Acad Sci USA* **103**, 17266–17271.
26. Dragovich, P. S., Zhou, R., Skalitzky, D. J., Fuhrman, S. A., Patick, A. K., Ford, C. E., Meador, J. W., III, Worland, S. T. (1999) Solid-phase synthesis of irreversible human rhinovirus 3C protease inhibitors. Part 1: optimization of tripeptides incorporating N-terminal amides. *Bioorg Med Chem* **7**, 589–598.
27. Matthews, D. A., Dragovich, P. S., Webber, S. E., Fuhrman, S. A., Patick, A. K., Zalman, L. S., Hendrickson, T. F., Love, R. A., Prins, T. J., Marakovits, J. T., Zhou, R., Tikhe, J., Ford, C. E., Meador, J. W., Ferre, R. A., Brown, E. L., Binford, S. L., Brothers, M. A., Delisle, D. M., Worland, S. T. (1999) Structure-assisted design of mechanism-based irreversible inhibitors of human rhinovirus 3C protease with potent antiviral activity against multiple rhinovirus serotypes. *Proc Natl Acad Sci USA* **96**, 11000–11007.
28. Thomas, J. B., Fall, M. J., Cooper, J. B., Rothman, R. B., Mascarella, S. W., Xu, H., Partilla, J. S., Dersch, C. M., McCullough, K. B., Cantrell, B. E., Zimmerman, D. M., Carroll, F. I. (1998) Identification of an opioid κ receptor subtype-selective N-substituent for (+)-(3*R*,4*R*)-dimethyl-4-(3-hydroxyphenyl)piperidine. *J Med Chem* **41**, 5188–5197.
29. Smith, R. A., Barbosa, J., Blum, C. L., Bobko, M. A., Caringal, Y. V., Dally, R., Johnson, J. S., Katz, M. E., Kennure, N., Kingery-Wood, J., Lee, W., Lowinger, T. B., Lyons, J., Marsh, V., Rogers, D. H., Swartz, S., Walling, T., Wild, H. (2001) Discovery of heterocyclic ureas as a new class of raf kinase inhibitors: identification of a second generation lead by a combinatorial chemistry approach. *Bioorg Med Chem Lett* **11**, 2775–2778.

Chapter 2

Chemoinformatics and Library Design

Joe Zhongxiang Zhou

Abstract

This chapter provides a brief overview of chemoinformatics and its applications to chemical library design. It is meant to be a quick starter and to serve as an invitation to readers for more in-depth exploration of the field. The topics covered in this chapter are chemical representation, chemical data and data mining, molecular descriptors, chemical space and dimension reduction, quantitative structure–activity relationship, similarity, diversity, and multiobjective optimization.

Key words: Chemoinformatics, QSAR, QSPR, similarity, diversity, library design, chemical representation, chemical space, virtual screening, multiobjective optimization.

1. Introduction

Library design is essentially a selection process, selecting a useful subset of compounds from a candidate pool. How to select this subset depends on the purpose of the library. For a simple probe of a local structure–activity relationship (SAR), medicinal chemists may be able to choose an excellent subset of representatives from a small pool of synthesizable compounds to achieve the goal without resorting to any sophisticated design tools. For complex applications of library though, design tools are indispensable for obtaining optimal results. Majority of the design tools used for library design fall into a field called chemoinformatics, a discipline that studies the transformation of data into information and information into knowledge for better decision making (1). Actually, the recent explosive development in chemoinformatics has mainly been stimulated by the ever-increasing applications of chemical library technologies in pharmaceutical industry.

J.Z. Zhou (ed.), *Chemical Library Design*, Methods in Molecular Biology 685,
DOI 10.1007/978-1-60761-931-4_2, © Springer-Science+Business Media, LLC 2011

Theoretically, there are 10^{60}–10^{100} compounds available to a small-molecule drug discovery program of any given drug target (2, 3). The purpose of a drug discovery program is to find a good compound that can modulate the function of the target while avoiding harmful side effects. It is not a trivial task to navigate even a small portion of this huge chemical space and locate a few optimal candidates with desirable properties. Therefore, a drug discovery program usually starts with the discovery of lead compounds followed by their optimizations, instead of the impossible task of sifting through the entire chemical space directly for a drug compound. Even this two-step divide-and-conquer approach cannot divide the chemical space small enough for manual identification of desirable compounds. Library design as a drug discovery technology faces the same "finding-a-needle-in-a-haystack" issues as the drug discovery itself. Computational tools are necessary for efficient navigations in the chemical space. Thus, chemoinformatic methods are developed to allow chemical data manipulations, chemoinformatic transformations, easy navigation in chemical space, predictive model building, etc. Chemoinformatics has played a very important role in the rapid development and widespread applications of chemical library technologies.

In this chapter, we will give a brief introduction to the basic concepts of chemoinformatics and their relevance to chemical library design. In Section **2**, we will describe chemical representation, molecular data, and molecular data mining in computer; we will introduce some of the chemoinformatics concepts such as molecular descriptors, chemical space, dimension reduction, similarity and diversity; and we will review the most useful methods and applications of chemoinformatics, the quantitative structure–activity relationship (QSAR), the quantitative structure–property relationship (QSPR), multiobjective optimization, and virtual screening. In Section **3**, we will outline some of the elements of library design and connect chemoinformatics tools, such as molecular similarity, molecular diversity, and multiple objective optimizations, with designing optimal libraries. Finally, we will put library design into perspective in Section **4**.

2. Chemoinformatics

Although still rapidly evolving, chemoinformatics as a scientific discipline is relatively mature. This section is meant to be introductory only. Interested readers are referred to various monographs on chemoinformatics for a deep understanding of the field (4–8).

2.1. Chemical Representation

The first task of chemoinformatics is to transform chemical knowledge, such as molecular structures and chemical reactions, into computer-legible digital information. The digital representations of chemical information are the foundation for all chemoinformatic manipulations in computer. There are many file formats for molecular information to be imported into and exported from computer. Some formats contain more information than others. Usually, intended applications will dictate which format is more suitable. For example, in a quantum chemistry calculation the molecular input file usually includes atomic symbols with three-dimensional (3D) atomic coordinates as the atomic positions, while a molecular dynamics simulation needs, in addition, atom types, bond status, and other relevant information for defining a force field.

Chemical representation can be rule-based or descriptive. Here we will give a short description of two popular file formats for molecular structures, MOLfiles (9) and SMILES (10–13), to illustrate how molecules are represented in computer. SMILES is a rule-based format while MOLfile is a more descriptive one.

A MOLfile usually contains a header block and a connection table (*see* **Fig. 2.1**). The header block consists of three lines

(a)

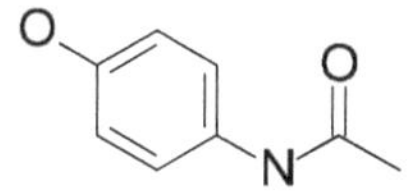

(b)

```
  SMMXDraw12120917342D

 11 11  0  0  0  0  0  0  0  0999 V2000
   12.3082   -7.1882    0.0000 C   0  0  0  0  0  0  0  0  0  0  0  0
   13.0242   -7.6016    0.0000 C   0  0  0  0  0  0  0  0  0  0  0  0
   13.7402   -7.1882    0.0000 C   0  0  0  0  0  0  0  0  0  0  0  0
   14.4562   -7.6016    0.0000 N   0  0  3  0  0  0  0  0  0  0  0  0
   15.1722   -7.1882    0.0000 C   0  0  0  0  0  0  0  0  0  0  0  0
   15.1722   -6.3615    0.0000 O   0  0  0  0  0  0  0  0  0  0  0  0
   15.8882   -7.6016    0.0000 C   0  0  0  0  0  0  0  0  0  0  0  0
   13.7402   -6.3615    0.0000 C   0  0  0  0  0  0  0  0  0  0  0  0
   13.0242   -5.9481    0.0000 C   0  0  0  0  0  0  0  0  0  0  0  0
   12.3082   -6.3615    0.0000 C   0  0  0  0  0  0  0  0  0  0  0  0
   11.5922   -5.9481    0.0000 O   0  0  0  0  0  0  0  0  0  0  0  0
  2  1  2  0  0  0  0
  3  2  1  0  0  0  0
  3  4  1  0  0  0  0
  4  5  1  0  0  0  0
  5  6  2  0  0  0  0
  5  7  1  0  0  0  0
  8  3  2  0  0  0  0
  9  8  1  0  0  0  0
 10  9  2  0  0  0  0
 10 11  1  0  0  0  0
  1 10  1  0  0  0  0
M  END
```

Header block

Counts line

Atom block

Connection Table (Ctab)

Bond block

Fig. 2.1. Illustrative example of a MOLFile for acetaminophen (also known as paracetamol). (**a**) Molecular structure of acetaminophen, commonly known as Tylenol. Tylenol is a widely used medicine for reducing fever and pain. (**b**) MOLFile for acetaminophen.

containing such information as molecular IDs, owner of the record, dates, and other miscellaneous information and comments. The connection table (CTab) contains the actual molecular structure information in several sections: a count line, an atom block, a bond block, and a property block. The count line includes number of atoms, number of bonds, number of atom lists, chiral flag for the molecule, and number of lines of additional property information in the property block. The atom block is made up of atom lines with each line containing atomic coordinates, atomic symbol, relative mass, charge, atomic stereo parity, valence, and other information. The bond block consists of bond lines for all bonds. Each bond line contains information about bond type, bond stereo, bond topology, and reacting center status. The property block consists of property lines. Most of the property lines start with a letter M followed by a property identifier. The usual properties appearing in property blocks are charges, radical status, isotope, Rgroup properties, 3D features, and other properties. The property block ends with an "M END" line.

The MOLfile format belongs to a general format definition for Chemical Table Files (CTFiles). CTFiles define file formats for various purposes. Particularly, multiple molecular entries can be stored in an SDFile format. Each molecular entry in an SDFile may consist of the MOLfile as described above and other data records associated with the molecule. Other important file formats of CTFiles definitions are RGFile for Rgroup files, rxnfile for reaction files, RDFile for multiple records of molecules and/or reactions along with their associated data, and XDFile for XML-based records of molecules and/or reactions along with their associated data. Interested readers are referred to Symyx's MDL white paper for a complete coverage of the CTFile formats in general and Molfile format in particular (9).

SMILES (Simplified Molecular Input Line Entry Systems) is a line notation system based on principles of molecular graph theory for entering and representing molecules and reactions in computer (10–13). It uses a set of simple specification rules to derive a SMILES string for a given molecular structure (or more precisely, a molecular graph). A simplified set of rules is as follows:

- Atoms are represented by their atomic symbols enclosed by square bracket, [ ], which can be dropped for the "organic" subset B, C, N, O, P, S, F, Cl, Br, and I. Hydrogen atoms are usually implicit.
- Bonds between adjacent atoms are assumed to be single unless specified otherwise; double and triple bonds are denoted as "=" and "#".
- Branches are specified by enclosing them in parentheses, which can be nested and stacked. The implicit connection

of a branch in a parenthesized expression is to the left of the string.

- Rings in cyclic structures are broken with a unique number attached to the two atoms at each break point. A single atom may involve in multiple ring breakages. In this case, it will have multiple numbers attached to it with each number corresponding to a single break point.
- Atoms in aromatic rings are denoted by lower case letters.
- Disconnected structures are separated by a period (.).

There are also rules specifying chiral centers, configurations around double bonds, charges, isotopes, etc. A complete list of specification rules can be found in the SMILES document at Daylight's web site (13). Even with this simplified subset of rules, SMILES strings can be derived for a lot of molecules. **Table 2.1** illustrates just a few of them.

Table 2.1
Illustrative SMILES: molecular structures and the corresponding SMILES strings are paired vertically. The numbered arrows on the three cyclic molecular structures are not part of the molecules. They are used to indicate the break points for deriving the corresponding SMILES strings (*see* text)

			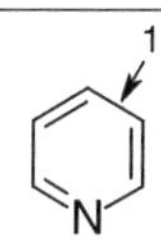
CCC	CC=C	CC#C	c1ccncc1
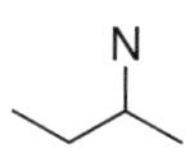	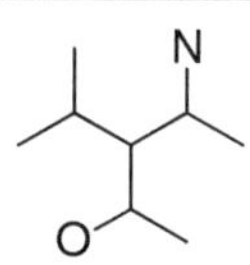	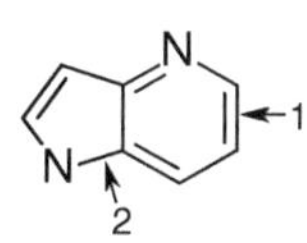	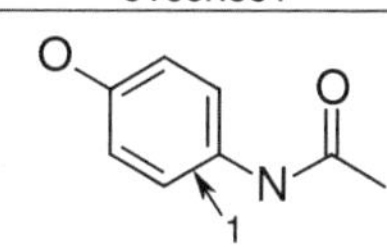
CCC(C)N	CC(C)C(C(C)N)C(C)O	c1cc2c(cc[nH]2)nc1	CC(=O)Nc1ccc(cc1)O

Note that a single molecule may correspond to many different, but equivalent, SMILES strings. For example, for a given asymmetric molecule, starting from a different asymmetric atom will lead to a different, but equally valid, SMILES string. These various SMILES are called isomeric SMILES. They can be converted to a unique form called canonical SMILES (11).

Daylight has extended SMILES rules to accommodate general descriptions of molecular patterns and chemical reactions (13). These SMILES extensions are called SMARTS and SMIRKS. SMARTS is a language for describing molecular patterns while SMIRKS defines rules for chemical reaction transformations.

SMILES strings are very concise and hence are suitable for storing and transporting a large number of molecular structures, while MOLfiles and its extension SDFiles have the option to store more complicated molecular data such as 3D molecular conformational information and biological data associated with the molecules. There are many other file formats not discussed here. Interested readers can find a list of file types at the following web site: http://www.ch.ic.ac.uk/chemime/.

2.2. Data, Databases, and Data Mining

Modern drug discovery is largely a data-driven process. There are tremendous amounts of data collected to facilitate decision making at almost every stage of the drug discovery process. Majority of the data are associated with molecules. These molecular data can be classified into two broad categories: physicochemical properties and biological assay data. Typical physicochemical properties for a molecule include molecular weight, number of heavy atoms, number of rings, number of hydrogen bond donors/acceptors, number of oxygen or nitrogen atoms, polar/nonpolar surface area, volume, water solubility, 1-octanol–water partition coefficient (CLogP), pK_a, and molecular stability. Most of these properties can be calculated while some are measured experimentally.

Biological data associated with small molecules come from a heterogeneous array of assays. Typical biological assay data include percentage inhibitions from high-throughput screening of binding assays against specific biological targets, biochemical binding constants, activity IC50 constants in cell-based assays, percentage inhibitions or binding constants against various CYP 450 proteins as first screening for metabolic liabilities, compound stabilities in human/animal microsome and hepatocytes, trans-membrane permeabilities (such as Caco-2 or PAMPA), dofetilide binding constants for finding potential hERG blockers (may cause prolongation of QT interval), genotoxicity data from assays like AMES tests, and various pharmacokinetic and pharmacodynamic data. Different biological assays vary greatly in experimental modes (biochemical, in vitro, in vivo, etc.), readout accuracies, and throughputs. Therefore, some types of data are abundant while others are only available very scarcely.

Computational models can be built based on experimental results for both physicochemical properties and biological assays. Thereby predicted physicochemical properties and biological assay data become available to compounds before their syntheses or to compounds without the data because of various experimental limitations such as cost or throughput. These computed data become an integral part of the molecular data.

Molecular data are usually stored in databases along with their corresponding molecular structures. Database is the central part of a typical chemoinformatics system that furthermore consists of interfaces and programs for capturing, storing, manipulating, and retrieving data. Careful data modeling for designing a robust chemoinformatics system integrating various heterogeneous molecular data is essential for the chemoinformatics system to deliver its designed functions with acceptable performances (14).

Data mining is to seek patterns among a given set of data. Mining molecular data to aid molecular design is one of the most important functions of a chemoinformatics system. Typical data mining tasks in drug discovery include subsetting libraries; identifying lead chemical series from HTS data (HTS hit triage); querying databases for similar compounds in terms of structural patterns, activity profiles across various biological targets, or property profiles across various physicochemical properties; and establishing quantitative structure–activity relationships (QSAR) or quantitative structure–property relationships (QSPR). In a general sense, drug design is an ideal field of applications for chemical data mining. Therefore, most of the drug design tools are actually chemical data mining tools.

2.3. Molecular Descriptors

To distinguish one molecule from another in computer and to establish various predictive QSAR/QSPR models for design purposes, molecules need to be projected into a chemical space of molecular characteristics. This projection is usually done through molecular descriptors. Given the diverse molecular characterizations, it is not an easy task to give a simple definition for all molecular descriptors. A formal definition of the molecular descriptor is given by Todeschini and Consonni as follows: *molecular descriptor* is the final result of a logic and mathematical procedure which transforms chemical information encoded within a symbolic representation of a molecule into a useful number or the result of some standardized experiment (15). Here the term "useful" has two meanings: the number can give more insight into the interpretation of the molecular properties and/or it is able to take part in a model for the prediction of some interesting property of other molecules.

Molecular descriptors vary greatly in both their origins and their applications. They come from both experimental measurements and theoretical computations. Typical molecular descriptors from experimental measurements include logP, aqueous solubility, molar refractivity, dipole moment, polarizability, Hammett substituent constants, and other empirical physicochemical properties. Notice that the majority of experimental descriptors are for entire molecules and come directly from experimental measurements. A few of them, such as various substituent constants, are for molecular fragments attached to certain molecular templates and they are derived from experimental results.

Theoretical molecular descriptors cover much broader varieties and usually are more readily available even though the complexity of their computational procedures may vary widely. Major classes of computed molecular descriptors include the following:

(i) Constitutional counts such as molecular weight, number of heavy atoms, number of rotatable bonds, number of rings, and number of aromatic rings.

(ii) 2D molecular properties such as number of hydrogen bond donor/acceptor and their strengths, number of polar atoms.

(iii) Topological descriptors from graph theory such as various graph-theoretic invariants of molecular graphs, 2D and 3D autocorrelations, various property-weighted graph-theoretic quantities.

(iv) Geometrical descriptors such as shape, radius of gyration, moments of inertia, volume, polar/nonpolar surface areas.

(v) Electrostatic properties such as dipole moment, partial atomic charges.

(vi) Fingerprints such as 2D fingerprints like Daylight fingerprints and UNITY fingerprints and 3D fingerprints like pharmacophore fingerprints.

(vii) Quantum chemical descriptors such as HOMO/LUMO energies, E-state values.

(viii) Predicted physicochemical properties such as calculated solubility, calculated logP, and various molecular properties from QSPR predictive models.

There are also various hybrid descriptors. For example, electrotopological descriptors are a hybrid of topological and electronic descriptors.

Applications of molecular descriptors are as diverse as their definitions. The important classes of applications include QSAR and/or QSPR, similarity, diversity, predictive models for virtual screening and/or data mining, data visualization. We will discuss briefly some of these applications in the next sections.

There are literally thousands of molecular descriptors available for various applications. We have only mentioned a few of them in previous paragraphs. Interested readers can find a more complete coverage of molecular descriptors in reference (15), which gives definitions for 3,300 molecular descriptors. Many software, or subroutines as an integral part of other programs, are available to generate various types of molecular descriptors. **Table 2.2** lists a few of these software.

2.4. Chemical Space and Dimension Reduction

Molecular descriptors for a given molecule can be considered as its coordinates in a multidimensional chemical space. Since

Table 2.2
A selected list of software for computing molecular descriptors[a]

Software name	Type of descriptors	Number of descriptors	Distributor (and/or author)	Reference
ADAPT	Topological, electronic, geometric, and some combination	>264	Peter Jurs, Penn State University	http://research.chem.psu.edu/pcjgroup/desccode.html
ADMET Predictor	Constitutional, functional group counts, topological, E-state, Moriguchi descriptors, Meylan flags, molecular patterns, electronic properties, 3D descriptors, hydrogen bonding, acid–base ionization, empirical estimates of quantum descriptors	297	Simulations Plus	http://www.simulations-plus.com/
ADRIANA.code	Global physicochemical descriptors, size and shape descriptors, atom property-weighted 2D and 3D autocorrelations and RDF, surface property-weighted autocorrelations	1,244	Molecular Networks	http://www.molecular-networks.com/products/adrianacode
CODESSA	Constitutional, topological, geometrical, electrostatic, surface property, quantum chemical, and thermodynamic descriptors	1,500	Alan R. Katritzky, Mati Karelson, and Ruslan Petrukhin, University of Florida	http://www.codessa-pro.com/index.htm
DRAGON	Constitutional descriptors, topological descriptors, walk and path counts, connectivity indices, information indices, 2D autocorrelations, edge adjacency indices, BCUT descriptors, topological charge indices, eigenvalue-based indices, Randic molecular profiles, geometrical descriptors, RDF descriptors, 3D-MoRSE descriptors, WHIM descriptors, GETAWAY descriptors, functional group counts, atom-lefted fragments, charge descriptors, molecular properties, 2D binary fingerprints, 2D frequency fingerprints	3,224	Talete srl, Milano, Italy	Reference for its web version, E-DRAGON: Tetko, I. V., et al. (2005) *J Comput Aid Mol Des* **19**, 453–463. http://www.talete.mi.it/products/dragon_description.htm

Table 2.2 (continued)

Software name	Type of descriptors	Number of descriptors	Distributor (and/or author)	Reference
JOELib	Counting, topological, geometrical, properties, etc.	68	J.K. Wegner, University of Tübingen, Germany	http://www.ra.cs.uni-tuebingen.de/software/joelib/index.html
MOE	Topological, structural keys, E-state, physical properties, surface area descriptors including CCG's VSA descriptors, etc.	>600	Chemical Computing Group	http://www.chemcomp.com/
Molconn-Z	Topological and E-state	>40	eduSoft	http://www.edusoft-lc.com/molconn/
Molgen-QSPR	Constitutional, topological, and geometrical	708	J. Braun, M. Meringer, and C. Rücker, University of Bayreuth, Germany	http://www.molgen.de/?src=documents/molgenqspr.html
PowerMV	Constitutional, atom pairs, fingerprints, BCUT, etc.	>1,000	J. Liu, J. Feng, A. Brooks, and S. Young, National Institute of Statistical Sciences, USA	http://nisla05.niss.org/PowerMV/
PreADMET	Constitutional, topological, geometrical, physicochemical, etc.	1,081	Bioinformatics & Molecular Design Research Center, South Korea	http://preadmet.bmdrc.org/index.php?option=com_content&view=frontpage&Itemid=1
Sarchitect	Constitutional, property-based, 2D topological, and 3D conformational descriptors	1,084	Strand Life Sciences, India	http://www.strandls.com/sarchitect/index.html
TAM	Topological	>20	M. Šaric-Medic. et al., University of Zagreb, Croatia	Vedrina, M., et al. (1997) *Computers & Chem* **21**, 355–361
TOPIX	Constitutional and topological	130	D. Svozil and H. Lohninger, Epina Software Labs, Austria	http://www.lohninger.com/topix.html

[a] *See* complete list at http://www.moleculardescriptors.eu/softwares/softwares.htm

value ranges for different descriptors may substantially differ for a given data set, it is desirable to scale (or normalize) descriptors selected before any mathematical manipulations. Another scenario for rescaling descriptors is to use weighting factors to differentiate important descriptors from unimportant ones. Therefore, a scaled individual descriptor is represented by one dimension in this multidimensional space and each molecule is represented by a single point in such a space. The distance between two molecules is often defined as their Euclidean distance in this space.

Chemical space so defined is highly degenerate because of the high redundancy of various molecular descriptors. For example, molecular weight is highly correlated with the number of heavy atoms. The high degeneracy along with the high dimensionality of the molecular descriptor space poses a real challenge to many applications of molecular descriptors. Therefore, dimension reduction of a chemical space is not only important to identifying key factors affecting the trends in various predictive models but also necessary for efficient mathematical manipulations during model developments.

It is evidently beneficial and easy to remove those trivial descriptors with constant or near-constant values across all molecules. To further eliminate duplication and redundancy of descriptors for a given data set, statistical methods, such as principal component analysis (PCA) (16), multidimensional scaling (MDS) (17), or nonlinear mapping (NLM) (18), can be very helpful for dimensionality reduction. PCA is a method of identifying patterns in a data set and expressing the data in such a way as to highlight their similarities and differences. It is able to find linear combinations of the variables (the so-called principal components) that correspond to directions of the maximal spread in the data. On the other hand, MDS is a method that represents measurements of similarity (or dissimilarity) among pairs of objects as distances between points of a low-dimensional multidimensional space. It preserves the original pairwise interrelationships as closely as possible. Finally, NLM tries to preserve distances between points as similar as possible to the actual distances in the original space. The NLM procedure for performing this transformation is as follows: compute interpoint distances in the original space; choose an initial configuration (generally random) in the display space (i.e., the target and low-dimensional space); calculate a mapping error from the distances in the two spaces; and modify iteratively the coordinates of points in the display space by means of a nonlinear procedure so as to minimize the mapping error. PCA is a linear method while both MDS and NLM are nonlinear methods. All these methods endeavor to optimally preserve information while reducing the dimensionality of the descriptor space (hence the mathematical complexity).

Reducing the dimensionality of the descriptor space not only facilitates model building with molecular descriptors but also makes data visualization and identification of key variables in various models possible. Notice that while a low dimension mathematically simplifies a problem such as model development or data visualization, it is usually more difficult to correlate trends directly with physical descriptors, and hence the data become less interpretable, after the dimension transformation. Trends directly linked with physical descriptors provide simple guidance for molecular modifications during potency/property optimizations.

2.5. Similarity and Diversity

Molecules with similar structures should behave similarly while it is more efficient to use a diverse set of compounds to cover a broad range of chemical space. Chemical similarity and diversity are interesting because even a fuzzy understanding of these concepts can aid the design of useful molecules. For example, similarity probe is essential to analogue designs during lead optimization while enough diversity of a chemical collection is critical to the successful lead generation through high-throughput screening (HTS) (19).

The quantification of molecular similarity generally involves three components: molecular descriptors to characterize the molecules, weighting factors to differentiate more important characteristics from less important ones, and the similarity coefficient to quantify the degree of similarity between pairs of molecules (20, 21). The first two components are related to the definition of chemical space as discussed in Section **2.4**. Therefore, it is natural to assume that structurally similar molecules should cluster together in a chemical space, and to define the similarity coefficient of a pair of molecules to be the distance between them in the chemical space. The shorter the distance is the more similar the pair is.

Because of the numerous choices for molecular descriptors, weighting factors, and similarity coefficients, there are many ways in which the similarities between pairs of molecules can be calculated. The most used molecular descriptors for defining similarity are probably the 2D fingerprints (22). The bit strings of the molecular fingerprints are used to calculate similarity coefficients. **Table 2.3** lists several selected similarity coefficients that can be used with various 2D fingerprints (23). The Tanimoto coefficient is the most popular one (22).

A related concept to similarity is dissimilarity. Dissimilarity can be considered as the opposite of similarity. It is also defined by the distance between two molecules in a chemical space. The larger the distance between the two molecules is the more dissimilar the pair is. Sometimes, dissimilarity is used interchangeably with diversity in literature even though there are subtle

Table 2.3
Selected similarity coefficients to be used with 2D fingerprints for molecule pair (A, B)

Coefficient	Expression[a]	Value range	Notes
Tanimoto	$\frac{c}{a+b+c}$	0.0–1.0	
Cosine	$\frac{c}{\sqrt{(a+b)(b+c)}}$	0.0–1.0	
Hamming	$a+b$	0.0–∞	This is a dissimilarity coefficient
Russell–Rao	$\frac{c}{a+b+c+d}$	0.0–1.0	
Forbes	$\frac{(a+b+c+d)c}{(a+c)(b+c)}$	0.0–∞	
Pearson	$\frac{cd-ab}{\sqrt{(a+c)(b+c)(a+d)(d+d)}}$	−1.0–1.0	
Simpson	$\frac{c}{\min\{(a+c),(b+c)\}}$	0.0–1.0	
Euclid	$\sqrt{\frac{c+d}{a+b+c+d}}$	0.0–1.0	

[a]a is the count of bits that is "on" in A string but "off" in B string; b is the count of bits that is "off" in A string but"on" in B string; c is the count of bits that is "on" in both A string and B string; d is the count of bits that is "off" in both A string and B string.

differences between diversity and dissimilarity. Diversity is a property of a molecular collection while dissimilarity can be defined for pairs of molecules as well.

Since diversity is a collective property, its precise quantification requires a mathematical description of the distribution of the molecular collection in a chemical space. When a set of molecules are considered to be more diverse than another, the molecules in this set cover more chemical space and/or the molecules distribute more evenly in chemical space. Historically, diversity analysis is closely linked to compound selection and combinatorial library design. In reality, library design is also a selection process, selecting compounds from a virtual library before synthesis. There are three main categories of selection procedures for building a diverse set of compounds: cluster-based selection, partition-based selection, and dissimilarity-based selection.

The cluster-based selection procedure starts with classifying compounds into clusters of similar molecules with a clustering algorithm followed by selection of representative(s) from each cluster (24). On the other hand, the partition-based selection procedure partitions chemical space into cells by dividing values of each dimension into various intervals and selects representative compounds from each cell (25). Because of the exponential dependence of cell numbers on dimensions of the chemical space, the partition-based selection procedure is only suitable for applications in a low-dimensional chemical space. Hence, most representative molecular descriptors need to be identified to form the chemical space, or the dimension reduction as described in Section **2.4** needs to be performed before the partition-based selection procedure can be used. In addition to the cell-based partitioning, statistical partitioning methods, such as decision tree method (26), are also used for classification. Finally, the dissimilarity-based selection procedure iteratively selects compounds that are as dissimilar as possible to those already selected (27). This method tends to select molecules with more complexity as well as a diverse set of chemical cores. For combinatorial library design, there is also an optimization-based selection procedure to select compounds from virtual libraries. It formulates the compound selection as an optimization problem with some quantitative measures of diversity (*see*, for example, reference (28)).

2.6. QSAR and QSPR

Building predictive QSAR and QSPR models is a cost-effective way to estimate biological activities, physicochemical properties such as partition coefficients and solubility, and more complicated pharmaceutical endpoints such as metabolic stability and volume of distribution. It seems to be reasonable to assume that structurally similar molecules should behave similarly. That is, similar molecules should have similar biological activities and physicochemical properties. This is the (Q)SAR/(Q)SPR hypothesis. Qualitatively, both molecular interactions and molecular properties are determined by, and therefore are functions of, molecular structures. Or

$$\text{Activity} = f_1(\text{mol structure/descriptors}) \quad [1]$$

and

$$\text{Property} = f_2(\text{mol structure/descriptors}) \quad [2]$$

There is a long history of efforts to find simple and interpretable f_1 and f_2 functions for various activities and properties (29, 30). The quest for predictive QSAR models started with Hammett's pioneer work to correlate molecular structures with chemical reactivities (30–32). However, the widespread applications of modern predictive QSAR and QSPR actually started with the seminal work of Hansch and coworkers on pesticides (29, 33, 34) and the developments of various powerful analysis tools, such as PLS (partial least squares) and neural networks, for multivariate analysis have fueled these widespread applications. Nowadays, numerous publications on guidelines, workflows, and common errors for building predictive QSAR and QSPR models, not to mention the countless papers of applications, are well documented in literature (35–41).

In principle, a valid QSAR/QSPR model should contain the following information (39): (i) a defined endpoint; (ii) an unambiguous algorithm; (iii) a defined domain of applicability; (iv) appropriate measures of goodness of fit, robustness, and predictivity; and (v) a mechanistic interpretation, if possible. Building predictive QSAR/QSPR models is a process from experimental data to model and to predictions. Collecting reliable experimental data (and subdividing the data into training set and testing set) is the first step of the model-building process. The second step of the process is usually to select relevant parameters (or molecular descriptors) that are most responsive to the variation of activities (or properties) in the data set. The third step is QSAR/QSPR modeling and model validation. Finally, the validated models are applied to make predictions. Usually, the second and the third, and sometimes the first, steps are repeated to select the best combination of parameter set and models (*see*, for example, reference (40)). Although majority of QSAR/QSPR models are built with molecular descriptors, there are parameter-free models. For example, the Free–Wilson method builds predictive QSAR/QSPR models for a series of substituted compounds without any molecular descriptors (42). Its drawback is that the Free–Wilson method requires a data set for almost all combinations of substituents at all substituted sites and the method is not applicable to molecular set of noncongeners.

It is interesting to note that various QSAR/QSPR models from an array of methods can be very different in both complexity and predictivity. For example, a simple QSPR equation with three parameters can predict logP within one unit of measured values (43) while a complex hybrid mixture discriminant analysis–random forest model with 31 computed descriptors can only predict the volume of distribution of drugs in humans within about twofolds of experimental values (44). The volume of distribution is a more complex property than partition coefficient. The former is a physiological property and has a much higher uncertainty in its experimental measurements while logP is a much simpler physicochemical property and can be measured more accurately. These and other factors can dictate whether a good predictive model can be built.

2.7. Multiobjective Optimization

The ultimate goal of a small-molecule drug discovery program is to establish an acceptable pharmacological profile for a drug candidate. To achieve this goal, usually many pharmacological attributes, or their numerous surrogates, of individual lead compounds need to be optimized either sequentially or in parallel. That is, drug discovery itself is a multiobjective optimization process (45). Furthermore, the multiobjective optimization is also involved in both various stages of the drug discovery process and many drug discovery enabling technologies. For example, to design libraries for lead generation or lead optimization, multiple physicochemical properties need to be optimized along with diversity and similarity (46–49). It is also a common practice to test multiple hypotheses in a single SAR/SPR run during lead optimization. The algorithms for solving these various multiobjective optimization problems can be quite similar even though the properties to be optimized are evaluated very differently, ranging from simple computations to complex in vivo experiments.

When optimizing multiple objectives, usually there is no best solution that has optimal values for all, and oftentimes competing, objectives. Instead, some compromises need to be made among various objectives. If a solution "A" is better than another solution "B" for every objective, then solution "B" is dominated by "A." If a solution is not dominated by any other solution, then it is a nondominated solution. These nondominated solutions are called Pareto-optimal solutions, and very good compromises for a multiobjective optimization problem can be chosen among this set of solutions. Many methods have been developed and continue to be developed to find Pareto-optimal solutions and/or their approximations (*see*, for example, references (50–52)). Notice that solutions in the Pareto-optimal set cannot be improved on one objective without compromising another objective.

Searching for Pareto-optimal solutions can be computationally very expensive, especially when too many objectives are to be optimized. Therefore, it is very appealing to convert a multiobjective optimization problem into a much simpler single-objective optimization problem by combining the multiple objectives into a single objective function as follows (53–55):

$$F(Obj_1, Obj_2, \ldots, Obj_n) = \sum_{i=1}^{n} w_i f_i(Obj_1) \quad [3]$$

where w_i is a weighting factor that reflects the relative importance of *i*th objective among all objectives. With this conversion, all algorithms used for single-objective optimizations can be applied to find optimal solutions as prescribed by equation [3]. Notice that both functional forms for $\{f_i\}$ and weighting factors w_i in equation [3] may be attenuated to achieve optimal results, when enough data are available for testing and validating (55).

It is a common practice in early drug discovery to select compounds by some very simple filters such as rule-of-five and "drug-likeness" (56, 57). For example, multiobjective optimization methods have been applied to design combinatorial libraries of "drug-like" compounds (53, 54).

2.8. Virtual Screening

Virtual screening (VS) has emerged as an important tool for drug discovery (58–67). The goal of VS is to separate active from inactive molecules in a compound collection and/or a virtual library through rapid in silico evaluations of their activities against a biological target. A full VS process generally involves three components: a library to be screened, an in silico "assay" to test the activities of molecules in the library, and a hit follow-up plan to experimentally verify the activities (*see* **Fig. 2.2**). The in silico "assay" is the core component of VS. The other two components are also very important for a successful VS campaign.

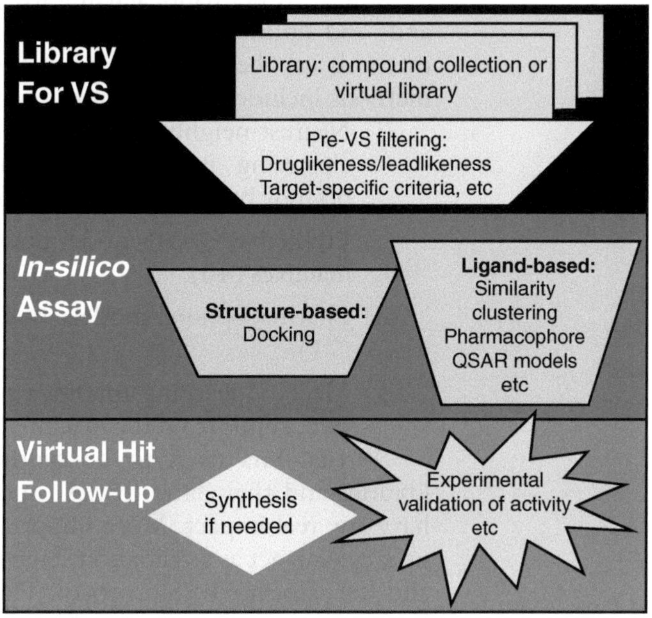

Fig. 2.2. Three components of a typical VS process: compound library, virtual "assay," and hit follow-up for virtual hits.

A compound library for VS could be a corporate compound collection, a public compound collection such as NCI's compound library (68), a collection of commercially available compounds, or a virtual library of synthesizable compounds. Nowadays, a corporate compound collection has a typical size of 10^6 compounds. More often than not, various filters are applied, for example, to filter out non-drug-like/non-lead-like compounds, thereby to reduce the library size for VS (56, 57, 64–67). Prefiltering becomes imperative for screening large virtual libraries within a reasonable period of time. It is obviously crucial to have target-relevant molecules in the library. Actually, library design in large part is to cover enough chemical space of these biologically relevant molecules.

Computational methods acting as in silico assay can be roughly classified into two major categories: structure based and ligand based (58–67). The structure-based methods require the knowledge of target structures. The most common structure-based approach is to dock each small molecule onto the active site of a target structure to determine its binding affinity (or docking score) to the target (69). A wide array of docking methods and their associated scoring functions are available for screening large libraries (70). The less-used methods in structure-based virtual screening include VS with pharmacophores built from target structures and the low-throughput free energy computations for ligand–receptor complexes via molecular dynamics or Monte Carlo simulations.

On the other hand, starting from knowledge about active ligands and optionally inactive compounds, various computational methods can be used to find related active compounds. These methods include the following:

(i) Nearest-neighbor methods such as similarity methods and clustering methods (assuming chemically similar compounds behave similarly biologically)

(ii) Predictive QSAR model built from actives and optionally inactives (40)

(iii) Pharmacophore models built from actives and inactives (71)

(iv) Machine learning methods such as classification, decision tree, support vector machine, and neural networks (72)

Virtual hits need to be synthesized for hits from virtual libraries and their bioactivities experimentally verified for VS to have any real impact. More importantly, these virtual hit follow-up steps can act as a validation stage for the computational models and the associated VS protocol. The results of experimental verification can be fed back to the in silico assay stage for building better predictive in silico models.

3. Library and Library Design

3.1. Compound Library for Drug Discovery

There are two major classes of libraries for drug discovery: diverse libraries for lead discovery and focused libraries for lead optimization. Lead discovery libraries emphasize diversity while lead optimization libraries prefer similar compounds. The purpose of lead discovery libraries is to find lead matter and to provide potential active compounds for further optimization. Without any prior knowledge about the active compounds for a given target, it is reasonable to start with a library of enough chemical space coverage to demarcate the biologically relevant chemical space for the target. Therefore, libraries for lead discovery often comprise diverse compounds with drug-like/lead-like properties. Lead matters without proper drug-likeness/lead-likeness properties might be trapped in a local and "unoptimizable" zone of the chemical space during lead optimization stage. On the other hand, the purpose of lead optimization libraries is to improve the activity and the property profile of the lead matter. With a lead compound, searching for better and optimized compounds is usually performed among similar compounds with limited diversity around the lead molecule in the chemical space.

There are three major sources for a typical corporate compound collection: project-specific compounds accumulated over a long period of time through medicinal chemistry efforts for various therapeutic projects, individual compounds from commercial sources, and compounds from combinatorial chemistry. In practice, compound collections are often divided into subsets, for example, the diverse subsets for general HTS and target-focused subsets (such as kinase libraries or GPCR libraries). For library design, diversity and similarity are generally built into the libraries of compounds to be synthesized and/or purchased (73).

Stimulated by the widespread applications of HTS technologies, combinatorial chemistry has provided a powerful tool for rapidly adding large number of compounds to corporate collections for many pharmaceutical companies. Virtual combinatorial library consists of libraries from individual reactions and compounds from a single reaction share a common product core (*see* **Fig. 2.3**). The number of compounds in a combinatorial library can grow rapidly with number of reaction components and numbers of reactants for individual components. For example, a full combinatorial library from a three-component reaction

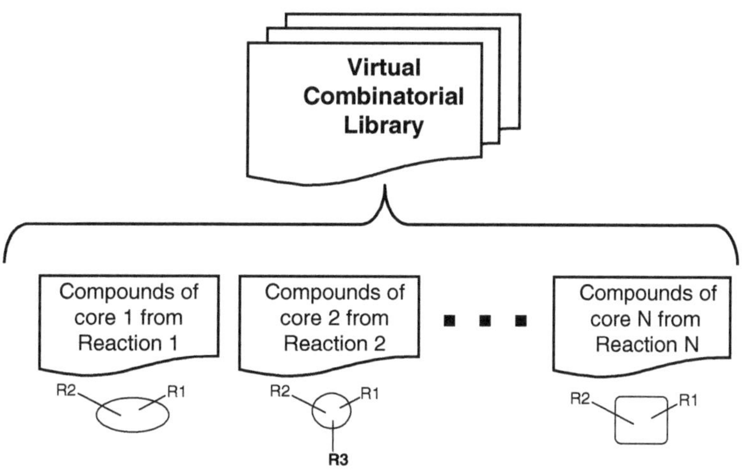

Fig. 2.3. Virtual combinatorial library is the start point for any combinatorial library design. It consists of libraries from individual reactions. Compounds from a given reaction share a unique product core.

with 200 reactants for each component would contain 8 million products. Virtual library can also be represented as a template with R-groups attached at various variation sites. This representation is also called Markush structure. Markush structure is the standard chemical structure often used in chemical patents. Template-based libraries can be considered as a generalization of the scenario shown in **Fig. 2.3** where the product cores of individual reactions are the templates. Notice that reaction-based virtual libraries have explicit chemistries for compound syntheses and therefore may include only those synthesizable compounds through careful selections of reactants while general template-based virtual libraries usually do not indicate chemistry accessibilities of the compounds.

3.2. Library Enumeration

The product structures of a combinatorial library can be formed from product core and structures of reactants or by attaching R-groups to the various variation sites of a template (*see* **Fig. 2.4**). Product formation is conventionally called product enumeration. It is accomplished by removing the leaving groups of reactants, a process also called clipping, followed by pasting the retained fragments at the variation sites of the product core or the template. For template-based enumerations, the R-groups, generated either by molecular fragmentation programs or from molecular clipping, are usually listed as part of the library definition. There are many automatic tools for library enumeration, either as standalone software or as subroutines of other application packages (*see*, for example, references (74–78)).

With product structures, many chemoinformatics tools can be applied to filter the virtual libraries and to select a few use-

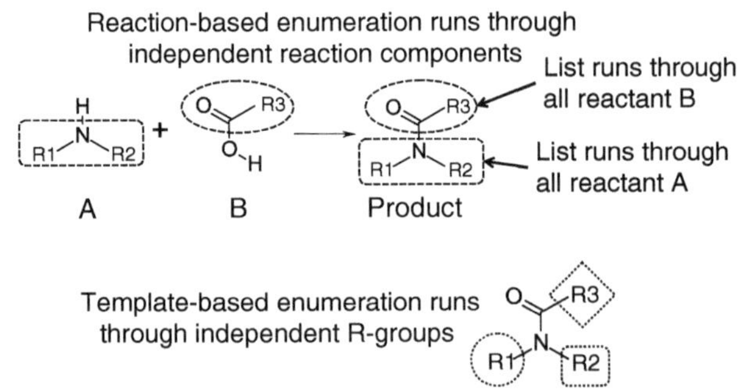

Fig. 2.4. Product enumerations of a combinatorial library. For reaction-based enumeration, individual groups of –N(R1)(R2) and –(CO)-R3 are replaced by corresponding molecular fragments from reactants A and B. For template-based enumeration, the R-groups R1, R2, and R3 are replaced by independent lists of molecular fragments. Note that some combinations of R1 and R2 may not exist in component A for reaction-based enumerations. The template-based product structure with R-groups is also called Markush structure and its enumeration is called Markush enumeration or Markush exemplification.

ful compounds for syntheses. These filtering tools include the various tools and methods discussed in the previous section (*see* Sections **2.5–2.8**). Multiobjective optimization algorithms can be used to design combinatorial libraries with optimal diversity/similarity, cost efficiency, and physical properties (46–47). Nevertheless, library design can be performed without the full enumeration of the entire virtual combinatorial libraries (79–80).

3.3. Library Design

Library design is a compound selection process that maximizes the number of compounds with desirable attributes while minimizing the number of compounds with undesirable characteristics. The "desirability" of compounds in a library is defined by the ultimate usage of the library and the cost efficiency for producing the library. Therefore, libraries for lead discovery demand sufficient diversity among compounds selected while lead optimization libraries usually contain compounds similar to those lead compounds. The diversity of a compound collection can be improved through inclusion of more diverse chemotypes/scaffolds and side chains (81–82). Diversity in chemotypes and scaffolds is usually derived from more reactions with novel chemistries. Diversity in side chains can be achieved by selecting more diverse reagents for a given reaction. It is well recognized that probability of finding effective ligand–receptor interactions decreases as a molecule becomes more complex (83). That is, relatively simple molecules from diverse chemotypes/scaffolds have a better chance than those complex molecules derived from diverse side chains to generate lead matter with more specific ligand–receptor interactions. Therefore, the current practice of building a diverse compound collection prefers more small libraries of many diverse novel chemistries to less large libraries of a few chemistries.

Another important consideration in designing a library is cost efficiency. Inexpensive reagents should always be favored as reactants for library production. Producing a library of a full combinatorial array is much more cost-effective than synthesizing cherry-picked singletons. Selections in a library design can be product based or reactant based. In a product-based design library, compounds are chosen purely based on their own properties. On the other hand, the reactant-based design chooses reactants, instead of library products directly, based on the collective properties of the associated products (47, 84). Reactant-based design generally leads to libraries of full combinatorial arrays. While costly, product-based design is more effective than reactant-based design in achieving optimal design objectives other than cost (47, 84). This seems to be obvious since limiting product choices to a subarray of a full combinatorial library will compromise other design objectives, unless the selected reactants are so dominant that the products derived from their combinations are superior to other products with respect to all objectives.

Frequently, library design involves simultaneous optimization of multiple objectives, among which diversity, similarity, and cost efficiency are three examples. Other typical properties include the "rule-of-five" properties (molecular weight, logP value, number of hydrogen bond donors, and total number of "N" and "O" atoms), polar surface area, and solubility. Complicated properties from predictive models can also be included. Library design in large part is actually a multiobjective optimization problem. Therefore, all methods discussed in Section **2.7** can be applied to library design.

To summarize, library design involves choices of diversity vs. similarity, product based vs. reactant based, and single objective vs. multiobjective optimizations. Chemoinformatics tools, such as various predictive models and chemoinformatics infrastructures, can be utilized to facilitate the selection process of library design.

4. Concluding Remarks

Library design has become an integral part of drug discovery process. Chemical library design underwent a transformation from a pure tool for supplying vast number of compounds to a power tool for generating quality leads and drug candidates. Although the controversy of how to define a best set of compounds for lead generation is not completely resolved, tremendous progress has been made to find biologically relevant subregions of the chemical space, particularly when confined to a target or a target family (*see*, for example, references (85, 86)). Providing biologically relevant compounds will continue to be one of the main goals of library design.

Since modern drug discovery is mainly a data-driven process and chemoinformatics is at the center of data integration and utilization, it is natural that majority of library design tools are chemoinformatics tools. Therefore, a deep understanding of chemoinformatics is necessary for taking full advantage of library technologies.

Though relatively mature, chemoinformatics is still an active field of intensive research. Numerous new methods and tools continue to be developed. Here we have selectively covered, without giving too many details, a few topics important to library design. Actually the interplays and costimulations of chemoinformatics with library design have been well documented in literature. We hope that the brief introduction in this chapter can serve as a guide for you to enter into the exciting field of chemoinformatics and its applications to chemical library design.

Acknowledgment

The chapter was prepared when the author was visiting with professor Andy McCammon's group. The author is very grateful to Professor Andy McCammon and his group for the exciting and stimulating scientific environment during the preparation of the chapter.

References

1. Brown, F. B. (1998) Chemoinformatics: what is it and how does it impact drug discovery. *Annu Rep Med Chem* **33**, 375–384.
2. Bohacek, R. S., McMartin, C., Guida, W. C. (1996) The art and practice of structure-based drug design: a molecular modeling perspective. *Med Res Rev* **16**, 3–50.
3. Walters, W. P., Stahl, M. T., Murcho, M. A. (1998) Virtual screening–an overview. *Drug Discov Today* **3**, 160–178.
4. Gasteiger, J. (ed.) (2003) *Handbook of Chemoinformatics: From Data to Knowledge*, Wiley-VCH, Weinhiem.
5. Bajorath, J. (ed.) (2004) *Chemoinformatics: Concepts, Methods, and Tools for Drug Discovery*, Humana Press, Totowa, NJ.
6. Oprea, T. I. (ed.) (2005) *Chemoinformatics in Drug Discovery*, Wiley-VCH, Weinheim.
7. Leach, A. R. and Gillet, V. J. (2007) *An Introduction to Chemoinformatics*, Springer, London.
8. Bunin, B. A., Siesel, B., Morales, G. A., Bajorath, J. (2007) *Chemoinformatics: Theory, Practice, & Products*, Springer, The Netherlands.
9. http://www.symyx.com/solutions/white_papers/ctfile_formats.jsp, last accessed February, 2010.
10. Weininger, D. (1988) SMILES, a chemical language and information system. 1. Introduction to methodology and encoding rules. *J Chem Inf Comput Sci* **28**, 31–36.
11. Weininger, D. (1989) SMILES, 2. Algorithm for generation of unique SMILES notation. *J Chem Inf Comput Sci* **29**, 97–101.
12. Weininger, D. (1990) SMILES, 3. Depict. Graphical depiction of chemical structures. *J Chem Inf Comput Sci* **30**, 237–243.
13. http://www.daylight.com/dayhtml/doc/theory/theory.smiles.html, last accessed February, 2010.
14. Simsion, G. C., Witt, G. C. (2001) *Data Modeling Essentials*, 2nd ed. Coriolis, Scottsdale, USA.
15. Todeschini, R., Consonni, V. (2009) *Molecular Descriptors for Chemoinformatics Vol. 1*, 2nd ed. Wiley-VCH, Weinheim, Germany.
16. Jolliffe, I. T. (2002) *Principal Component Analysis*, 2nd ed. Springer, New York.
17. Borg, I. and Groenen, P. J. F. (2005) *Modern Multidimensional Scaling: Theory and Applications*, 2nd ed. Springer, New York.
18. Domine, D., Devillers, J., Chastrette, M., Karcher, W. (1993) Non-linear mapping for structure-activity and structure-property modeling. *J Chemometrics* **7**, 227–242.
19. Wermuth, C. G. (2006) Similarity in drugs: reflections on analogue design. *Drug Discov Today* **11**, 348–354.
20. Willett, P. (2000) Chemoinformatics–similarity and diversity in chemical libraries. *Curr Opin Biotech* **11**, 85–88.
21. Maldonado, A. G., Doucet, J. P., Petitjean, M., Fan, B. -T. (2006) Molecular similarity and diversity in chemoinformatics: from theory to applications. *Mol Divers* **10**, 39–79.
22. Willett, P. (2006) Similarity-based virtual screening using 2D fingerprints. *Drug Discov Today* **11**, 1046–1053.
23. Holliday, J. D., Hu, C. -Y., Willett, P. (2002) Grouping of coefficients for the calculation of inter-molecular similarity and dissimilarity using 2D fragment bitstrings. *Comb Chem High Throughput Screening* **5**, 155–166.
24. Dunbar J. B. (1997) Cluster-based selection. *Perspect Drug Discov Des* **7/8**, 51–63.
25. Mason J. S., Pickett S. D. (1997) Partition-based selection *Perspect Drug Discov Des* **7/8**, 85–114.
26. Rusinko, A. III, Farmen, M. W., Lambert, C. G. et al. (1999) Analysis of a large structure/biological activity dataset using recursive partitioning. *J Chem Inf Comput Sci* **39**, 1017–1026.
27. Lajiness, M. S. (1997) Dissimilarity-based compound selection techniques. *Perspect Drug Discov Des* **7/8**, 65–84.
28. Pickett, S. D., Luttman, C., Guerin, V., Laoui, A., James, E. (1998) DIVSEL and

COMPLIB–strategies for the design and comparison of combinatorial libraries using pharmacophore descriptors. *J Chem Inf Comput Sci* **38**, 144–150.
29. Hansch, C., Hoekman, D., Gao, H. (1996) Comparative QSAR: toward a deeper understanding of chemicobiological interactions. *Chem Rev* **96**, 1045–1074.
30. Jaffé, H. H. (1953) A reexamination of the Hammett equation. *Chem Rev* **53**, 191–261.
31. Hammett, L. P. (1935) Some relations between reaction rates and equilibrium. *Chem Rev* **17**, 125–136.
32. Hammett, L. P. (1937) The effect of structure upon the reactions of organic compounds. Benzene derivatives. *J Am Chem Soc* **59**, 96–103.
33. Hansch, C., Maloney, P. P., Fujita, T., Muir, R. M. (1962) Correlation of biological activity of phenoxyacetic acids with Hammett substituent constants and partition coefficients. *Nature* **194**, 178–180.
34. Hansch, C. (1993) Quantitative structure-activity relationships and the unnamed science. *Acc Chem Res* **26**, 147–153.
35. Livingstone, D. J. (2004) Building QSAR models: a practical guide, in (Cronin, M. T. D., Livingstone, D. J. eds.) *Predicting Chemical Toxicity and Fate*. CRC Press, Boca Raton, FL, 2004, pp. 151–170.
36. Walker, J. D., Dearden, J. C., Schultz, T. W., Jaworska, J., Comber M. H. I. (2003) in (Walker, J. D. ed.) *QSARs for New Practitioners, in QSARs for Pollution Prevention, Toxicity Screening, Risk Assessment, and Web Applications*. SETAC Press, Pensacola, FL, pp. 3–18.
37. Walker, J. D., Jaworska, J., Comber, M. H. I., Schultz, T. W., Dearden, J. C. (2003) Guidelines for developing and using quantitative structure–activity relationships. *Environ Toxicol Chem* **22**, 1653–1665.
38. Cronin, M. T. D., Schultz, T. W. (2003) Pitfalls in QSAR *J Theoret Chem (Theochem)* **622**, 39–51.
39. OECD Principles for the Validation of (Q)SARs, http://www.oecd.org/dataoecd/33/37/37849783.pdf, last accessed February, 2010.
40. Tropsha, A., Golbraikh, A. (2007) Predictive QSAR modeling workflow, model applicability domains, and virtual screening. *Curr Pharmaceut Design* **13**, 3494–3504.
41. Dearden, J. C., Cronin, M. T. D., Kaiser, K. L. E. (2009) How not to develop a quantitative structure-activity or structure-property relationship (QSAR/QSPR). *SAR and QSAR in Environ Res* **20**, 241–266.
42. Free, S. M., Wilson, J. W. (1964) A mathematical contribution to structure-activity studies. *J Med Chem* **7**, 395–399.
43. Xing, L., Glen, R. C. (2002) Novel methods for the prediction of logP, pK_a, and logD. *J Chem Inf Comput Sci* **42**, 796–805.
44. Lombardo, F., Obach, R. S., et al. (2006) A hybrid mixture discriminant analysis-random forest computational model for the prediction of volume of distribution of drugs in human. *J Med Chem* **49**, 2262–2267.
45. Nicolaou, C. A., Brown, N., Pattichis, C. S. (2007) Molecular optimization using computational multi-objective methods *Curr Opin Drug Discov Develop* **10**, 316–324.
46. Gillet, V. J., Willett, P., Bradshaw, J., Green, D. V. S. (1999) Selecting combinatorial libraries to optimize diversity and physical properties. *J Chem Inf Comput Sci* **39**, 169–177.
47. Brown, R.D., Hassan, M., Waldman, M. (2000) Combinatorial library design for diversity, cost efficiency, and drug-like characters. *J Mol Graph Model* **18**, 427–437.
48. Gillet, V. J., Khatib, W., Willett, P., Fleming, P. J., Green, D. V. S. (2002) Combinatorial library design using a multiobjective genetic algorithm. *J Chem Inf Comput Sci* **42**, 375–385.
49. Chen, G., Zheng, S., Luo, X., Shen, J., Zhu, W., Liu, H., Gui, C., Zhang, J., Zheng, M., Puah, C.M., Chen, K., Jiang, H. (2005) Focused combinatorial library design based on structural diversity, drug likeness and binding affinity score. *J Comb Chem* **7**, 398–406.
50. Eichfelder, G. (2008) *Adaptive Scalarization Methods in Multiobjective Optimization*, Springer-Verlag, Berlin, Germany.
51. Abraham, A., Jain, L., Goldberg, R. (eds.) (2005) *Evolutionary Multiobjective Optimization: Theoretical Advances and Applications*, Springer-Verlag, London, UK.
52. Van Veldhurizen, D. A., Lamont, G. B. (2000) Multiobjective evolutionary algorithms: analyzing the state-of-the-art. *Evol Comput* **8**, 125–147.
53. Gillet, V. J., Willett, P., Bradshaw, J., Green, D. V. S. (1999) Selecting combinatorial libraries to optimize diversity and physical properties. *J Chem Inf Comput Sci* **39**, 169–177.
54. Zheng, W., Hung, S. T., Saunders, J. T., Seibel, G. L. (2000) PICCOLO: a tool for combinatorial library design via multicriterion optimization. *Pac Symp Biocomput* **5**, 585–596.
55. A multi-endpoint optimization tool with a graphics user interface developed at Pfizer–La

Jolla by Zhou, J. Z., Kong, X., Mattaparti, S, et al. (unpublished).
56. Lipinski, C. A., Lombardo, F., Dominy, B. W., Feeney, P. J. (1997) Experimental and computational approaches to estimate solubility and permeability in drug discovery and development settings. *Adv Drug Deliv Rev* **23**, 3–25.
57. Gillet, V. J., Willett, P., Bradshaw, J. (1998) Identification of biological activity profiles using substructural analysis and genetic algorithms. *J Chem Inf Comput Sci* **38**, 165–179.
58. Walter, W. P., Stahl, M. T., Murcko, M. A. (1998) Virtual screening–an overview. *Drug Discov Today* **3**, 160–178.
59. Bajorath, J. (2002) Integration of virtual and high-throughput screening. *Nat Rev Drug Discov* **1**, 882–894.
60. Reddy, A. S., Pati, S. P., Kumar, P. P., Pradeep, H. N., Sastry, G. N. (2007) Virtual screening in drug discovery–a computational perspective. *Curr Prot Pept Sci* **8**, 329–351.
61. Klebe, G. (ed.) (2000) *Virtual Screening: An Alternative or Complement to High Throughput Screening?* Kluwer Academic Publishers, Boston.
62. Alvarez, J., Shoichet, B. (ed.) (2005) *Virtual Screening in Drug Discovery*, Taylor & Francis, Boca Raton, USA.
63. Varnek, A., Tropsha, A. (ed.) (2008) *Chemoinformatics: An Approach to Virtual Screening*, RSC, Cambridge, UK.
64. Rishton, G. M. (1997) Reactive compounds and *in vitro* false positives in HTS. *Drug Discov Today* **2**, 382–384.
65. Walters, W. P., et al. (1998) Can we learn to distinguish between 'druglike' and 'nondrug-like' molecules? *J Med Chem* **41**, 3314–3324.
66. Sadowski, J., Kubinyi, H. A. (1998) A scoring scheme for discriminating between drugs and nondrugs. *J Med Chem* **41**, 3325–3329.
67. Rishton, G. M. (2003) Nonleadlikeness and leadlikeness in biochemical screening. *Drug Discov Today* **8**, 86–96.
68. http://dtp.nci.nih.gov/docs/3d_database/Structural_information/structural_data.html, last accessed February, 2010.
69. Kuntz, I. D. (1992) Structure-based strategies for drug design and discovery. *Science* **257**, 1078–1082.
70. Kitchen, D. B., Decornez, H., Furr, J. R., Bajorath, J. (2004) Docking and scoring in virtual screening for drug discovery: methods and applications. *Nat Rev Drug Discov* **3**, 935–949.
71. Sun, H. (2008) Pharmacophore-based virtual screening. *Curr Med Chem* **15**, 1018–1024.
72. Melville, J. L., Burke, E. K., Hirst, J. D. (2009) Machine learning in virtual screening. *Comb Chem High Throughput Screening* **12**, 332–343.
73. Harper, G., Pickett, S. D., Green, D. V. S. (2004) Design of a compound screening collection for use in high throughput screening. *Comb Chem High Throughput Screening* 7, 63–70.
74. Schüller, A., Hähnke, V., Schneider, G. (2007) SmiLib v2.0: a Java-based tool for rapid combinatorial library enumeration *QSAR. Comb Sci* **26**, 407–410.
75. Pipeline Pilot distributed by Accelrys Inc. can be used to enumerate libraries defined either by reactions or by Markush structures: http://accelrys.com/resource-center/case-studies/enumeration.html, last accessed February, 2010.
76. CombiLibMaker is software distributed by Tripos Inc.: http://tripos.com/data/SYBYL/combilibmaker_072505.pdf, last accessed February, 2010.
77. Yasri, A., Berthelot, D., Gijsen, H., Thielemans, T., Marichal, P., Engels, M., Hoflack, J. (2004) REALISIS: a medicinal chemistry-oriented reagent selection, library design, and profiling platform. *J Chem Inf Comput Sci* **44**, 2199–2206.
78. (a) Peng, Z., Yang, B., Mattaparti, S., Shulok, T., Thacher, T., Kong, J., Kostrowicki, J., Hu, Q., Na, J., Zhou, J. Z., Klatte, K., Chao, B., Ito, S., Clark, J., Coner, C., Waller, C., Kuki, A. PGVL Hub: an integrated desktop tool for medicinal chemists to streamline design and synthesis of chemical libraries and singleton compounds, in (Zhou, J. Z., ed.) *Chemical Library Design*. Humana Press, New York, Chapter 15.
78. (b) Truchon, J. -F. GLARE: a tool for product-oriented design of combinatorial libraries, in (Zhou, J. Z., ed.) *Chemical Library Design*. Humana Press, New York, Chapter 17.
78. (c) Lam, T. H., Bernardo, P. H., Chai, C. L. L., Tong, J. C. CLEVER – a general design tool for combinatorial libraries, in (Zhou, J. Z., ed.) *Chemical Library Design*. Humana Press, New York, Chapter 18.
79. Shi, S., Peng, Z., Kostrowicki, J., Paderes, G., Kuki, A. (2000) Efficient combinatorial filtering for desired molecular properties of reaction products. *J Mol Graph Model* **18**, 478–496.
80. Zhou, J. Z., Shi, S., Na, J., Peng, Z., Thacher, T. (2009) Combinatorial library-based design with basis products. *J Comput Aided Mol Des* **23**, 725–736.

81. Grabowski, K., Baringhaus, K. -H., Schneider, G. (2008) Scaffold diversity of natural products: inspiration for combinatorial library design. *Nat Prod Rep* **25**, 892–904.
82. Stocks, M. J., Wilden, G. R. H, Pairaudeau, G., Perry, M. W. D, Steele, J., Stonehous, J. P. (2009) A practical method for targeted library design balancing lead-like properties with diversity. *ChemMedChem* **4**, 800–808.
83. Hann, M. M., Leach, A. R., Harper, G. (2001) Molecular complexity and its impact on the probability of finding leads for drug discovery. *J Chem Inf Comput Sci* **41**, 856–864.
84. Gillet, V. J. (2002) Reactant- and product-based approaches to the design of combinatorial libraries. *J Comput Aided Mol Des* 16:371–380.
85. Balakin, K. V., Ivanenkov, Y. A., Savchuk, N. P. (2009) Compound library design for targeted families, in (Jacoby, E. ed.) *Chemogenomics*. Humana Press, New York, pp 21–46.
86. Xi, H., Lunney, E. A. (2010) The design, annotation and application of a kinase-targeted-library, in (Zhou, J. Z. ed.) *Chemical Library Design*. Humana Press, New York, Chapter 14.

Chapter 3

Molecular Library Design Using Multi-Objective Optimization Methods

Christos A. Nicolaou and Christos C. Kannas

Abstract

Advancements in combinatorial chemistry and high-throughput screening technology have enabled the synthesis and screening of large molecular libraries for the purposes of drug discovery. Contrary to initial expectations, the increase in screening library size, typically combined with an emphasis on compound structural diversity, did not result in a comparable increase in the number of promising hits found. In an effort to improve the likelihood of discovering hits with greater optimization potential, more recent approaches attempt to incorporate additional knowledge to the library design process to effectively guide the search. Multi-objective optimization methods capable of taking into account several chemical and biological criteria have been used to design collections of compounds satisfying simultaneously multiple pharmaceutically relevant objectives. In this chapter, we present our efforts to implement a multi-objective optimization method, MEGALib, custom-designed to the library design problem. The method exploits existing knowledge, e.g. from previous biological screening experiments, to identify and profile molecular fragments used subsequently to design compounds compromising the various objectives.

Key words: Multi-objective molecular library design, multi-objective evolutionary algorithm, selective library design, MEGALib.

1. Introduction

Drug discovery can be seen as the quest to design small molecules exhibiting favourable biological effects in vivo. Such molecules need to balance a combination of multiple properties including binding affinity to the pharmaceutical target, appropriate pharmacokinetics, limited (or no) toxicity (1, 2). The lack of consideration of the multitude of properties in the early stages of lead identification and optimization frequently hinders subsequent efforts

J.Z. Zhou (ed.), *Chemical Library Design*, Methods in Molecular Biology 685,
DOI 10.1007/978-1-60761-931-4_3, © Springer Science+Business Media, LLC 2011

for drug discovery (3). Indeed, one of the common causes for lead compounds to fail in the later stages of drug discovery is the lack of consideration of multiple objectives at the early stage of optimization of candidate compounds (4).

Traditional molecular library design (MLD) methods, modelled after the standard experimental drug discovery procedures, ignored the multi-objective nature of drug discovery and focussed on the design of libraries taking into account a single criterion. Often, the focus has been on maximizing library diversity in an effort to select compounds representative of the entire possible population (5) or in designing compound collections exploring a well-defined region of the chemical space defined by similarity to known ligands (6). The resulting molecular libraries, typically synthesized using combinatorial chemistry which enables the synthesis of large numbers of compounds and screened via high-throughput screening systems, revealed that simply synthesizing and screening large numbers of diverse (or similar) compounds may not increase the probability of discovering promising hits (7). Instead, due to the multi-objective nature of drug discovery, other factors, such as absorption, distribution, metabolism, excretion, toxicity (ADMET), selectivity and cost, molecular screening libraries need to be carefully planned and a number of design objectives must be taken into account (8). In recent times, MLD efforts have been exploring the use of multi-objective optimization (MOOP) techniques capable of designing libraries based on a number of properties simultaneously (9).

1.1. Multi-objective Optimization Basics

Problems that require the accommodation of multiple objectives, such as molecular library design, are widely known as multi-objective problems (MOP) or 'vector' optimization problems (10). In contrast to single-objective problems where optimization methods explore the feasible search space to find the single best solution, in multi-objective settings, no best solution can be found that outperforms all others in every criterion (3). Instead, multiple 'best' solutions exist representing the range of possible compromises of the objectives (11). These solutions, known as non-dominated, have no other solutions that are better than them in all of the objectives considered. The set of non-dominated solutions is also known as the Pareto-front or the trade-off surface. **Figure 3.1** illustrates the concept of non-dominated solutions and the Pareto-front in a bi-objective minimization problem.

MOPs are often characterized by vast, complex search spaces with various local optima that are difficult to explore exhaustively, largely due to the competition among the various objectives. In order to decrease the complexity of the search landscape, MOPs have traditionally been simplified, either by ignoring all objectives but one or by aggregating them. Multi-objective optimization (MOOP) methods enable the simultaneous optimization of

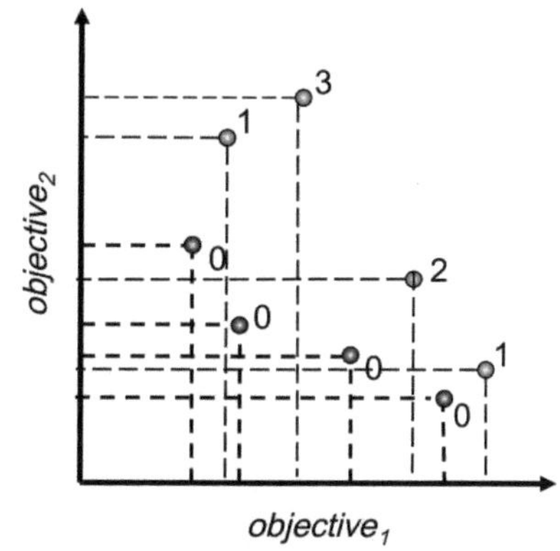

Fig. 3.1. A MOP with two minimization objectives and a set of solutions represented as circles. The rank of each solution (number next to *circle*) is based on the number of solutions that dominate it (i.e. are better) in both objectives. The area defined by the *dashed lines* of each solution contains the solutions that dominate it. Non-dominated solutions are labelled "0". Point (0, 0) represents the ideal solution to this problem.

several objectives by considering numerous dependent properties to guide the search. Pareto-based MOOP methods produce a set of solutions representing various compromises among the objectives and allow the user to choose the solutions that are most suitable for the task. The challenge facing these methods is to ensure the convergence of well-dispersed solutions to guarantee the effective coverage of the true optimal front (11). The major benefit of MOOP methods is that local optima corresponding to one objective can be avoided by consideration of all the objectives, thereby escaping single objective dead ends.

1.2. Evolutionary Algorithms

Evolutionary algorithms (EAs) have been used extensively for MOPs with several multi-objective optimization EAs (MOEA) cited in the literature (12, 13). EA-based algorithms use populations of individuals evolved through a set of genetic operators such as reproduction, mutation, crossover and selection of the fittest for further evolution (11). In the case of single objectives, selection of solutions involves ranking the individual solutions according to their fitness and choosing a subset. MOEAs extend traditional EAs by adding a Pareto-ranking component to enable the algorithm to handle multiple objectives simultaneously. MOEAs are particularly attractive since their population-based approach enables the exploration of multiple search space regions and thus the identification of numerous Pareto-solutions in a single run. EAs impose no constraints on the morphology of the search space, and thus, are suitable for complex, multi-modal search spaces with various local optima such as the ones typically found in MOPs (9). **Figure 3.2** outlines the main steps of an MOEA algorithm.

Generate initial population P
Evaluate solutions in P against objectives O_{1-n}
Assign Pareto-rank to solutions
Assign efficiency value to solutions based on Pareto-rank
While Not Stop Condition:
    Select parents $P_{parents}$ in proportion to efficiency values
    Generate population $P_{offspring}$ by reproduction of $P_{parents}$
        Mutation on individual parents
        Crossover on pairs of parents
    Evaluate solutions in $P_{offspring}$ against objectives O_{1-n}
    Merge P, $P_{offspring}$ to create P_{new}
    Assign Pareto-rank to solutions
    Assign efficiency value to solutions based on Pareto-rank

Fig. 3.2. The MOEA algorithm.

1.3. Multi-objective Molecular Library Design Applications

Typical multi-objective molecular library design approaches use the weighted-sum-of-objective-functions method that combines the multiple objectives into a composite one via a weighted-average transformation (14). Representative methods include SELECT (15) which combines diversity and drug-likeness criteria to design libraries via an EA-based optimization method and PICCOLO (16) which combines various objectives including reagent diversity, product novelty, similarity to known ligands and pharmacokinetics into a single one and uses simulated annealing (11) to search for optimal solutions. Alternatively, the method described by Bemis and Murcko enumerated a large virtual library of compounds and applied a set of filters, including predictive models for target-specific activity and drug-likeness thresholds on chemical properties, to generate a compound library satisfying multiple objectives (17).

In more recent years, Pareto-based methods have also been used for molecular library design. MoSELECT employs the multi-objective genetic algorithm (MOGA) (12) to simultaneously handle multiple objectives such as diversity, physicochemical properties and ease of synthesis (7), and MoSELECT II incorporates library size (i.e. number of compounds) and configuration (i.e. number of reagents at each position) as additional objectives (18). A multi-objective incremental construction method, generating libraries based on a supplied scaffold and a set of reagents, was proposed in reference (5). The method relies on the selection of appropriate reagents based on the similarity of the virtual molecules they produce to the set of query molecules. The multiple similarities calculated for the virtual products are subjected to Pareto-ranking that is subsequently used for reagent selection.

This chapter describes our work in developing an MOEA algorithm specifically designed to address the problem of multi-objective library design given available knowledge, including results from initial rounds of screening. The next sections describe the algorithm in detail and present the software implemented.

A sample application of the method focussing on designing a selective library of compounds for secondary screening is also presented. The chapter concludes with a set of notes for a user to avoid common mistakes and make better use of the method.

2. Materials

2.1. Multi-objective Optimization Software

1. NSisDesign: A molecular library design application program, part of the NSisApps0.8 software suite (19), was developed and used. The program is capable of generating a collection of chemical designs of a given size produced by combining building blocks from a fragment collection supplied at run time. The designs produced represent compromises of a number of objectives also supplied at run time.
2. Molecular Fitness Assessment Software:
   a. Fuzzee (20), a molecular similarity method based on a fuzzy, property-based molecular representation.
   b. OEChem (21), a chemoinformatics toolkit used to calculate chemical structure properties such as molecular weight, hydrogen bond donors and acceptors.

2.2. Molecular Building Block Preparation Software

1. NSisFragment, a molecular fragmentation and substructure mining tool part of the NSisUtilities0.9 software suite (19) was used. The tool is able to extract fragments from molecular graphs in a variety of ways including frequent subgraph mining (22) and the RECAP chemical bond type identification and cleaving technique (23). The fragments contain information about their attachment points and the type of bond cleaved at each attachment point.
2. NSisProfile, a chemical fragment characterization and profiling tool from the NSisUtilities0.9 software suite. The tool characterizes supplied molecular fragments with respect to chemical structure characteristics, e.g. molecular weight, hydrogen bond donors and acceptors, complexity (24) and number of rotatable bonds. When supplied with molecular libraries annotated with biological screening information, the tool matches fragments and molecules, prepares lists of molecules containing each fragment and annotates fragments with properties related to the molecules containing them, for example, average IC_{50} values for a specific assay.

2.3. Datasets

1. Dataset 1, a set of well-known estrogen receptor (ER) ligands, contains five compounds, three with increased selectivity to ER-β and two with increased selectivity to ER-α.

LIGAND	RBA ER-α	RBA ER-β	Selectivity ratio (β/α)
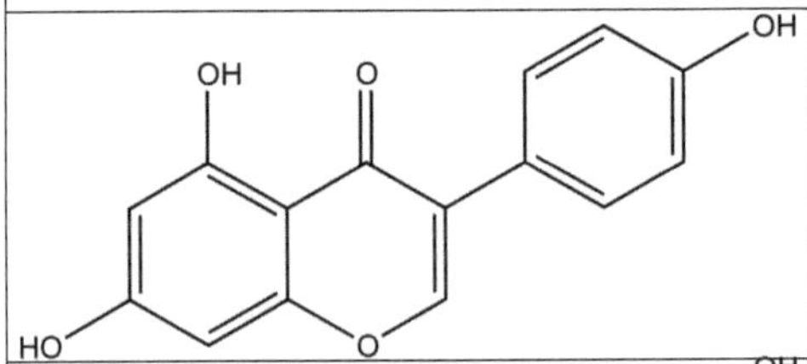	0.17	13	76
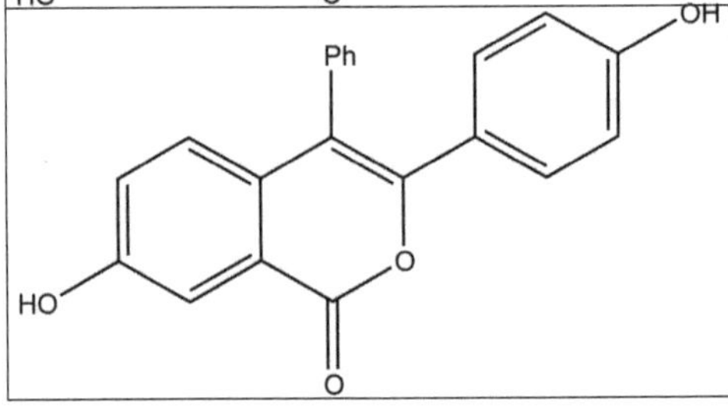	32.2	6.4	0.2

Fig. 3.3. Ligands and their relative binding affinity (RBA) to estrogen receptors α and β (25).

Figure 3.3 shows two of the molecules used, representative of the two sets used.

2. Dataset 2 is an ER-inhibitor dataset obtained from PubChem (26). The dataset consists of 86,098 compounds tested on both ER-α (Bioassay 629: HTS of Estrogen Receptor-alpha Coactivator Binding inhibitors, Primary Screening) and ER-β (Bioassay 633: HTS for Estrogen Receptor-beta Coactivator Binding inhibitors, Primary Screening).

3. Methods

Recently, we proposed the Multi-objective Evolutionary Graph Algorithm (MEGA), an optimization algorithm designed for the evolution of chemical structures satisfying multiple constraints (9). The technique combines evolutionary techniques with graph data structures to directly manipulate graphs and perform a global search for promising molecule designs. MEGA supports the use of problem-specific knowledge and local search techniques with an aim to improve both performance and scalability. Initial applications of the algorithm to the problem of de novo design showed that the technique is able to produce a diverse collection of equivalent solutions and, thus, support the drug discovery process (9). Based on our experiences we have designed a custom version of the original algorithm, termed MEGALib, to meet the requirements of multi-objective library design. The method focusses on designing the best possible products, i.e. chemical structures, for the problem under investigation and makes no attempt to minimize the number of reagents used; its main applications to date have been in designing small, focussed molecular libraries for secondary screening. This section initially describes MEGALib followed by a detailed overview of the methodology used to prepare the fragment collection and the computational objectives required by the algorithm. The later part of the section thoroughly describes an application of MEGALib to the problem of designing a selective library of compounds.

3.1. Multi-objective Library Design Algorithm Description

1. MEGALib input. MEGALib requires the supply of a set of molecular building blocks, the implemented objectives to be used for scoring molecules, a set of attributes controlling evolutionary operations, including mutation and crossover methods and probabilities, and hard filters for solution elimination. User input indicating the size of the designed library is also supplied. MEGALib operates on two population sets, the normal, working population and the secondary population or the Pareto-archive. The size of the two populations is also supplied by the user.
2. Initial working population generation. The first phase of the algorithm generates the initial population by combining pairs of building blocks from the collection supplied by the user and initiates the external archive of solutions intended to store the secondary population. The virtual synthesis step operates by taking into account the weight associated with each building block, if one is provided. Specifically, a roulette-like method selects building blocks via a probabilistic mechanism that assigns higher selection probability to those having a higher weight (11). To synthesize a member of the initial population the algorithm selects a core building block and attaches to each of its attachment points a building block with matching attachment point bond type. The algorithm repeats the above process until the number of initial population members reaches a multiple of the user-defined working population size, by default five times more, in order to avoid problems with insufficient working population size resulting from the elimination of solutions by the application of filtering in step 4 below. It is worth noting that the algorithm uses graph-based chromosomes corresponding to chemical structures to avoid the information loss associated with the encoding of more complex structures into simpler ones (9).

3. Solution fitness assessment. The population is then subjected to fitness assessment through application of the available objectives, a process that results in the generation of a list of scores for each individual.
4. Hard filter elimination. The list of solution scores is used for the elimination of solutions with values outside the range allowed by the corresponding active hard filters defined by the user.
5. Working population update. This step combines the two populations, working and secondary, to update the working population pool. This step is eliminated from the first iteration of the algorithm since the secondary, archive population is empty.
6. Pareto-ranking. The individuals' list of scores is subjected to a Pareto-ranking procedure to set the rank of each individual. According to this procedure the rank of an individual is set to the number of individuals that dominate it incremented by 1, thus, non-dominated individuals are assigned rank order 1.
7. Efficiency score calculation. The algorithm then proceeds to calculate an efficiency score for each individual using a methodology that operates both in parameter and objective space. The methodology employs an elaborate niching mechanism that performs diversity analysis of the population based on the genotype, i.e. the chromosome graph structure, and subsequently assigns an efficiency score that takes into account both the Pareto-rank and the diversity analysis outcome (9). The current implementation of the diversity analysis uses the Wards agglomerative clustering technique (27) and atom-type descriptors (28). The resulting Wards cluster tree is processed with the Kelley cluster level selection method (27) to produce a set of natural clusters. The results from clustering are subsequently used in the preparation of the efficiency score of individuals, which consists of its Pareto-rank and the cluster assignment.
8. Secondary population update. Efficiency scores are initially used to update the Pareto-archive. The current Pareto-archive is erased and a subset of the current working population that favours individuals with high efficiency score, i.e. low domination rank and high chromosome graph diversity, takes its place. Note that the size of the secondary population selected is limited by a user-supplied parameter. The secondary population mechanism has been designed specifically to preserve good solutions, non-dominated or dominated but substantially structurally unique, from all

iterations from getting lost due to working population size limitations.

9. Parent selection. Following the update of the Pareto-archive, MEGALib checks for the termination conditions and terminates if they have been satisfied. If this is not the case the process moves to select the parent subset population from the combined population set using a variation of the roulette method (11) operating on the dual-valued efficiency scores of the candidate solutions. Specifically, the selection method is applied on the clusters rather than the entire population. The process picks one solution from each cluster starting from the largest cluster and proceeding to clusters containing fewer compounds (9). The process traverses the set of clusters until the number of parents is selected. The parent selection method can be fine-tuned via user-supplied parameters to favour the parameter space or the objective space. Favouring the objective space amounts to selecting non-dominated solutions from each cluster. This method only proceeds to select dominated solutions when all non-dominated have been selected. Favouring the parameter space focusses on selecting solutions from all clusters by applying the roulette-like, weighted selection method to each cluster.

10. Offspring generation. The parents are then subjected to mutation and crossover according to the probabilities indicated by the user. MEGALib evolves solutions through a set of fragment-based operations inspired by mutation and crossover techniques. Mutation processes include insertion, removal and exchange of fragments. For fragment insertion, an attachment point is first chosen and a fragment from the weighted fragment collection is chosen and attached. For the fragment removal and exchange operations RECAP (23) is used to break the molecule into two disconnected parts and either remove or replace one of them with a fragment from the fragment collection. Note that fragment weights influence the probability of selection of a fragment for the insertion and exchange operations. Also note that the exchange fragment operation involves building blocks with attachment points of compatible bond types. Crossover takes place by identifying and cleaving a RECAP-type bond in each of two parents and recombining the resulting fragments to generate offspring. In a manner similar to the exchange fragment operation described above, this type of crossover is restricted to breaking specific bond types and combining fragments with compatible bond types in order to produce reasonable chemical designs.

11. New working population generation. The new working population is formed by merging the original working population and the newly produced mutants and crossover children. The process then iterates as shown in **Fig. 3.4**.

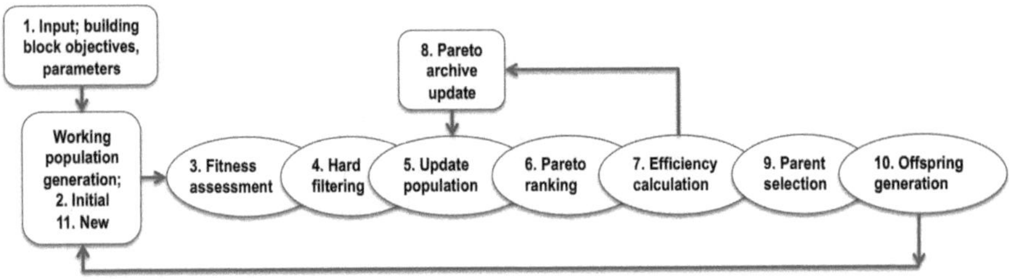

Fig. 3.4. The MEGALib algorithm.

Upon termination of the process the algorithm selects a compound set from the working population equal to the user-supplied library size as the library proposed. The selection of the library members is performed in a manner identical to the parent selection method described previously.

The algorithm exploits existing knowledge through the inclusion of multiple, problem-specific objectives, the use of bond-type information when evolving molecules and the exploitation of the weights associated with the building blocks provided which result in favouring those with an increased weight, i.e. having 'privileged' status.

3.2. Fragment Collection Preparation

The building block collection required by MEGALib consists of information-rich reagents, e.g. chemical fragments annotated with information on attachment points and bond types, as well as weights that designate their privileged, or not, status. The building blocks may be prepared via the application of the NSis-Fragment and NSisProfile tools, described previously, on a set of compounds with biological property information. The building blocks may also be obtained using other means by following the detailed advanced programmer interface (API) provided by the toolkit. For example, commercially available reagents may be appropriately annotated with information about attachment points, reaction types and privileged status by expert chemists.

3.3. Computational Objective Encoding

Fitness scores required for the application of MEGALib rely on the encoding of computational objective scorers that measure, or predict, molecular attributes.The main use of such scorers is to guide the optimization process, i.e. to direct the search towards interesting regions of the chemical space. Additionally, objective scorers may be used as hard filters to remove solutions with fitness values outside a predefined allowed range provided by the user. Objectives used in this manner are typically referred to as secondary while objectives used to guide optimization are considered primary.

MEGALib can use a wide range of molecular scorers provided that they have been encoded inline with a well-defined available API that allows smooth integration with the algorithm. The set of scorers available by the current implementation includes the following:

(a) Binding affinity scorers: MEGALib provides an interface that facilitates the encoding of objectives based on the predicted binding affinity of a designed molecule to a target protein. The implementation uses the docking program Glamdock and the ChillScore scoring method recently developed by Tietze and Apostolakis (29) to dock the designed molecules into the binding site of a receptor site provided by the user interactively, in real time. The interaction score of the best solution is used as an objective function. Settings for docking correspond to the slow settings described in Tietze and Apostolakis (29).

(b) Molecular similarity scorers: MEGALib encodes molecular similarity to a collection of user-supplied molecules as a distinct objective. The method uses the Fuzzee tool from the $Chil^2$ molecular modelling platform (20) which operates on abstractions of molecular graphs that replace atoms with molecular features to produce the so-called feature graphs. The actual similarity is calculated in a pair-wise manner by first aligning the feature graphs of two molecules, identifying common features, and then applying the Tanimoto similarity measure (30). In the event of similarity to a set of compounds, the average value of the pair-wise similarities is used.

(c) Chemical structure scorers: A list of chemical structure objectives, including molecular weight, number of hydrogen bond donors and acceptors, rotatable bonds and complexity is also available in the current implementation of MEGALib. Typically, chemical structure scorers are used as secondary objectives to constrain the search space by filtering out solutions such as those not conforming to the Rule-of-Five (31) or those estimated to be highly complex (24).

3.4. Selective Library Design Case Study

Designing selective libraries implies taking into consideration more objectives than just collecting compounds from various structural classes (32). The sample case study described in this section involves the application of MEGALib to design a library of compounds potentially exhibiting selectivity to one of two related but distinct pharmaceutical targets, namely ER-β over ER-α. The example given is meant to highlight the steps to be followed to produce a library satisfying multiple criteria.

A single collection of 51,123 building blocks was used for all the tests performed. The building blocks were obtained via fragmentation of Dataset 2 described previously with the fragmentation tool NSisFragment. The resulting fragments were profiled using the NSisProfile tool against the properties of the molecules that contain them as found in the Pubchem Assays 629 and 633 and weights corresponding to the values of the properties have been recorded. For the purposes of this application, a property-specific weight of a fragment is the average value of the property for the molecules that contain the fragment. Note that known ligands were not included in the fragmentation and building block generation process to favour the design of structurally different chemical designs.

The experimental settings used population size 100 and 1,000 generations. Runs were performed using both mutation and crossover. Mutation probability was set at 0.25 and crossover at 1.0. The maximum Pareto-archive set was set to 1,000. The desired library size was set at 250. Parent selection was set to balance between the diversity in parameter and objective space.

Two ligand-based objectives that measured the average similarity of a query molecule to known ligands were used. Similarity was calculated using the tool Fuzzee (20). The two objectives measured shape and property-based similarity of a given query molecule to the set of ER-α-selective and ER-β-selective ligands in Dataset 1. The experiments aimed at designing a library of molecules exploring the selectivity potential between the two ERs, with an emphasis on designs more similar to compounds selective to ER-β, and so the algorithm was set so as to maximize average similarity to the ER-β ligand set and minimize the average similarity to the ER-α ligand set. The search was constrained by imposing limits on the acceptable similarity values of the new designs to the two objectives. Specifically, minimum acceptable similarity to ER-β was set to 0.5 and maximum acceptable similarity to ER-α was also set to 0.5. Additionally, a set of hard filters based on chemical structure objectives was applied in order to remove potentially problematic designs from further consideration in line with step 4 of the MEGALib algorithm. The set of hard filters included limitations in the number of hydrogen bond donors and acceptors, and molecular weight, in line with the Rule-of-Five (31).

Progress monitoring of the MEGALib execution was performed by calculating the quality of the Pareto-approximation using quantitative measures in a post-processing step taking place after each generation. Specifically, the performance measures encoded included the calculation of the Pareto-approximation set hypervolume (13), the spacing measure (11) and the chromosome/structural diversity. The latter was calculated by averaging the Euclidean distances of each solution to all other solutions in the proposed set, using atom-pair descriptors (28) of the molecules involved. All three measures were calculated using code developed in-house for this reason. To avoid the extraction of misleading conclusions obtained through chance results a total of five runs were performed with identical input parameter settings but different initial population sets resulting from alternative initial population generation settings. The assessment of the results obtained from the five runs indicated similar performance with respect to the hypervolume, spacing and chromosome diversity and no major deviations. The results presented in the figures below correspond to one of the five runs and are representative of the set of results produced. Time requirements for the execution of the runs were sufficiently reasonable. A typical run of MEGALib executed, with population 250 and 1,000 iterations, took approximately 6 h on a normal PC.

The resulting library consisted of 250 compounds representing different compromises between the two conflicting objectives supplied. **Figure 3.5** presents a plot of the Pareto-approximation proposed by the software library (circles connected by line). Each of the remaining circles represents a solution from the initial population set after the hard filtering process. The *x*-axis represents similarity to ER-α ligands and the *y*-axis dissimilarity (1-similarity)

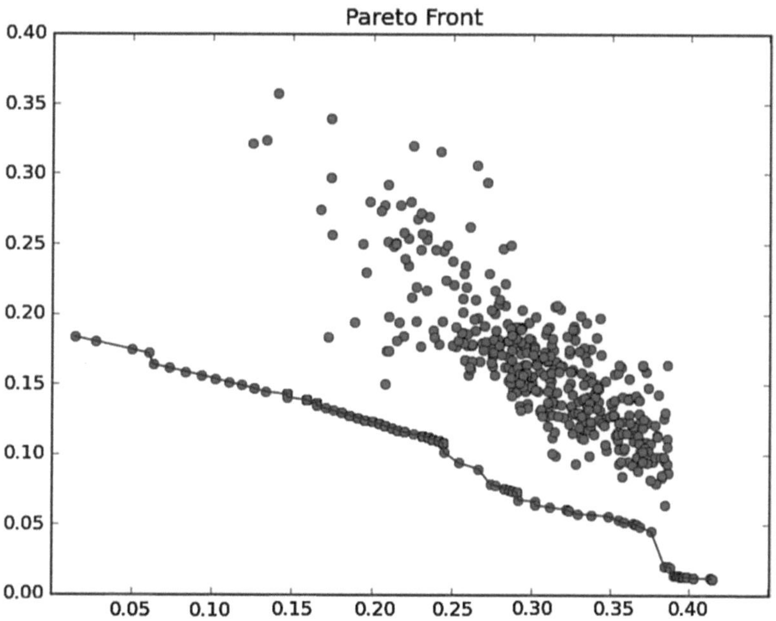

Fig. 3.5. Pareto-approximation formed by the designed library. The non-connected circles represent the initial population set. The *x*-axis represents shape similarity to ER-α ligands and the *y*-axis shape dissimilarity to ER-β ligands. Both objectives were minimized.

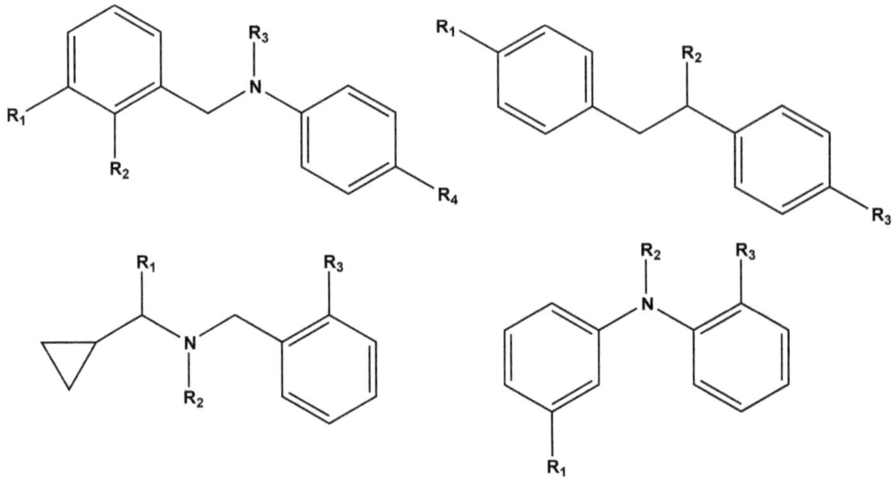

Fig. 3.6. Scaffolds representative of the compounds in the library designed using MEGALib.

to ER-β ligands; thus, the problem has been transformed to a bi-objective minimization problem with the ideal solution at point (0, 0).

Figure 3.6 presents a small subset from the collection of the scaffolds found in the compounds of the designed library. Each scaffold gave rise to one or more compounds of the designed library with varying performance to the objectives of the experiment through different substitutions on the various attachment points indicated as R groups. Consequently, the resulting library was sufficiently diverse indicating that MEGALib has been successful in identifying and preserving the structural diversity of the designed compounds.

4. Notes

1. Designing focussed vs diverse libraries. The scope and diversity of the library designed by MEGALib can be controlled using the user-supplied parameters required by the algorithm primarily by the choice of objectives and building block pool. Diverse libraries may be designed by formulating population diversity as one of the objectives of MEGALib. To this end the Wards clustering method combined with the Kelley cluster level selection described in Section **3** may be used. Additional objectives ensure that the set of diverse molecules produced will meet, for example, drug-likeness criteria. Focussed libraries are meant for a specific target (or related targets) and therefore objectives encoding target-specific information must be used (17). The use of a carefully

selected building block set consisting of fragments privileged for the specific target as well as objectives based on similarity to one or more known ligands can guide the search to generate a custom library for the target. The sample application presented in this chapter belongs to the latter library design category.

2. Types of objectives. MEGALib is agnostic to the type of objectives used. It is sufficient to prepare a computational method implementing a specific objective with an interface strictly in line with the NSisDesign API to enable its use by MEGALib during execution time. While this provides great flexibility to the user it is worth noting that special consideration must be given when preparing objectives to ensure their quality and reliability to facilitate the search. Typically, objectives based on noisy data or models of questionable quality may impede the algorithmic search and should only be used to provide general guidance to the search or as loose hard filters. Similarly, the use of highly correlated objectives should be avoided since their presence is not beneficial and may instead result in degraded computational performance.
3. Hard filtering. The use of multiple and/or strict sets of hard filters may cause problems especially in the initial iterations of the execution of MEGALib since they may reduce the population below the size required for subsequent operations and/or decrease greatly the working population diversity. The current algorithm implementation checks whether the solutions passing through the hard filters satisfy the population size indicated by the user. In the event that this is not the case eliminated solutions are selected and added to the working population. The solution 'recovery' step sorts the eliminated solutions according to the number of filters they failed and selects a large enough subset to add to the working population in a quasi-random fashion favouring each time the least problematic individuals.
4. Performance issues. The performance of MEGALib is largely dependent on the computational cost of the objectives used for the fitness assessment of the population. Certain objectives, such as those based on docking, require substantial execution time while others, such as those based on chemical structure or comparisons to known ligands, are less costly.
5. Pareto-archive size. MEGALib, as well as other MOEAs, has the ability to generate a large number of equivalent solutions for a given MOP. Consequently, the size of the Pareto-archive may increase to several thousands or even more depending on the number of iterations, the size of the working population, the number of building blocks, etc. An overly large archive, even though theoretically able to hold all promising solutions from all iterations, in practice

imposes a significant performance cost during execution time mostly due to the clustering step invoked by the niching mechanism. Extensive experimentation has shown that limiting the size of the archive using a user-supplied parameter available in the current implementation and a cluster-based elimination of solutions is able to maintain population diversity and reduce the computational cost to reasonable times.

6. Niching mechanism. Care must be exercised when sampling from clusters to accommodate the likely presence of singleton and under-represented clusters often found when the population size is small or particularly diverse. Such clusters may cause problems during selection, for example, when attempting to sample from singleton clusters. To avoid this type of problem MEGA implements appropriate rules, such as allowing only simple selection from singleton clusters (9).
7. Repair mechanism. Following the virtual synthesis step that takes place during parent solution evolution a repair mechanism is applied to ensure that the resulting offspring are valid molecules with respect to valences. Briefly, in its current implementation the mechanism identifies atoms with valence problems and attempts to repair them either by removing hydrogens attached to the atom or by downgrading atom bonds to a lower order, i.e. converts a double bond to single or a triple to double. If such action is not possible or sufficient to fix the problem, the offspring is discarded (9).
8. Parent selection method. Typical settings of the MEGALib algorithm use the parent selection method favouring the parameter space, i.e. selecting solutions from clusters using the roulette-like method described in Section **3**. This setting has been experimentally proven to preserve graph chromosome diversity and ensure that a variety of different promising subgraphs (scaffolds/chemotypes) survive long enough in the evolution cycle to contribute to the solution search.

References

1. Ekins, S., Boulanger, B., Swaan, P. W., Hupcey, M. A. (2002) Towards a new age of virtual ADME/TOX and multidimensional drug discovery. *J Comput Aided Mol Des* **16**, 381–401.
2. Agrafiotis, D. K., Lobanov, V. S., Salemme, F. R. (2002) Combinatorial informatics in the post-genomics era. *Nat Rev Drug Discov* **1**, 337–346.
3. Baringhaus, K. -H., Matter, H. (2004) Efficient strategies for lead optimization by simultaneously addressing affinity, selectivity and pharmacokinetic parameters, in (Oprea, T., ed.) *Chemoinformatics in Drug Discovery*. Wiley-VCH, Weinheim, Germany, pp. 333–379.
4. Nicolaou, C. A., Brown, N., Pattichis, C. S. (2007) Molecular optimization using computational multi-objective methods. *Curr Opin Drug Discov Dev* **10**, 316–324.
5. Soltanshahi, F., Mansley, T. E., Choi, S., Clark, R. D. (2006) Balancing focused

combinatorial libraries based on multiple GPCR ligands. *J Comput Aided Mol Des* **20**, 529–538.
6. Gillet, V. J., Willet, P., Fleming, P. J., Green, D. V. (2002) Designing focused libraries using MoSELECT. *J Mol Graph Model* **20**, 491–498.
7. Gillet, V. J., Khatib, W., Willett, P., Fleming, P. J., Green, D. V. (2002) Combinatorial library design using a multiobjective genetic algorithm. *J Chem Inf Comput Sci* **42**, 375–385.
8. Agrafiotis, D. K. (2000) Multiobjective optimization of combinatorial libraries. *Mol Divers* **5**, 209–230.
9. Nicolaou, C. A., Apostolakis, J., Pattichis, C. S. (2009) De novo drug design using multiobjective evolutionary graphs. *J Chem Inf Model* **49**, 295–307.
10. Coello Coello, C. A. (2002) Evolutionary multiobjective optimization: a critical review, in (Sarker, R., Mohammadian, M., Yao, X. eds.) *Evolutionary Optimization.* New York: Springer **48**, pp. 117–146.
11. Yann, C., Siarry, P. (eds.) (2004) *Multiobjective Optimization: Principles and Case Studies*, Springer, Berlin, Germany.
12. Fonseca, C. M., Fleming, P. J. (1998) Multiobjective optimization and multiple constraint handling with evolutionary algorithms. I: a unified formulation. *IEEE Trans Syst Man Cybernet* **28**, 26–37.
13. Zitzler, E., Thiele, L. (1999) Multiobjective evolutionary algorithms: a comparative case study and the strength Pareto approach. *IEEE Trans Evol Comput* **3**, 257–271.
14. Gillet, V. J. (2004) Designing combinatorial libraries optimized on multiple objectives in methods in molecular biology, in (Bajorath, J., ed.) *Chemoinformatics: Concepts, Methods, and Tools for Drug Discovery.* Humana Press, Totowa, NJ, **275**, pp. 335–354.
15. Gillet, V. J., Willett, P., Bradshaw, J., Green, D. V. S. (1999) Selecting combinatorial libraries to optimize diversity and physical properties. *J Chem Inf Comput Sci* **39**, 169–177.
16. Zheng, W., Hung, S. T., Saunders, J. T., Seibel, G. L. (2000) PICCOLO: a tool for combinatorial library design via multicriterion optimization. *Pac Symp Biocomput* **5**, 585–596.
17. Bemis, A. G. W., Murcko, M. A. (1999) Designing libraries with CNS activity. *J Med Chem* **42**, 4942–4951.
18. Wright, T., Gillet, V. J., Green, D. V., Pickett, S. D. (2003) Optimizing the size and configuration of combinatorial libraries. *J Chem Inf Comput Sci* **43**, 381–390.
19. Noesis Chemoinformatics, Ltd. http://www.noesisinformatics.com (accessed August 12, 2009).
20. MoDest. http://www.chil2.de (accessed June 30, 2009).
21. OpenEye, Inc. http://www.eyesopen.com (accessed July 3, 2009).
22. Nicolaou, C. A., Pattichis, C. S. (2006) Molecular substructure mining approaches for computer-aided drug discovery: a review. Proceedings of the 2006 ITAB Conference, October 26–28, Ioannina, Greece.
23. Lewell, X. O., Budd, D. B., Watson, S. P., Hann, M. M. (1998) RECAP – Retrosynthetic Combinatorial Analysis Procedure: a powerful new technique for identifying privileged molecular fragments with useful applications in combinatorial chemistry. *J Chem Inf Comput Sci* **38**, 511–522.
24. Barone R., Chanon, M. (2001) A new and simple approach to chemical complexity. Application to the synthesis of natural products. *J Chem Inf Comput Sci* **41**, 269–272.
25. Angelis, M. D., Stossi F., Waibel M., Katzenellenbogen, B. S., Katzenellenbogen, J. A. (2005) Isocoumarins as estrogen receptor beta selective ligands: isomers of isoflavone phytoestrogens and their metabolites. *Bioorg Med Chem* **13**, 6529–6542.
26. Wheeler, D. L., et al. (2006) Database resources of the National Center for Biotechnology Information. *Nucleic Acids Res* **34**, 173–180.
27. Wild, D. J., Blankley, C. J. (2000) Comparison of 2D fingerprint types and hierarchy level selection methods for structural grouping using wards clustering. *J Chem Inf Comput Sci* **40**, 155–162.
28. Kearsley, S. K., Sallamack, S., Fluder, E. M., Andose, J. D., Mosley, R. T., Sheridan, R. P. (1996) Chemical similarity using physiochemical property descriptors. *J Chem Inf Comput Sci* **36**, 118–127.
29. Tietze, S., Apostolakis, J. (2007) GlamDock: development and validation of a new docking tool on several thousand protein-ligand complexes. *J Chem Inf Model* **47**, 1657–1672.
30. Willet, P., Barnard, J. M., Downs, G. M. (1998) Chemical similarity searching. *J Chem Inf Comput Sci* **39**, 983–996.
31. Lipinski, C. A., Lombardo, F., Dominy, B. W., Feeney, P. J. (1997) Experimental and computational approaches to estimate solubility and permeability. Drug discovery and development settings. *Adv Drug Discovery Rev* **23**, 3–25.
32. Prien, O. (2005) Target-family-oriented focused libraries for kinases – conceptual design aspects and commercial availability. *ChemBioChem* **6,** 500–505.

Chapter 4

A Scalable Approach to Combinatorial Library Design

Puneet Sharma, Srinivasa Salapaka, and Carolyn Beck

Abstract

In this chapter, we describe an algorithm for the design of lead-generation libraries required in combinatorial drug discovery. This algorithm addresses simultaneously the two key criteria of diversity and representativeness of compounds in the resulting library and is computationally efficient when applied to a large class of lead-generation design problems. At the same time, additional constraints on experimental resources are also incorporated in the framework presented in this chapter. A computationally efficient scalable algorithm is developed, where the ability of the deterministic annealing algorithm to identify clusters is exploited to truncate computations over the entire dataset to computations over individual clusters. An analysis of this algorithm quantifies the trade-off between the error due to truncation and computational effort. Results applied on test datasets corroborate the analysis and show improvement by factors as large as ten or more depending on the datasets.

Key words: Library design, combinatorial optimization, deterministic annealing.

1. Introduction

In recent years, combinatorial chemistry techniques have provided important tools for the discovery of new pharmaceutical agents. Lead-generation library design, the process of screening and then selecting a subset of potential drug candidates from a vast library of similar or distinct compounds, forms a fundamental step in combinatorial drug discovery (1). Recent advances in high-throughput screening such as using micro/nanoarrays have given further impetus to large-scale investigation of compounds. However, combinatorial libraries often consist of extremely large collections of chemical compounds, typically several million. The time and cost of associated experiments makes it practically

J.Z. Zhou (ed.), *Chemical Library Design*, Methods in Molecular Biology 685,
DOI 10.1007/978-1-60761-931-4_4, © Springer Science+Business Media, LLC 2011

impossible to synthesize each and every combination from such a library of compounds. To overcome this problem, chemists often work with virtual combinatorial libraries (VCLs), which are combinatorial databases containing enumeration of all possible structures of a given pharmacophore with all available reactants. A subset of lead compounds from this VCL is selected which is used for physical synthesis and biological target testing. The selection of this subset is based on a complex interplay between various objectives, which is cast as a combinatorial optimization problem. The main goal of this optimization problem is to identify a subset of compounds that is representative of the underlying vast library as well as manageable, where these lead compounds can be synthesized and subsequently tested for relevant properties, such as activity and bioaffinity. The combinatorial nature of the selection problem makes it impractical to exhaustively enumerate each and every possible subset of obtaining the optimal solution. For example, to select 30 lead compounds from a set of 1,000, there are approximately 3×10^{25} different possible combinations. Selection based on enumeration is thus impractical and requires numerically efficient algorithms to solve the constrained combinatorial optimization problem.

2. Issues in Lead-Generation Library Design

In addition to the computational complexity that arises due to the combinatorial nature of the problem, any algorithm that aims to address the lead-generation library design problem must address the following key issues:

Diversity versus representativeness: The most widely used method to obtain a lead-generation library involves maximizing the diversity of the overall selection (2, 3), based on the premise that the more diverse the set of compounds, the better the chance to obtain a lead compound with desired characteristics. Such a design strategy suffers from an inherent problem that using diversity as the sole criterion may result in a set where a large number of lead compounds disproportionately represent outliers or singletons (4, 5). However, from a drug discovery point of view, it is desirable for the lead-generation library to more proportionally represent all the compounds, or at least to quantify how representative each lead compound is in order to allot experimental resources. A maximally diverse subset is of little practical significance because of its limited pharmaceutical applications. Therefore, representativeness should be considered as a lead-generation library design criterion along with diversity (6, 7).

Design constraints: In addition to diversity and representativeness, other design criteria include confinement, which quantifies the degree to which the properties of a set of compounds lie in a prescribed range (8), and maximizing the activity of the set of compounds against some predefined targets. Activity is usually measured in terms of the quantitative structure of the given set. Additionally, the cost of chemical compounds and experimental resources is significant and presents one of the main impediments in combinatorial diagnostics and drug synthesis. Different compounds require different experimental supplies which are typically available in limited quantities. The presence of these multiple (and often conflicting) design objectives makes the library design a multiobjective optimization problem with constraints.

3. Basic Problem Formulation and Modifications

Basic formulation: The problem of selecting lead compounds for lead-generation library design can be stated in general as follows:

Given a distribution of N compounds, x_i, in a descriptor space Ω, find the set of M lead compounds, r_j, that solves the following minimization problem:

$$\min_{r_j, 1 \leq j \leq M} \sum_{i=1}^{N} p(x_i) \left\{ \min_{1 \leq j \leq M} d(x_i, r_j) \right\} \qquad [1]$$

Here, Ω represents the chemical property space corresponding to the VCL, $d(x_i, r_j)$ represents an appropriate distance metric between the lead compound r_j and the compound x_i, $p(x_i)$ is the relative weight that can be attached to compound x_i (if all compounds are of equal importance, then the weights $p(x_i) = \frac{1}{N}$ for each i), and M is typically much smaller than N. That is, this problem seeks a subset of M lead compounds r_j in a descriptor space such that the average distance of a compound x_i from its nearest lead compound is minimized. Alternatively, this problem can also be formulated as finding an optimal partition of the descriptor space ω into M clusters $\{R_j\}$ and assigning to each cluster a lead compound r_j such that the following cost function is minimized:

$$\sum_{j=1}^{M} \sum_{x_i \in R_j} d(x_i, r_j) p(x_i)$$

Incorporating diversity and representativeness: One drawback of the basic formulation is that all the lead compounds are weighted equally. However, design constraints often require distinguishing them from one another to reflect different aspects of the clusters. For example, when addressing the issue of representativeness in the lead-generation library, the lead compounds that represent larger clusters need to be distinguished from those that represent outliers.

We incorporate representativeness into the problem formulation by specifying an additional *relative weight* parameter $\lambda_j, 1 \leq j \leq M$ for each lead compound. This parameter λ_j quantifies the size of the cluster represented by the compound r_j, and it is proportional to the number of the compounds in that cluster. Thus, the resulting library design will associate lead compounds that represent outliers with low values of λ and the lead compounds that represent the majority members with corresponding high values. In this way, the algorithm can be used to identify distinct compounds through property vectors r_j in the descriptor space Ω that denote the *jth* lead compound and at the same time determine how representative each lead compound is. For instance, $\lambda_j = 0.2$ implies that lead compound r_j represents 20% of all compounds in the VCL. The following modified optimization problem adequately describes the diversity goals in the basic formulation as well as the representativeness through the relative weights λ_j:

$$\min_{r_j,\lambda_j,1\leq j\leq M} \sum_i p(x_i)\left\{\min_{1\leq j\leq M} d(x_i, r_j)\right\} \qquad [2]$$

such that $\sum_{j=1}^{M} \lambda_j = 1$

where λ_j is the fraction of compounds in VCL that are nearest to (represented by) the lead compound r_j.

Incorporating constraints on experimental resources: Experiments associated with compounds with different properties often require different experimental resources. The constraints on availability of these resources can vary depending on their respective handling costs and time. These constraints can be incorporated in the selection problem by associating appropriate weights to lead compounds. For instance, consider a VCL that is classified into q types of compounds corresponding to q types of experimental supplies required for testing. More specifically, the *jth* lead compound can avail only W_{jn} amount of the *nth* experimental resource $(1 \leq n \leq q)$. The modified optimization problem is then given by (9, 10)

$$\min_{r_j} D = \sum_n \sum_i p_n(x_i^n)\left\{\min_j d(x_i^n, r_j)\right\} \qquad [3]$$

such that $\lambda_{jn} = W_{jn} 1 \le j \le M, 1 \le n \le q$

where $p_n(x_i^n)$ is the weight of the compound location x_i^n, which requires the *nth* type of supply.

4. Computational Issues

Problem formulations [1–3] for designing lead-generation library under different constraints belong to a class of combinatorial resource allocation problems, which have been widely studied. They arise in many different applications such as minimum distortion problems in data compression (11), facility location problems (12), optimal quadrature rules and discretization of partial differential equations (13), locational optimization problems in control theory (9), pattern recognition (14), and neural networks (15). Combinatorial resource allocation problems are nonconvex and computationally complex and it is well documented (16) that most of them have many local minima that riddle the cost surface. Therefore, the main computational issue is developing an efficient algorithm that avoids local minima. Due to the large size of VCLs, and the combinatorial nature of the problem, the issue of algorithm scalability takes central importance. Since the number of computations to be performed by the lead-generation library design algorithm scales up exponentially with an increase in the amount of data, most algorithms become prohibitively slow and expensive (computationally) for large datasets.

4.1. Deterministic Annealing Algorithm

The main drawback of most popular algorithms that address the basic combinatorial resource allocation problem [1], such as Lloyd's or K-means algorithms (11, 17), is that they are extremely sensitive to initialization step in their procedures and typically get trapped in local minima. Other algorithms such as simulated annealing that actively try to avoid local minima are often computationally inefficient. Other drawbacks of these algorithms mainly stem from the lack of flexibility to incorporate various constraints on the resource locations discussed in **Section 3**. The deterministic annealing (DA) algorithm (18) overcomes these drawbacks; this algorithm is heuristically based on law of minimum free energy in statistical chemistry that models similar combinatorial problems occurring in nature. The DA algorithm is versatile in terms of accommodating constraints on resource locations while simultaneously it is designed to be insensitive to the initialization step and to avoid local minima.

The central concept of the DA algorithm is based on developing a homotopy from an appropriate convex function to the nonconvex cost function; the local minima of cost function at every step of homotopy serves as the initialization for the subsequent step. Since minimization of the initial convex function yields a global minimum, this procedure is independent of initialization. The heuristic is that the global minimum is tracked as the initial convex function deforms into the actual nonconvex cost function via the homotopy. Accordingly, the DA algorithm solves the following multiobjective optimization problem:

$$\min_{r_j} \min_{p(r_j|x_i)} \underbrace{D - T_k H}_{:=F}$$

over iterations indexed by k, where T_k is a parameter called temperature which tends to zero as k tends to infinity. The cost function F is called free energy, where this terminology is motivated by statistical chemistry (18). Here the distortion

$$D = \sum_{i=1}^{N} p(x_i) \sum_{j=1}^{M} d(x_i, r_j) p(r_j|x_i)$$

which is similar to the cost function in equation [1] is the "weighted average distance" of a lead compound r_j from a compound x_i in the VCL. This formulation associates each x_i to every r_j through the weighting parameter $p(r_j|x_i)$ and thus diminishes the sensitivity of algorithm to initialization of locations r_j. The more uniformly (or randomly) these weights are distributed, the more insensitive is the algorithm with respect to the initialization. The term $H = -\sum_{i,j} p(y_j|x_i) \log p(y_j|x_i)$ is the entropy of the weights $\{p(y_j|x_i)\}$ that quantifies their uniformity (or randomness). The annealing parameter T_k defines the homotopy from the convex function $-H$ to the nonconvex function D. Clearly, for large values of T_k, we mainly attempt to maximize the entropy. As T_k is lowered, we trade entropy for the reduction in distortion, and as T_k approaches zero, we minimize D directly to obtain a hard (nonrandom) solution, where $p(y_j|x_i)$ is either 0 or 1 for each pair (i,j). Minimizing the free energy term F with respect to the weighting parameter $p(r_j|x_i)$ is straightforward and gives the Gibbs distribution

$$p(r_j|x_i) = \frac{e^{-d(x_i,r_j)/T_k}}{Z_i}, \text{ where } Z_i := \sum_j e^{-d(x_i,r_j)/T_k} \qquad [4]$$

Note that the weighting parameters $p(r_j|x_i)$ are simply radial basis functions, which clearly decrease in value exponentially as r_j and x_i move farther apart. The corresponding minimum of F is obtained by substituting for $p(r_j|x_i)$ from equation [4]:

$$\widehat{F} = -T_k \sum_i p(x_i) \log Z_i \qquad [5]$$

To minimize $\widehat{F}$ with respect to the lead compounds $\{r_j\}$, we set the corresponding gradients equal to zero, i.e., $\frac{\partial \widehat{F}}{\partial r_j} = 0$; this yields the corresponding implicit equations for the locations of lead compounds:

$$r_j = \sum_i p(x_i|r_j)x_i, 1 \le j \le M, \text{where}$$
$$p(x_i|r_j) = \frac{p(x_i)p(r_j|x_i)}{\sum_k p(x_k)p(r_j|x_k)} \qquad [6]$$

Note that $p(x_i|r_j)$ denotes the posterior probability calculated using Bayes' rule and the above equations clearly convey the *centroid* aspect of the solution.

The DA algorithm consists of minimizing $\widehat{F}$ with respect to $\{r_j\}$ starting at high values of T_k and then tracking the minimum of $\widehat{F}$ while lowering T_k. At each k,

1. Fix $\{r_j\}$ and use equation [4] to compute the new weights $\{p(r_j|x_i)\}$.
2. Fix $\{p(r_j|x_i)\}$ and use equation [6] to compute the lead compound locations $\{r_j\}$.

4.2. A Scalable Algorithm

As noted earlier, one of the major problems with combinatorial optimization algorithms is that of scalability, i.e., the number of computations scales up exponentially with an increase in the amount of data. In the DA algorithm, the computational complexity can be addressed in two steps – first by reducing the number of iterations and second by reducing the number of computations at every iteration. The DA algorithm, as described earlier, exploits the phase transition feature (18) in its process to decrease the number of iterations (in fact in the DA algorithm, typically the temperature variable is decreased exponentially which results in few iterations). The number of computations per iteration in the DA algorithm is $O(M^2 N)$, where M is the number of lead compounds and N is the total number of compounds in the underlying VCL. In this section, we present an algorithm that requires fewer computations per iteration. This amendment becomes necessary in the context of the selection problem in combinatorial chemistry as the sizes of the dataset are so large that DA is typically too slow and often fails to handle the computational complexity. We exploit the features inherent in the DA algorithm that, for a given temperature, the farther an individual compound is from a cluster, the lower is its influence on the cluster (as is evident from equation [4]). That is, if two clusters are far apart, then they have very small interaction between them. Thus, if we ignore the effect of a separated cluster on the remaining compound locations, the resulting error will not be significant (*see* **Fig. 4.1**). Ignoring the effects of separated regions (i.e., groups

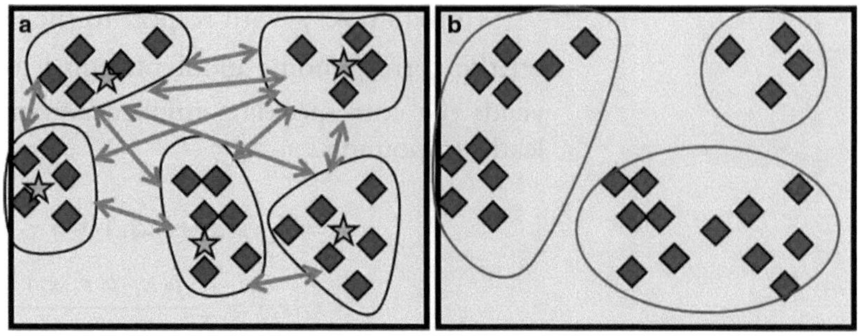

Fig. 4.1. (**a**) Illustration depicting the different clusters in the dataset, together with the interaction between each pair of points (and *clusters*). (**b**) Separated regions determined after characterizing intercluster interaction and separation.

of clusters) on one another will result in a considerable reduction in the number of computations since the points that constitute a separated region will not contribute to the distortion and entropy computations for the rest. This computational saving increases as the temperature decreases since the number of separated regions, which are now smaller, increases as the temperature decreases.

4.2.1. Cluster Interaction and Separation

In order to characterize the interaction between different clusters, it is necessary to consider the mechanism of cluster identification during the process of the DA algorithm. As the temperature (T_k) is reduced after every iteration, the system undergoes a series of phase transitions (*see* (18) for details). In this annealing process, at high temperatures that are above a pre-computable critical value, all the lead compounds are located at the centroid of the entire descriptor space, thereby there is only one distinct location for the lead compounds. As the temperature is decreased, a critical temperature value is reached where a phase transition occurs, which results in a greater number of distinct locations for lead compounds and consequently finer clusters are formed. This provides us with a tool to control the number of clusters we want in our final selection. It is shown (18) for a square Euclidean distance $d(x_i, r_j) = \|x_i - r_j\|^2$ that a cluster R_i splits at a critical temperature T_c when twice the maximum eigenvalue of the posterior covariance matrix, defined by $C_{x|r_j} = \sum_i p(x_i)p(x_i|r_j)(x_i - r_j)(x_i - r_j)^T$, becomes greater than the temperature value, i.e., when $T_c \leq 2\lambda_{\max}\left[C_{x|r_i}\right]$. This is exploited in the DA algorithm to reduce the number of iterations by jumping from one critical temperature to the next without significant loss in performance. In the DA algorithm, the lead location r_j is primarily determined by the compounds near it since far-away points exert small influence, especially at low temperatures. The association probabilities $p(r_j|x_i)$ determine the level of interaction between the cluster R_j and the data-point x_i. This interaction decays exponentially with the increase in the distance between r_j and x_i. The total interaction exerted by all the data-points in a given space determines the relative weight of each cluster, $p(r)_j := \sum_i^N p(x_i, r_j) = \sum_i^N p(r_j|x_i)p(x_i)$, where $p(r_j)$ denotes the weight of cluster R_j. We define the level of interaction that data-points in cluster R_i exert on cluster R_j by $\varepsilon_{ji} = \sum_{x \in R_i} p(r_j|x)p(x)$. The higher this value is, the more interaction exists between clusters R_i and R_j. This gives us an effective way to characterize the interaction between various clusters in a dataset. In a probabilistic framework, this interaction can also be interpreted as the probability of transition from R_i to R_j. Consider the $m \times n$ matrix $m \geq n$

$$A = \begin{pmatrix} \sum\limits_{x \in R_1} p(r_1|x)p(x) & \cdots & \sum\limits_{x \in R_m} p(r_1|x)p(x) \\ \sum\limits_{x \in R_1} p(r_2|x)p(x) & \cdots & \sum\limits_{x \in R_m} p(r_1|x)p(x) \\ \vdots & \ddots & \vdots \\ \sum\limits_{x \in R_1} p(r_m|x)p(x) & \cdots & \sum\limits_{x \in R_m} p(r_1|x)p(x) \end{pmatrix}$$

In a probabilistic framework, this matrix is a finite-dimensional Markov operator, with the term $A_{j,i}$ denoting the transition probability from region R_i to R_j. The higher the transition probability, the greater is the amount of interaction between the two regions. Once the transition matrix is formed, the next step is to identify regions, that is, groups of clusters, which are separate from the rest of the data. The separation is characterized by a quantity which we denote by ε. We say a cluster (R_j) is ε-separate if the level of its interaction with each of the other clusters $(A_{j,i}, i = 1, 2, \ldots, n, i \neq j)$ is less than ε. The value ε is used to partition the descriptor space into separate regions for reduced and scalable computational effort, and it quantifies the increase in the distortion cost function of the proposed scalable algorithm with respect to the DA algorithm.

4.2.2. Trade-Off Between Error in Lead Compound Location and Computation Time

As was discussed in **Section 4.2**, the greater the number of separate regions we use, the smaller the computation time for the scalable algorithm. At the same time, a greater number of separate regions results in a higher deviation in the distortion term of the proposed algorithm from the original DA algorithm. This trade-off between reduction in computation time and increase in distortion error is systematically addressed in the following. For any pair (r_j, V), where r_j is a lead compound and V is a subset of the descriptor space Ω, we define

$$G_j(V) := \sum_{x_i \in V} x_i p(x_i) p(r_j|x_i), \qquad [7]$$
$$H_j(V) := \sum_{x_i \in V} p(x_i) p(r_j|x_i)$$

Then, from the DA algorithm, the location of the lead compound (r_j) is determined by $r_j = \frac{G_j(\Omega)}{H_j(\Omega)}$. Since the cluster Ω_j is separated from all the other clusters, the lead compound location r'_j will be determined in the scalable algorithm by

$$r'_j = \frac{\sum_{x_i \in \Omega_j} x_i p(x_i) p(r_j|x_i)}{\sum_{x_i \in \Omega_j} p(x_i) p(r_j|x_i)} = \frac{G_j(\Omega_j)}{H_j(\Omega_j)} \quad [8]$$

We obtain the component-wise difference between r_j and r'_j by subtracting terms. Note that we use the symbols $\prec$ and $\succ$ for component-wise operations. On simplifying, we have

$$\left| r'_j - r_j \right| \prec= \frac{\max\left(G_j(\Omega_j^c) H_j(\Omega_j), G_j(\Omega_j) H_j(\Omega_j^c) \right)}{H_j(\Omega_j) H_j(\Omega)}, \quad [9]$$

where $\Omega_j^c = \Omega \backslash \Omega_j$

Denoting the cardinality of Ω by N and $M_j^c = \frac{1}{N} \sum_{x_i \in \Omega_j^c} x_i$, we note that

$$G_j(\Omega_j^c) \leq \left(\sum_{x_i \in \Omega_j^c} x_i \right) H_j(\Omega_j^c) = N M_j^c H_j(\Omega_j^c) \quad [10]$$

We have assumed $x \succ= 0$ without any loss of generality since the problem definition is independent of translation or scaling factors. Thus,

$$\begin{aligned} \left| r'_j - r_j \right| \prec= & \frac{\max\left(N M_j^c H_j(\Omega_j), G_j(\Omega_j) \right) H_j(\Omega_j^c)}{H_j(\Omega_j) H_j(\Omega)} \\ & = \max\left(N M_j^c, \frac{G_j(\Omega_j)}{H_j(\Omega_j)} \right) \left(\frac{H_j(\Omega_j^c)}{H_j(\Omega)} \right) \end{aligned} \quad [11]$$

then dividing through by N and using $M = \frac{1}{N} \sum_{x_i \in \Omega} x_i$ gives

$$\frac{\left| r'_j - r_j \right|}{MN} \prec= \max\left(\frac{M_j^c}{M}, \frac{M_j}{M} \right) \eta_j, \text{where } \eta_j = \frac{\sum\limits_{k \neq j} \varepsilon_{kj}}{\sum\limits_{k} \varepsilon_{kj}} \quad [12]$$

and ε_{kj} is the level of interaction between cluster Ω_j and Ω_k. For a given dataset, the quantities M, M_j,and M_j^c are known a priori. For the error in lead compound location $\left| r'_j - r_j \right| / M$ to be less than a given value δ_j (where $\delta_j > 0$), we must choose η_j such that

$$\eta_j \leq \frac{\delta_j}{N \max\left(\frac{M_j^c}{M}, \frac{M_j}{M}\right)} \qquad [13]$$

4.2.3. Scalable Algorithm

1. Initiate the DA algorithm and determine lead compound locations together with the weighting parameters.
2. When a split occurs (phase transition), identify individual clusters and use the weights $(p(r_j|x))$ to construct the transition matrix.
3. Use the transition matrix to identify separated clusters and group them to form separated regions. Ω_k will be separated from Ω_j if the entries $A_{j,k}$ and $A_{k,j}$ are less than a chosen ε_{jk}.
4. Apply the DA to each region, neglecting the effect of separate regions on one another.
5. Stop if the terminating criterion (such as maximum number of lead compounds (M) or maximum computation time) is met, otherwise go to 2.

Identification of separate regions in the underlying data provides us with a tool to efficiently scale the DA algorithm. In the DA algorithm, at any iteration, the number of computations is $M^2\ N$. In the proposed scalable algorithm, the number of computations at a given iteration is proportional to $\sum_{k=1}^{s} M_k^2 N_k$, where $N_k \left(N = \sum_{k=1}^{s} N_k\right)$ is the number of compounds and M_k is the number of clusters in the *kth* region. Thus, the scalable algorithm saves computations at each iteration. This savings increases as temperature decreases since corresponding values of N_k decrease. Moreover, since the scalable algorithm can run these *s* DA algorithms in *parallel*, it will result in additional potential savings in computational time.

5. Simulation Results

5.1. Design for Diversity and Representativeness

As a first step, a fictitious dataset (VCL) was created to present the "proof of concept" for the proposed optimization algorithm. The VCL was specifically designed to simultaneously address the issue of diversity and representativeness in the lead-generation library design. This dataset consists of few points that are outliers while most of the points are in a single cluster. Simulations were carried out in MATLAB. The results for dataset 1 are shown in **Fig. 4.2**. The pie chart in **Fig. 4.2** shows the relative weight of each lead compound. As was required, the algorithm gave larger weights at locations which had larger numbers of similar compounds. At the same time, it should be noted that the key issue of diversity is not

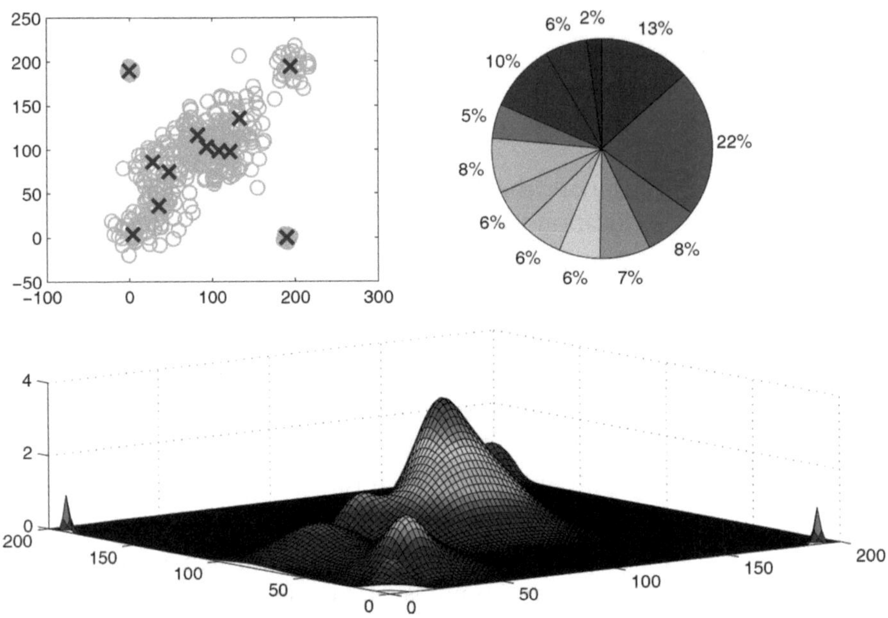

Fig. 4.2. Simulation results for dataset 1. (**a**) The locations $x_i, 1 \leq i \leq 200$, of compounds (*circles*) and $r_j, 1 \leq j \leq 10$, of lead compounds (*crosses*) in the 2-d descriptor space. (**b**) The weights λ_j associated with different locations of lead compounds. (**c**) The given weight distribution $p(x_i)$ of the different compounds in the dataset. Reprinted ("adapted" or "in part") with permission from *Journal of Chemical Information and Modeling*. Copyright 2008 American Chemical Society.

compromised. This is due to the fact that the algorithm inherently recognizes the natural clusters in the VCL. As is seen from the figure, the algorithm identifies all clusters. The two clusters which were quite distinct from the rest of the compounds are also identified albeit with a smaller weight. As can be seen from the pie chart, the outlier cluster was assigned a weight of 2%, while the central cluster was assigned a significant weight of 22%.

5.2. Scalability and Computation Time

In order to demonstrate the computational savings, the algorithm was tested on a suite of synthesized datasets. The first set was obtained by identifying ten random locations in a square region of size 400×400. These locations were then chosen as the cluster centers. Next, the size of each of these clusters was chosen and all points in the cluster were generated by a normal distribution of randomly chosen variance. A total of 5,000 points comprised this dataset. All the points were assigned equal weights (i.e., $p(x_i) = \frac{1}{N}$ for all $x_i \in \Omega$). **Figure 4.3** shows the dataset and the lead compound locations obtained by the original DA algorithm. The crosses denote the lead compound locations (r_j) and the pie chart gives the relative weight of each lead compound (λ_j).

The algorithm starts with one lead compound at the centroid of the dataset. As the temperature is reduced, the cluster is split and separate regions are determined at each such split.

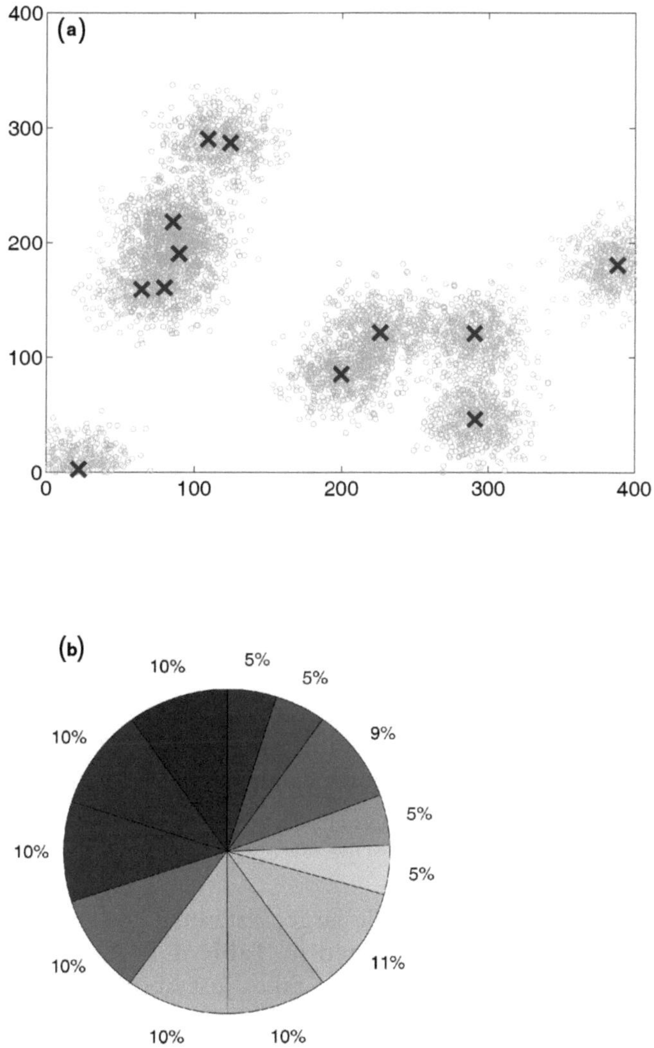

Fig. 4.3. (**a**) Locations $x_i, 1 \leq i \leq 5,000$, of compounds (*circles*) and $r_j, 1 \leq j \leq 12$, of lead compounds (*crosses*) in the 2-d descriptor space determined from the original algorithm. (**b**) Relative weights λ_j associated with different locations of lead compounds. Reprinted ("adapted" or "in part") with permission from *Journal of Chemical Information and Modeling*. Copyright 2008 American Chemical Society.

Figure 4.4a shows the four separate regions identified by the algorithm (as described in **Section 4.2.1**) at the instant when 12 lead compound locations have been identified. **Figure 4.4b** shows a comparison between the two algorithms. Here the crosses represent the lead compound locations (r_j) determined by the original DA algorithm and the circles represent the locations (r'_j) determined by the proposed scalable algorithm. As can be seen from the figure, there is little difference between the locations obtained by the two algorithms. The main advantage of the scalable algorithm is in terms of computation time and its ability to

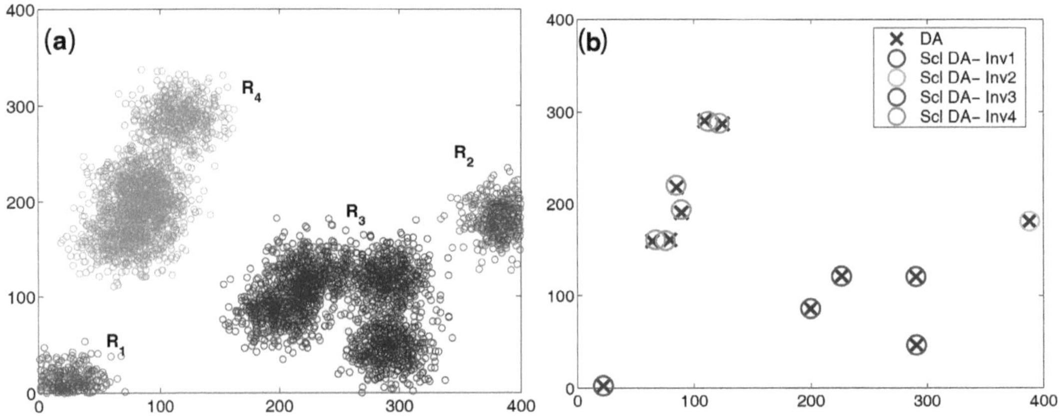

Fig. 4.4. (**a**) Separated regions R_1, R_2, R_3, and R_4 as determined by the proposed algorithm. (**b**) Comparison of lead compound locations r_j and r'_j. Reprinted ("adapted" or "in part") with permission from *Journal of Chemical Information and Modeling.* Copyright 2008 American Chemical Society.

Table 4.1
Comparison between the original and proposed algorithm

Algorithm	Distortion	Computation time (s)
The original DA	300.80	129.41
Proposed algorithm	316.51	21.53

Reprinted ("adapted" or "in part") with permission from *Journal of Chemical Information and Modeling.* Copyright 2008 American Chemical Society

handle larger datasets. The results from the two algorithms are presented in **Table 4.1**. As can be seen, the proposed scalable algorithm takes just about 17% of the time used by the original (nonscalable) algorithm and results in only a 5.2% increase in distortion; this was obtained for $\varepsilon = 0.005$. Both the algorithms were terminated when the number of lead compounds reached 12. The computation time for the scalable algorithm can be further reduced (by changing ε), but at the expense of increased distortion.

5.2.1. Further Examples

The scalable algorithm was applied to a number of different datasets. Results for three such cases have been presented in **Fig. 4.5**. The dataset in Case 2 is comprised of six randomly chosen cluster centers with 1,000 points each. All the points were assigned equal weights (i.e., $p(x_i) = \frac{1}{N}$ for all $x_i \in \Omega$). **Figure 4.5a** shows the dataset and the eight lead compound locations obtained by the proposed scalable algorithm. The dataset in Case 3 is also comprised of eight randomly chosen cluster locations with 1,000 points each. Both the algorithms were executed till they identified eight lead compound locations in the underlying dataset. Case 4 is comprised of two cluster centers with 2,000

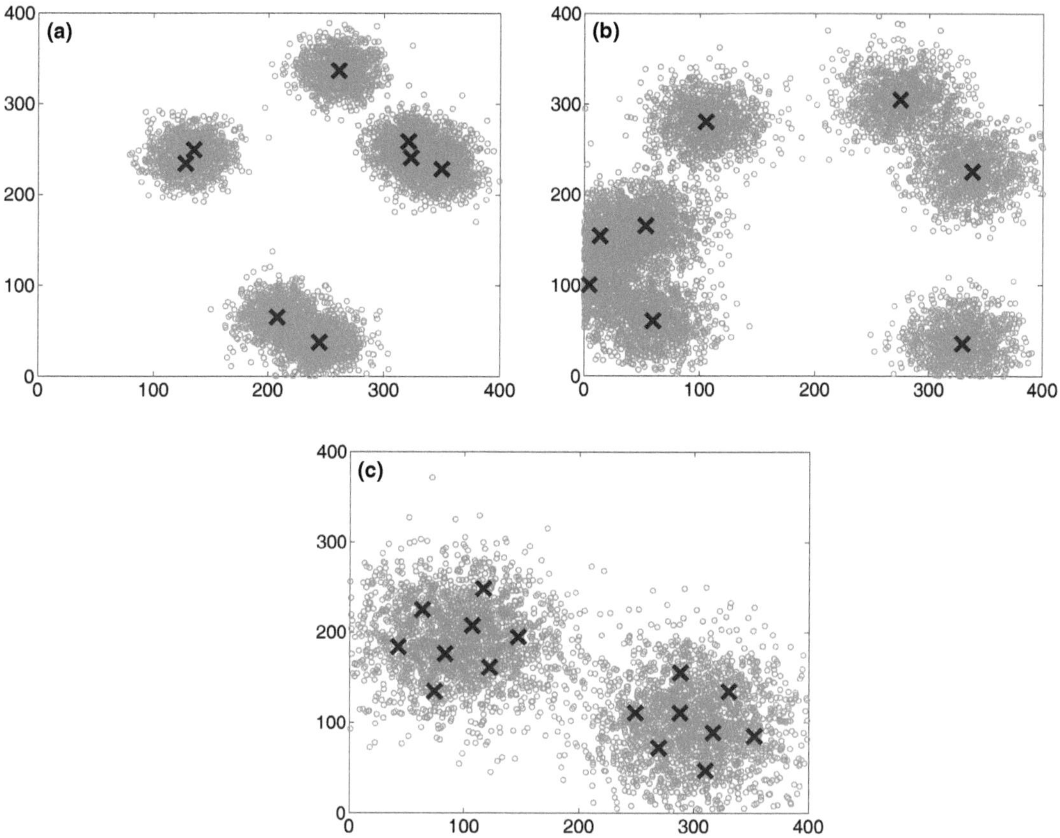

Fig. 4.5. (**a**, **b**, **c**) Simulated dataset with locations x_i of compounds (*circles*) and lead compound locations r_j (*crosses*) determined by the algorithm. Reprinted ("adapted" or "in part") with permission from *Journal of Chemical Information and Modeling*. Copyright 2008 American Chemical Society.

Table 4.2
Distortion and computation times for different datasets

Case	Algorithm	Distortion	Computation time (s)
Case 2	The original DA	290.06	44.19
	Proposed algorithm	302.98	11.98
Case 3	The original DA	672.31	60.43
	Proposed algorithm	717.52	39.77
Case 4	The original DA	808.83	127.05
	Proposed algorithm	848.79	41.85

Reprinted ("adapted" or "in part") with permission from *Journal of Chemical Information and Modeling*. Copyright 2008 American Chemical Society

points each. Both the algorithms were executed till they identified 16 lead compound locations. Results for the three cases have been presented in **Table 4.2**.

It should be noted that both the algorithms were terminated after a specific number of lead compound locations had been identified. The proposed algorithm took far less computation time when compared to the original algorithm while maintaining less than 5% error in distortion.

5.3. Drug Discovery Dataset

This dataset is a modified version of the test library set (19). Each of the 50,000 members in this set is represented by 47 descriptors which include topological, geometric, hybrid, constitutional, and electronic descriptors. These molecular descriptors are computed using the Chemistry Development Kit (CDK) Descriptor Calculator (20, 21). These 47-dimensional data were then normalized and projected onto a two-dimensional space. The projection was carried out using Principal Component Analysis. Simulations were completed on this two-dimensional dataset. The proposed scalable algorithm was used to identify 25 lead compound locations from this dataset (*see* **Fig. 4.6**). The algorithm gave higher weights at locations which had larger numbers of similar compounds. Maximally diverse compounds are identified with a very small weight. The original version of the algorithm could not complete the computations for this dataset (on a 512 MB RAM 1.5 GHz Intel Centrino processor).

5.4. Additional Constraints on Lead Compounds

As was discussed in **Section 3**, the multiobjective framework of the proposed algorithm allows us to incorporate additional constraints in the selection problem. In this section, we have addressed two such constraints, namely the experimental resources constraint and the exclusion/inclusion constraint.

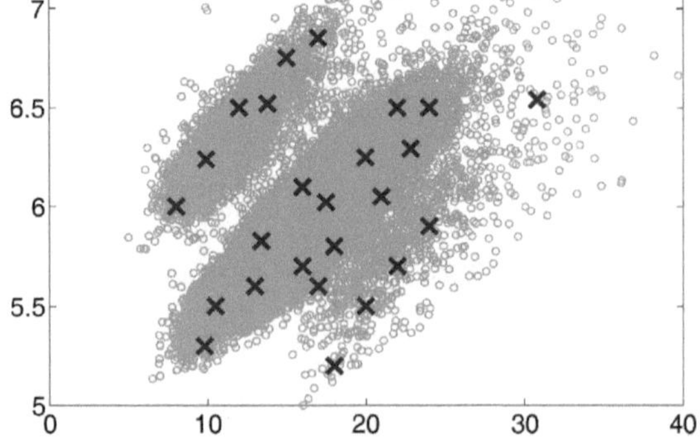

Fig. 4.6. Choosing 25 lead compound locations from the drug discovery dataset. Reprinted ("adapted" or "in part") with permission from *Journal of Chemical Information and Modeling*. Copyright 2008 American Chemical Society.

5.4.1. Constraints on Experimental Resources

In this dataset, the VCL is divided into three classes based on the experimental supplies required by the compounds for testing, as shown in **Fig. 4.7a** by different symbols. It contains a total of 280 compounds with 120 of the first class (denoted by circles), 40 of the second class (denoted by squares), and 120 of the third class (denoted by triangles). We incorporate experimental supply constraints into the algorithm by translating them into direct constraints on each of the lead compounds. With these experimental supply constraints, the algorithm was used to select 15 lead compound locations (r_j) in this dataset with capacities (W_{jn}) fixed for

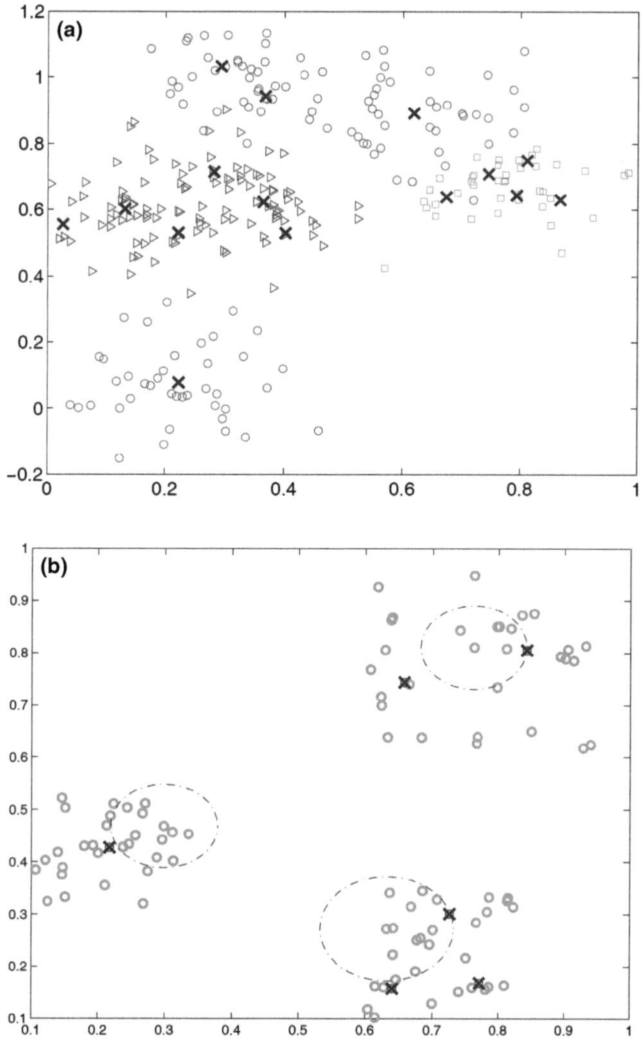

Fig. 4.7. (**a**) Simulation results with constraints on experimental resources. (**b**) Simulation results with exclusion constraint. The locations $x_i, 1 \leq i \leq 90$, of compounds (*circles*) and $r_j, 1 \leq j \leq 6$, of lead compounds (*crosses*). *Dotted circles* represent undesirable properties. Reprinted ("adapted" or "in part") with permission from *Journal of Chemical Information and Modeling.* Copyright 2008 American Chemical Society.

each class of resource. The crosses in **Fig. 4.7a** represent the selection from the algorithm in the wake of the capacity constraints for different types of compounds. As can be seen from the selection, the algorithm successfully addressed the key issues of diversity and representativeness together with the constraints that were placed due to experimental resources.

5.4.2. Constraints on Exclusion and Inclusion of Certain Properties

There may arise scenarios where we would like to inhibit selection of compounds exhibiting properties within certain prespecified ranges. This constraint can be easily incorporated in the cost function by modifying the distance metric used in the problem formulation. Consider a case in a 2-d dataset where each point x_i has an associated radius (denoted by χ_{ij}). The selection problem is the same, but with the added constraint that all the selected lead compounds (r_j) must be at least χ_{ij} distance removed from x_i. The proposed algorithm can be modified to solve this problem by defining the distance function, given by $d(x_i, r_j) = \left(\|x_i - r_j\| - \chi_{ij}\right)^2$, which penalizes any selection (r_j) which is in close proximity to the compounds in the VCL. For the purpose of simulation, a dataset was created with 90 compounds ($x_i, i = 1, \ldots, 90$). The dotted circle around the locations x_i denotes the region in the property space that is to be avoided by the selection algorithm. The objective was to select six lead compounds from this dataset such that the criterion of diversity and representativeness is optimally addressed in the selected subset. The selected locations are represented by crosses. From **Fig. 4.7b**, note that the algorithm identifies the six clusters under the constraint that none of the cluster centers are located in the undesirable property space (denoted by dotted circles).

6. Conclusions

In this chapter, we proposed an algorithm for the design of lead-generation libraries. The problem was formulated in a constrained multiobjective optimization setting and posed as a resource allocation problem with multiple constraints. As a result, we successfully tackled the key issues of diversity and representativeness of compounds in the resulting library. Another distinguishing feature of the algorithm is its scalability, thus making it computationally efficient as compared to other such optimization techniques. We characterized the level of interaction between various clusters and used it to divide the clustering problem with huge data size into manageable subproblems with small size. This resulted in significant improvements in the computation time and enabled the algorithm to be used on larger sized datasets. The trade-off between computation effort and error due to truncation is also characterized, thereby giving an option to the end user.

References

1. Gordon, E. M., Barrett, R. W., Dower, W. J., Fodor, S. P. A., Gallop, M. A. (1994) Applications of combinatorial technologies to drug discovery. 2. Combinatorial organic synthesis, library screening strategies, and future directions. *J Med Chem* **37**(10), 1385–1401.
2. Blaney, J., Martin, E. (1997) Computational approaches for combinatorial library design and molecular diversity analysis. *Curr Opin Chem Biol* **1**, 54–59.
3. Willett, P. (1997) Computational tools for the analysis of molecular diversity. *Perspect Drug Discov Design*, **7/8**, 1–11.
4. Rassokhin, D. N., Agrafiotis, D. K. (2000) Kolmogorov-Smirnov statistic and its applications in library design. *J Mol Graph Model* **18**(4–5), 370–384.
5. Lipinski, C. A., Lomabardo, F., Dominy, B. W., Feeny, P. J. (1997) Experimental and computational approaches to estimate solubility and permeability in drug discovery and development setting. *Adv Drug Del Review* **23**, 2–25.
6. Higgs, R. E., Bemis, K. G., Watson, I. A., Wikel, J. H. (1997) Experimental designs for selecting molecules from large chemical databases. *J Chem Inf Comput Sci* **37**, 861–870.
7. Clark, R. D. (1997) Optisim: an extended dissimilarity selection method for finding diverse representative subsets. *J Chem Inf Comput Sci* **37**(6), 1181–1188.
8. Agrafiotis, D. K., Lobanov, V. S. (2000) Ultrafast algorithm for designing focussed combinatorial arrays. *J Chem Inf Comput Sci* **40**, 1030–1038.
9. Salapaka, S., Khalak, A. (2003) Constraints on locational optimization problems. *Proceedings of the IEEE Control and Decisions Conference.* Maui, HI, 9–12 December 2003, pp. 1741–1746.
10. Sharma, P., Salapaka, S., Beck, C. (2008) A scalable approach to combinatorial library design for drug discovery. *J Chem Inf Model* **48**(1), 27–41.
11. Gersho, A., Gray, R. (1991) *Vector Quantization and Signal Compression.* Kluwer, Boston, Massachusetts.
12. Drezner, Z. (1995) Facility location: a survey of applications and methods. *Springer Series in Operations Research*, Springer, New York.
13. Du, Q., Faber, V., Gunzburger, M. (1999) Centroidal Voronoi tessellations: applications and algorithms. *SIAM Rev* **41**(4), 637–676.
14. Therrien, C. W. (1989) *Decision, Estimation and Classification: An Introduction to Pattern Recognition and Related Topics*, 1st ed. Wiley, New York.
15. Haykin, S. (1998) *Neural Networks: A Comprehensive Foundation*, Prentice Hall, Englewoods Cliffs, NJ.
16. Gray, R., Karnin, E. D. (1982) Multiple local minima in vector quantizers. *IEEE Trans Inform Theor* **28**, 256–361.
17. Lloyd, S. P. (1982) Least squares quantization in PCM. *IEEE Trans Inform Theory* **28(2)**, 129–137.
18. Rose, K. (1998) Deterministic annealing for clustering, compression, classification, regression and related optimization problems. *Proc IEEE* **86**(11), 2210–2239.
19. Mcmaster hts lab competition. HTS data mining and docking competition. http://hts.mcmaster.ca/downloads/82bfbeb4-f2a4-4934-b6a8-804cad8e25a0.html (accessed June 2006).
20. Guha, R. (2006) Chemistry Development Kit (CDK) descriptor calculator GUI (v 0.46). http://cheminfo.informatics.indiana.edu/rguha/code/java/cdkdesc.html (accessed October 2006).
21. Steinbeck, C., Hoppe, C., Kuhn, S., Floris, M., Guha, R., Willighagen, E. L. (2006) Recent developments of the Chemistry Development Kit (CDK) – an open-source JAVA library for chemo and bioinformatics. *Curr Pharm Des* **12**(17), 2110–2120.

Chapter 5

Application of Free–Wilson Selectivity Analysis for Combinatorial Library Design

Simone Sciabola, Robert V. Stanton, Theresa L. Johnson, and Hualin Xi

Abstract

In this chapter we present an application of in silico quantitative structure–activity relationship (QSAR) models to establish a new ligand-based computational approach for generating virtual libraries. The Free–Wilson methodology was applied to extract rules from two data sets containing compounds which were screened against either kinase or PDE gene family panels. The rules were used to make predictions for all compounds enumerated from their respective virtual libraries. We also demonstrate the construction of R-group selectivity profiles by deriving activity contributions against each protein target using the QSAR models. Such selectivity profiles were used together with protein structural information from X-ray data to provide a better understanding of the subtle selectivity relationships between kinase and PDE family members.

Key words: QSAR, Free–Wilson, MLR, virtual libraries, combinatorial chemistry, protein kinase, PDE, enzyme inhibition, enzyme selectivity, docking.

1. Introduction

Combinatorial chemistry has become an essential tool in the pharmaceutical industry for identifying new leads and optimizing the potency of potential lead candidates while reducing the time and costs associated with producing effective and competitive new drugs. By speeding up the process of chemical synthesis, it is now possible to generate large diverse compound libraries to screen for novel bioactivities. At the same time improvements in high-throughput screening (HTS) allow selectivity panels for

J.Z. Zhou (ed.), *Chemical Library Design*, Methods in Molecular Biology 685,
DOI 10.1007/978-1-60761-931-4_5, © Springer Science+Business Media, LLC 2011

gene families or diverse off-target activity to be regularly run against all compounds of interest. Unfortunately, despite these significant synthetic and screening efforts, only few novel lead candidates have been identified for optimization, resulting in increased interest in the use of computational techniques for the design of focused combinatorial libraries rather than simply diverse ones. An additional benefit of these libraries is that they can be used to probe enzyme specificity by analyzing the activity of diverse groups of intrafamily proteins using in silico methods. In this respect, protein kinases (PKs) and phosphodiesterases (PDEs) represent two well-known examples of enzyme superfamilies which have been heavily pursued by both pharmaceutical companies and academic groups because of their mechanistic role in many diseases, thus providing us with a large amount of structural and biological data to be used for developing and validating new in silico methodologies.

Protein kinases (1, 2) catalyze the transfer of the terminal phosphoryl group of ATP to specific hydroxyl groups of serine, threonine, or tyrosine residues of their protein substrates. Because protein kinases have profound effects on a cell, their activity is highly regulated by the binding of activator/inhibitor proteins or small molecules or by controlling their location in the cell relative to their substrates. Intracellular phosphorylation by protein kinases, triggered in response to extracellular signals, provides a mechanism for the cell to switch on or off many diverse processes (3). Deregulated kinase activity is a frequent cause of disease, particularly cancer, since kinases regulate many aspects that control cell growth, movement, and death. Drugs which inhibit specific kinases are being developed to treat many diseases and several are currently in clinical use. These include (1) Gleevec (Imatinib) (4) for chronic myeloid leukemia (CML), (2) Sutent (Sunitinib) (5), a multitargeted receptor tyrosine kinase for the treatment of renal cell carcinoma (RCC) as well as imatinab-resistant gastrointestinal stromal tumor (GIST), (3) Iressa (Gefitinib) (6), and Erlotinib (Tarceva) (7) for non-small cell lung cancer (NSCLC). Previously, studies have shown how molecular specificity varies widely among known inhibitors (8), and this variation is not dictated by the general chemical scaffold of an inhibitor (e.g., EGFR inhibitors, belonging to the quinazoline/quinoline class, range from highly specific to quite promiscuous) or by the primary, intended kinase target toward which the particular inhibitor was initially optimized (e.g., compounds considered tyrosine kinase inhibitors also bind to Ser-Thr kinases and vice versa). Moreover, with over 500 kinases in the human genome, selectivity is a daunting task and predicting selectivity based on the protein-binding site or ligand pharmacophores is extremely challenging given the high degree of homology across the kinase protein family, particularly in the active site region.

The second gene family we included in our study is the PDE superfamily of enzymes that degrade cyclic adenosine monophosphate (cAMP) and cyclic guanosine monophosphate (cGMP) (9–11). Both cAMP and cGMP are intracellular second messengers that play a key role in mediating cellular responses to various hormones and neurotransmitters (12, 13), and their intracellular concentration is tightly regulated at the level of synthesis (by the catalytic reaction of adenylyl cyclase and guanylyl cyclase) as well as degradation (by binding to cyclic nucleotide phosphodiesterases). PDEs are involved in a wide array of pharmacological processes, including proinflammatory mediator production and action, ion channel function, muscle contraction, learning, differentiation, apoptosis, glycogenolysis, and gluconeogenesis (14), and have become recognized as important drug targets for the treatment of various diseases, such as heart failure, depression, asthma, inflammation, and erectile dysfunction (13, 15–17).

Since the early discovery of multiple phosphodiesterase isoforms and their potential use as therapeutic targets (18), the biological and functional understanding around PDEs has expanded from what was understood to be a family of three isozymes (19) toward a total of 21 human PDE genes falling into 11 families with over 60 isoforms (10, 12, 15, 20–24). Selective inhibitors for each of the multiple PDE forms can offer an opportunity for desired therapeutic intervention and would be an extremely useful tool in drug discovery efforts for a medicinal chemist. Although there are distinct differences in the full-length structure of the PDEs, not surprisingly the catalytic domain that shares a common function across different isoforms has a more conserved structure, making the design of highly selective PDE inhibitors a difficult challenge. As for kinases, PDE inhibition has potential therapeutic utility but care must be taken in the rational design of active inhibitors to avoid unwanted off-target PDE inhibition.

Over the past few years, Pfizer has focused on developing selectivity screening platforms (25) to provide high-quality data against a diverse range of PKs and PDEs, which has been used to guide therapeutic projects by analyzing structure–activity relationship (SAR) and identifying potential off-target liabilities of compounds within a chemical series. The integration of this highly valuable data together with appropriate computational methods can speed up the overall lead discovery process by allowing the optimization of property-based design within a homologous series. However, the success of such studies depends on the choice of an appropriate molecular characterization, through the use of informative descriptors.

In the chapter, we report a successful application of the Free–Wilson (26–30) methodology to model structure–activity/selectivity relationships. The Fujita–Ban (31–34) modification of Free–Wilson coupled with multiple linear regression analysis (MLR) was used to model the selectivity profiles of different chemical series in our in-house kinase and PDE screening panel. Overall, reliable estimations for R-group activity contributions against each protein in the data set were observed and used for enumerating focused virtual libraries to predict more selective inhibitors. When an external test set of cherry-picked compounds was used to test the validity of the in silico models, a strong correlation of experimental versus predicted inhibition values was found. Lastly, the availability of X-ray structures in the public domain for both PKs and PDEs allowed us to further validate our QSAR models by combining the information from the Free–Wilson approach with the three-dimensional (3D) structural knowledge of the target, providing more insight into specific enzyme selectivity.

2. Methods

2.1. Assay Conditions

All of the kinase assays are performed in a 384-well format using either a radioactive or Caliper protocol (25). In all assays, 5 μL of 5× concentration compound in 3.75% DMSO is added to the plates. 10 μL of 2.5× enzyme in 1.25× kinase buffer (optimized for each individual kinase) is then added, followed by a 15-min preincubation at room temperature. 10 μL of a 2.5× mixture of peptide substrate (optimized for each individual kinase) and ATP in 1.25× kinase buffer are then added to initiate the reaction. Each assay is run at the experimentally determined Michaelis–Menten constant (K_m) concentration of ATP for the relevant kinase with an incubation time that was determined to be within the linear reaction time. Reactions are stopped by the addition of EDTA to a final concentration of 20 mM. Detection of phosphorylated substrate is achieved using either a radioactive method or a nonradioactive mobility shift assay format (Caliper). In the radioactive assay, tracer amounts of γ-$^{(33)}$P-labeled ATP are included in the reaction, and biotinylated peptide substrates are used. After the reactions are stopped, 25 μL is transferred to streptavidin-coated Flashplates^{TM} (Perkin Elmer). Plates are washed with 50 mM Hepes and soaked for 1 h with 500 μM unlabeled ATP before reading in a TopCount. Alternatively, for the mobility shift assay, reactions are stopped within the assay plates followed by detection of fluorescently labeled substrates on a Caliper LC3000 using a 12-sipper chip and conditions that were optimized for each kinase.

The PDE assays are performed in a 384-well format using a radioactive protocol where the enzymatic activities were assayed by using ^{3}H-cAMP or ^{3}H-cGMP as substrates to a final concentration of 20 nM. The catalytic domain of PDEs was incubated with a reaction mixture of 50 mM Tris·HCl, pH 7.5, 1.3 mM $MgCl_2$, 1 mM DTT, and ^{3}H-cAMP or ^{3}H-cGMP at room temperature on an orbital shaker for 30 min. Compounds to be tested are submitted to the assay at a concentration of 4 mM in 100% DMSO. Compounds are initially diluted in 50% DMSO/water. Subsequent dilutions are in 15% DMSO/water to achieve 5× the desired assay concentration. Each well receives 10 μL drug or DMSO vehicle, 20 μL ^{3}H-cAMP or ^{3}H-cGMP, and 20 μl enzyme (diluted 1:1,000 in assay buffer). The incubation is terminated by the addition of 25 μL of PDE SPA beads (0.2 mg/well). The reaction product ^{3}H-cAMP or ^{3}H-cGMP was precipitated out by $BaSO_4$ while unreacted ^{3}H-cAMP or ^{3}H-cGMP remained in supernatant. After centrifugation, the radioactivity in the supernatant was measured in a liquid scintillation counter after a 500 min delay. The enzymatic properties were analyzed by the steady-state kinetics. The nonlinear regression of the Michaelis–Menten equation as well as Eadie–Hofstee plots was analyzed to obtain the values of K_M, V_{max}, and k_{cat}. For measurement of IC_{50}, ten concentrations of inhibitors were used at the substrate concentration of <1/10 K_M and the suitable enzyme concentration. All measurements were repeated three times.

2.2. Data Sets

In the kinase case study, 975 compounds based on chemical series belonging to four different chemotypes (**Table 5.1**) were screened against 45 protein kinases, selected to provide maximal coverage across subfamilies within the kinome (25). The data set consists of 388 compounds with the Diaminopyrimidine core (R1=77, R2=183), 312 with the Pyrrolopyrazole core (R1=124, R2=87), 181 with the Pyrrolopyrimidine core (R1=8, R2=169), and 94 sharing the Quinazoline core (R1=19, R2=5, R3=37, R4=33). Due to changes in the panel over time, not all the compounds in the study were screened against each individual kinase, giving rise to an incomplete combinatorial matrix. However, these chemical series were selected trying to consistently meet the criteria of having a high number of compound per kinase assay (**Fig. 5.1**). Percent inhibition data at two compound concentrations, 1 μM and 10 μM, was first transformed into pIC_{50} (35) and then combined to give a single pIC_{50} value by applying the following equations:

$$pIC_{50}@1\,\mu M = -\log\left(10^{-6} \times \frac{100 - (\text{percent inhibition}@10\,\mu M)}{\text{percent inhibition}@10\,\mu M}\right)$$

$$pIC_{50}@10\,\mu M = -\log\left(10^{-5} \times \frac{100 - (\text{percent inhibition}@10\,\mu M)}{\text{percent inhibition}@10\,\mu M}\right)$$

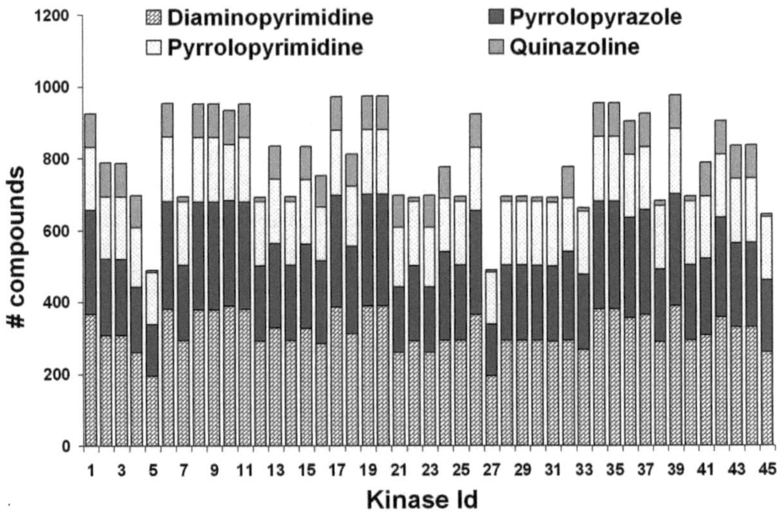

Fig. 5.1. Number of compounds tested against each of the 45 protein kinases in the in-house selectivity panel. Histogram bars are subdivided according to the compound's frequency in the four kinase chemical series.

$$pIC_{50}^{Calc} = \begin{cases} pIC_{50}@1\mu M, & Inhib@10\mu M > 99\% \\ pIC_{50}@10\mu M, & Inhib@1\mu M < 5\% \\ \frac{pIC_{50}@1\mu M + pIC_{50}@10\mu M}{2}, & 5\% \leq Inhib@1\mu M \leq 99\% \end{cases}$$

The reported block function was adopted to improve the overall correlation between calculated and experimental pIC_{50} at the two different concentrations. As reported previously (25, 36), in the lower range of inhibition, below 5%, a stronger correlation between pIC_{50}^{Calc} computed at 10 μM concentration and experimental pIC_{50}^{Calc} was found, when compared to 1 μM. An opposite trend was present in the upper range of inhibition (above 99%), where pIC_{50}^{Calc} computed at 1 μM concentration tended to correlate better with experiment than that at 10 μM. For inhibition values between the previously defined cut-offs, we used the average pIC_{50}.

The second data set consists of 1,505 total compounds sharing a unique chemotype (Pyrazolopyrimidine) tested in two different PDE biochemical assays (PDE2 and PDE10). Four sites of substitutions (R1=62, R2=157, R3=872, R4=339) were allowed to change around the Pyrazolopyrimidine core substructure (**Table 5.1**). Although not all the compounds in the study were tested against both PDEs (1,357 and 1,346 compounds tested, respectively, in PDE2 and PDE10), a large number of compounds (1,198) were in common between the two assays, providing us with a wealth of data to be used for studying their selectivity profiles. Different from the kinase data set, all the

Table 5.1
2D depiction for the five chemical series. R-positions represent sites which were allowed to change within a given library while X-positions indicate not changing chemical matter whose structure cannot be disclosed

Protein family	Chemical series	2D depiction	Number of Compounds	R-groups
Kinases	Diaminopyrimidine	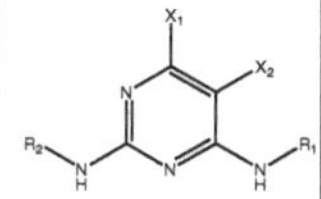	388	R1 = 77 R2 = 183
	Pyrrolopyrazole	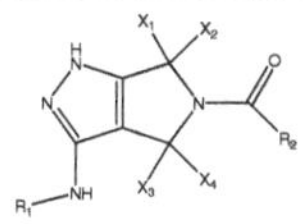	312	R1 = 124 R2 = 87
	Pyrrolopyrimidine	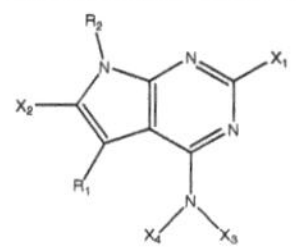	181	R1 = 8 R2 = 169
	Quinazoline	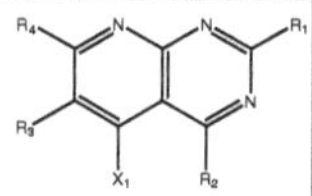	94	R1 = 19 R2 = 5 R3 = 37 R4 = 33
PDEs	Pyrazolopyrimidine	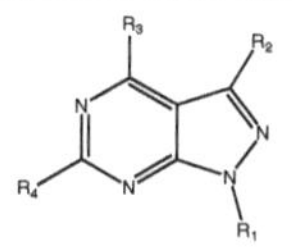	1505	R1 = 62 R2 = 157 R3 = 872 R4 = 339

PDE compounds were tested for IC_{50}, therefore, no data transformation was required in the case. The negative logarithm of IC_{50} was used as a dependent variable in the model-building process.

2.3. Free–Wilson (F–W)

The Free–Wilson approach was the first mathematical technique to be developed for the quantitative prediction of the structure–activity relationships for a series of chemical analogs (26). The basic idea behind this methodology is that the biological activity of a molecule can be described as the sum of the activity contributions of specific substructures (parent fragment and the corresponding substituents). It does not require any substituent parameters or descriptors to be defined; only the activity is needed. The underlying assumption in Free–Wilson modeling is that the contribution of each substituent to the biological activity is additive and constant, regardless of the structural variation on the other sites of substitution in the rest of the molecule. The classical Free–Wilson linear model is expressed by the following equation:

$$\text{BioActivity} = \sum_{\text{ij}} \alpha_{ij}{}^{*} R_{ij} + \mu$$

where the constant term μ (activity value of the unsubstituted compound) is the overall average of biological activities and α_{ij} is the R-group contribution of substituent R_i in position j. If substituent R_i is in position j, then $R_{ij} = 1$, otherwise $R_{ij} = 0$. This gives rise to a set of equations that can be potentially solved by MLR, where α_{ij} are the regression coefficients, R_{ij} the independent variables, and μ the intercept. Unfortunately, MLR cannot be applied directly to the resulting structural matrix due to a linear dependence on its columns (34). One way to get around these dependencies is to use the Fujita–Ban modification where the activity contribution of each substituent is relative to H and the constant term μ, obtained by the least-squares method, is a theoretically predicted activity value of the unsubstituted compound itself (all R-groups set to H) (31). Kubinyi et al. have shown that the original Free–Wilson and the Fujita–Ban modifications are linearly related, with the latter approach being a linear transformation of the classical Free–Wilson model (34). Additionally, the Fujita–Ban model leads to a number of important advantages. First, no complex transformation of the structural matrix is required and only the removal of one column for each site of substitution is necessary to move from the structural matrix to the Fujita–Ban matrix. Second, the matrix is not changed by the addition or elimination of a compound. Third, in the Fujita–Ban model the constant term μ in the linear equation is derived theoretically by applying the least-squares method and therefore not markedly influenced by the addition or elimination of a compound. In consideration of these advantages, the Fujita–Ban modification of the Free–Wilson mathematical model was implemented for the analysis reported here.

2.4. F–W Model Building and Validation

The Fujita–Ban modification of the Free–Wilson methodology was applied to the structural matrices of descriptors corresponding to each chemical series/biochemical assay combination analyzed in this study and individual QSAR models were built. The first step consisted of generating the R-groups by fragmenting all the compounds within each series, thus obtaining the initial structural matrices. After that, compounds with correlated R-groups and outlier compounds whose R-groups did not occur in other compounds were removed from the data set as the activity contribution for these R-groups could not be estimated. Then the remaining structural matrix was rearranged into independent blocks where R-groups from one block would not cross over with other blocks, and statistical analysis was applied to each block separately to estimate activity contribution for each R-group. Furthermore, blocks whose R-group activity contributions could not be estimated due to a lack in R-group crossovers were further eliminated. This block separation and compound removal procedure maximized the total number of R-group activity contributions that could be estimated.

The relationship between the enzyme inhibition data and the chemical structures was analyzed using MLR, a multivariate regression method able to quantitatively model the relationship between two or more explanatory variables and a response variable by fitting a linear equation to the observed data. An MLR model was first built independently for each series/biochemical assay combination. The quality of the models both in terms of fitting the experimental data and predicting the activity for new compounds through cross-validation techniques was assessed by computing the squared Pearson correlation coefficient (r^2_{corr}) between predicted and actual activities together with the associated standard error of correlation (STE):

$$r^2_{\text{corr}} = \frac{\left[\sum_{i\in\text{test}} \left(y_i^{\text{pred}} - \bar{y}_i^{\text{pred}}\right)\left(y_i^{\text{act}} - \bar{y}_i^{\text{act}}\right)\right]^2}{\sum_{i\in\text{test}} \left(y_i^{\text{pred}} - \bar{y}_i^{\text{pred}}\right)^2 \sum_{i\in\text{test}} \left(y_i^{\text{act}} - \bar{y}_i^{\text{act}}\right)^2}$$

$$\text{STE} = \sqrt{\frac{1}{n-2}\left[\sum_{i\in\text{test}} \left(y_i^{\text{pred}} - \bar{y}_i^{\text{pred}}\right)^2 - \frac{\left[\sum_{i\in\text{test}} \left(y_i^{\text{pred}} - \bar{y}_i^{\text{pred}}\right)\left(y_i^{\text{act}} - \bar{y}_i^{\text{act}}\right)\right]^2}{\sum_{i\in\text{test}} \left(y_i^{\text{act}} - \bar{y}_i^{\text{act}}\right)^2}\right]}$$

Here, y_i^{pred} is the predicted activity for the ith test set compound, y_i^{act} is its measured activity, $\bar{y}_i^{\text{pred}}$ and $\bar{y}_i^{\text{act}}$ are the average of the predicted and measured activity values, respectively, and n is the sample size.

The squared Pearson correlation coefficients for the linear models built upon the Diaminopyrimidine, Pyrrolopyrazole, Pyrrolopyrimidine, and Quinazoline series across the 45 protein kinases are, respectively, in the range of $r^2_{\text{fitting}} = 0.82 \leftrightarrow 0.95$ (average $r^2_{\text{fitting}} = 0.87$), $r^2_{\text{fitting}} = 0.73 \leftrightarrow 0.93$ (average $r^2_{\text{fitting}} = 0.85$), $r^2_{\text{fitting}} = 0.36 \leftrightarrow 0.99$ (average $r^2_{\text{fitting}} = 0.80$), and $r^2_{\text{fitting}} = 0.46 \leftrightarrow 0.97$ (average $r^2_{\text{fitting}} = 0.76$). For the PDE case study, the correlation coefficients for the Pyrazolopyrimidine series when tested in the PDE2 and PDE10 biochemical assays are, respectively, $r^2_{\text{fitting}} = 0.94 \pm 0.17$ and $r^2_{\text{fitting}} = 0.92 \pm 0.18$. The highly significant correlation between experimental and calculated pIC_{50}

confirmed the basic assumption of the Free–Wilson method for this set of biological data, which is the additivity of R-group effects.

The models predictivity was evaluated using standard Leave-One-Out (LOO) analysis as "internal validation" technique. LOO is a cross-validation procedure that works by building reduced models (models for which one object at a time is removed) and using them to predict the *Y*-variables of the object held out. Results obtained by applying LOO validation to the kinase and PDE data sets are shown in **Figs. 5.2** and **5.3**, respectively. In general, the predicted pIC_{50} is in agreement with the calculated pIC_{50} derived from experimental data. In the Diaminopyrimidine series, taking all 45 kinase models together, 6,712 LOO estimations were carried out giving a global correlation coefficient $r^2_{\mathrm{corr,CV}} = 0.90$ and a standard error of the predicted pIC_{50} value in the regression STE = 0.35. Similar results

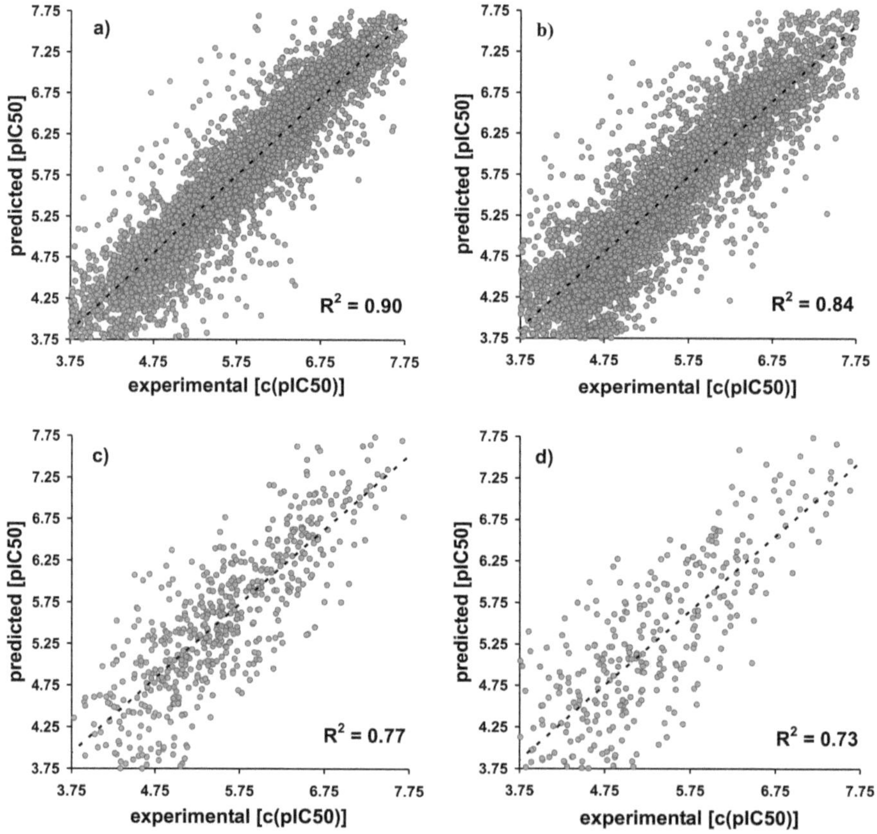

Fig. 5.2. Leave-One-Out cross-validation results reported as predicted vs. experimental pIC_{50} values for the four kinase chemical series. In general, model prediction of pIC_{50} is in good agreement with experimental pIC_{50} derived from percent of inhibition, with a global correlation coefficient $r^2_{\mathrm{corr,CV}} = 0.90$ for Diaminopyrimidine (**a**), $r^2_{\mathrm{corr,CV}} = 0.84$ for the Pyrrolopyrazole (**b**), $r^2_{\mathrm{corr,CV}} = 0.77$ for the Pyrrolopyrimidine (**c**), and $r^2_{\mathrm{corr,CV}} = 0.73$ for the Quinazoline (**d**) series.

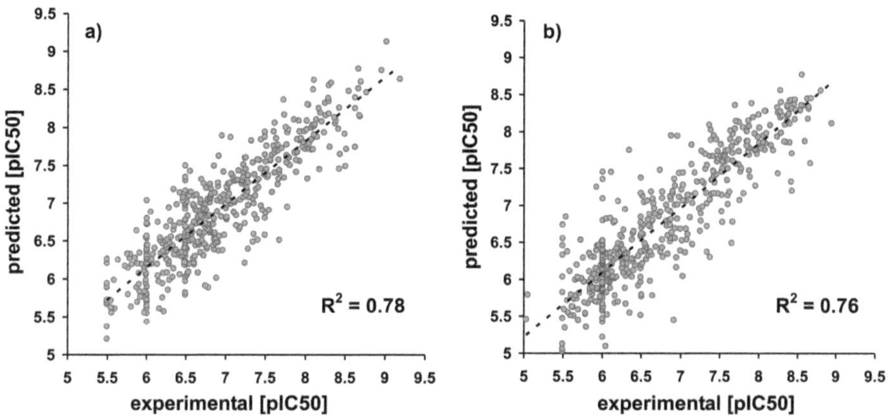

Fig. 5.3. Leave-One-Out cross-validation results for the Pyrazolopyrimidine series tested in the biochemical assays PDE2 (**a**) and PDE10 (**b**).

were obtained for the Pyrrolopyrazole series, where LOO estimations of 5,413 objects gave an overall correlation coefficient $r^2_{\mathrm{corr,CV}} = 0.85$ (STE = 0.47), the Pyrrolopyrimidine series with $r^2_{\mathrm{corr,CV}} = 0.77$ (STE = 0.53 based on 650 LOO estimations), and the Quinoline series where 707 LOO estimations resulted in $r^2_{\mathrm{corr,CV}} = 0.73$ (STE = 0.64). The same cross-validation protocol was carried out in the case of Pyrazolopyrimidine series obtaining the following correlation coefficients: $r^2_{\mathrm{corr,CV}} = 0.78$ (STE = 0.38, 485 LOO estimations) and $r^2_{\mathrm{corr,CV}} = 0.76$ (STE = 0.46, 473 LOO estimations) when tested in the PDE2 and PDE10 assays, respectively (**Fig. 5.3**).

Since Free–Wilson models use the presence or absence of distinct R-group fragments as the basic variables in regression, the derived model coefficients can be treated as a quantitative estimate of the activity contribution of each R-group. Assuming the additive assumption holds, then these R-group contributions can be used to make reliable predictions for all the enumerated compounds in a virtual library, where all R-group fragments are crossed with each other.

2.5. Virtual Library Space Analysis

After model building and validation, the R-groups within each chemical series were exhaustively combined with each other and their pIC_{50} contributions from the F–W QSAR models used to predict the final activity of the compounds enumerated in the virtual library. This step represents one of the key advantages of using F–W methodology over standard descriptors-based QSAR techniques that is the deconvolution of the biological activity of a molecule into its components (parent fragment plus the corresponding substituents). Indeed, due to experimental and synthetic limitations, typically only a small number of compounds can be synthesized and screened against a given biochemical assay.

As a result, many compounds with desired potency and selectivity profiles could potentially be missed. By using high-quality QSAR models, the activity and selectivity of compounds in the virtual library can be reliably estimated, thus, greatly expanding the chemical space coverage and increasing the chance of finding compounds with attractive biological properties.

To demonstrate this, we enumerated the full virtual library for the five chemical series shown in **Table 5.1**. We obtained 861 compounds for the Diaminopyrimidine series, 1,764 compounds for the Pyrrolopyrazole series, 598 for the Pyrrolopyrimidine series, 2,370 for the Quinazoline series, and 214,486 for the Pyrazolopyrimidine series, using only those R-groups from the existing compounds for which the activity contribution could be estimated across the 45 protein kinase (first four chemical series) and the two PDE assays (Pyrazolopyrimidine). We then calculated their selectivity profile using the QSAR models derived from Free–Wilson analysis. Among the existing compounds in the kinases series, 27 of them (17 Diaminopyrimidines, 1 Pyrrolopyrazole, 6 Pyrrolopyrimidine, 3 Quinazoline) met our selectivity criteria ($pIC_{50} > 5.3$ against no more than 5 kinases on the panel). In the full virtual library, however, 111 additional compounds (57 Diaminopyrimidines, 8 Pyrrolopyrazoles, 31 Pyrrolopyrimidine, 15 Quinazoline) were predicted to be selective. In the PDE series, the library expansion provided with a greater enrichment in the number of compounds potentially selective, moving from three selective compounds in the original library ($pIC_{50} \geq 7$ in one assay and $pIC_{50} \leq 5.3$ in the second assay) to 4,103 selective compounds in the virtual space.

We have also noticed an increase in the number of kinases selectively targeted upon the expansion of the inhibitor's chemical space, suggesting that such a procedure would also be suitable as a tool for exploring potential "Target Hopping." Indeed, when applied to our data set, existing selective compounds from the Diaminopyrimidine, Pyrrolopyrazole, Pyrrolopyrimidine, and Quinazoline series targeted 14, 5, 7, and 3 protein kinases, respectively. However, after complete enumeration of the virtual libraries, 28, 19, 31, and 12 protein kinases were predicted to be selectively inhibited by compounds in the four series, respectively. This shows how series originally developed for a specific kinase could be turned into selective inhibitors for other kinases by exploiting different R-group combinations.

2.6. R-Group Selectivity Profiles

The objective of this analysis was to gain knowledge from the R-group contributions as determined by the Free–Wilson methodology. Only R-groups for which a coefficient could be determined across the 45 kinases in the panel and the 2 PDE biochemical assays reported in this study were taken into account. For the Diaminopyrimidine series, this resulted in 36 R1- and 26

R2-group structures, giving rise to two different matrices containing 36×45 R1- and 26×45 R2-group contributions. In the Pyrrolopyrazole series, a total of 60 R1- and 35 R2-group structures were available for analysis, leading to two coefficient matrices of 60×45 R1- and 35×45 R2-group contributions. Analysis of the R-group structures for the Pyrrolopyrimidine and Quinazoline series resulted in two coefficient matrices of 3×45 R1- and 57×45 R2-group contributions for the former series and four coefficient matrices of 4×45 R1-, 2×45 R2-, 15×45 R3-, and 11×45 R4-group contributions for the latter. In the Pyrazolopyrimidine PDE series, a total of 5 R1-, 79 R2-, 543 R3-, and 3 R4-group structures were available for analysis, leading to four coefficient matrices of 5×2 R1-, 79×2 R2-, 543×2 R3-, and 3×2 R4-group contributions.

The main objective in this R-group selectivity analysis was to detect whether small changes in structure could give rise to large variations in activity. This was achieved by computing all pairwise structural similarities between R-groups at each substitution site (using a combination of structural descriptors (37, 38) and Tanimoto as similarity measure), then keeping only R-group pairs with Tanimoto similarity greater than 0.8. Afterward, each surviving R-group pair was assigned a profile resulting from the difference in the original coefficients profiles for the R-groups being compared. This produced one selectivity map for each R-group position within each different chemical series. **Figures 5.4** and **5.5** show a few snapshots of this data transformation for the Diaminopyrimidine and Pyrazolopyrimidine series, reported as heat maps where each R-group pair/assay combination is assigned a color ranging from white ($\Delta pIC_{50} = 0$) to red ($\Delta pIC_{50} \geq 2$).

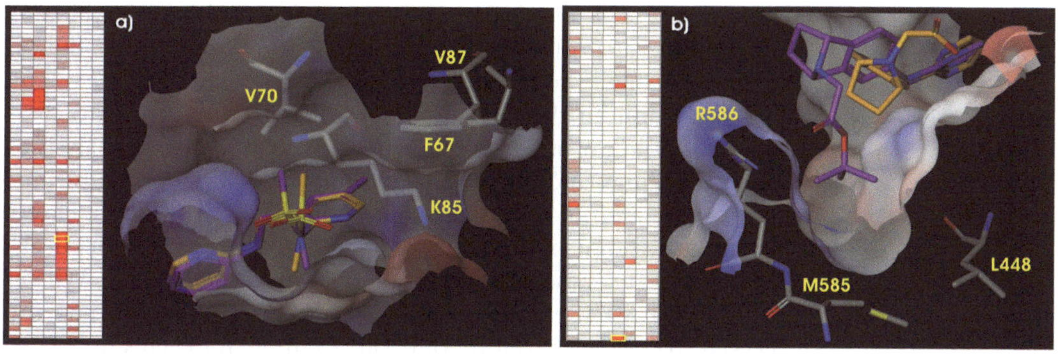

Fig. 5.4. Structural models for binding site interactions of Diaminopyrimidine series. Selectivity maps are shown next to each binding site model. ΔpIC_{50} for the specific R-group pair/assay combination is highlighted in *yellow* (R-group combinations are reported as rows and protein kinase assays as columns within the heat map). (**a**) R-group^{A} (*orange*) and R-group^{B} (*violet*) at site R1 of Diaminopyrimidine docked into the crystal structure of GSK3β (1O9U). The extra methyl in R-group^{B} is responsible for its increased activity contribution. (**b**) Position R2 of Diaminopyrimidine in protein kinase PAK4 (2CDZ). R-group^{B} (*violet*) undergoes a 45° rotation in order to orient the *tert*-butyloxy tail toward the buried lipophilic pocket made up residues R586, M585, and L448.

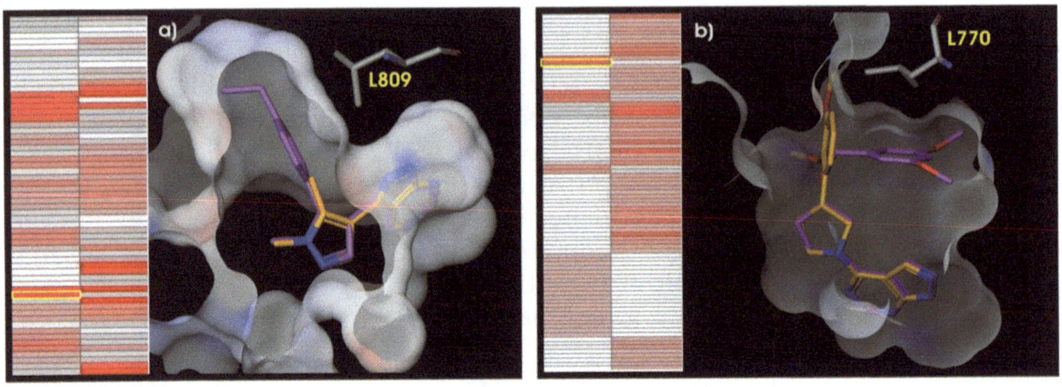

Fig. 5.5. Structural models for binding site interactions of Pyrazolopyrimidine series. Selectivity maps are shown next to each binding site model. ΔpIC_{50} for the specific R-group pair/assay combination is highlighted in *yellow* (R-group combinations are reported as rows and protein PDE assays as columns within the heat map). (**a**) R-group^{A} (*orange*) and R-group^{B} (*violet*) at site R2 of Pyrazolopyrimidine docked into the in-house PDE2 crystal structure. The extra phenethyl moiety in R-group^{B} makes an extended hydrophobic interaction with residue L809 and it is responsible for the observed increased in activity. (**b**) Position R3 of Pyrazolopyrimidine in the PDE2 crystal structure. The presence of two extra atoms linker in R-group^{B} (*violet*) determines its different binding mode compared to R-group^{A}. The 1,3-dimethoxy benzene portion of R-group^{B} undergoes a 90° rotation in order to orient itself toward a buried lipophilic pocket and interacting directly with the side chain of residue L770.

To provide more insight into kinase/PDE selectivity and to analyze the variations in pIC_{50} based upon small structural changes at the R-group level, we combined the information from the Free–Wilson approach with the 3D structural knowledge of the target. This analysis was made possible by the availability of numerous in-house as well as public protein kinase and phosphodiesterase crystal structures. In this respect, a structure-based study was carried out for each R-group/protein combination using an internal core-docking workflow (39), which consists of a protocol specifically designed for screening multiple combinatorial libraries against a family of proteins and relies on the common alignment of all the available protein X-ray structures.

Although all the virtual compounds were docked into their corresponding protein crystal structures, an exhaustive analysis of these dockings and the interpretation of the R-group contributions contained in each of the individual selectivity heat maps is beyond the scope of this study. Our objective here was spot checking the ligand-based results obtained through the Free–Wilson analysis to see if they were consistent with the known enzyme crystal structure; therefore, only one example for each site of substitution for the Diaminopyrimidine kinase series and the Pyrazolopyrimidine PDE series is shown here (**Figs. 5.4** and **5.5**).

Starting with the R1 position of Diaminopyrimidine, structural poses for R-group^{A} and R-group^{B}, as described in **Table 5.2**, were analyzed after docking into protein active site of kinase GSK3β (PDB entry: 1O9U). A variation in pIC_{50} of 1.8

Table 5.2
R-group/kinase contributions from Free–Wilson selectivity maps

Core	Site	R-group^A	R-group^B	Protein	F-W$^{\Delta pIC50}$ (R^B–R^A)
Diaminopyrimidine	R1	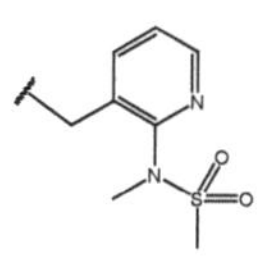	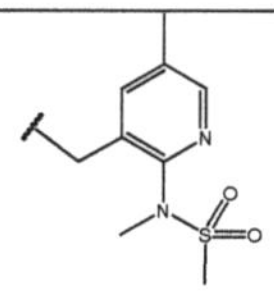	gsk3β (1O9U)	+1.8
	R2	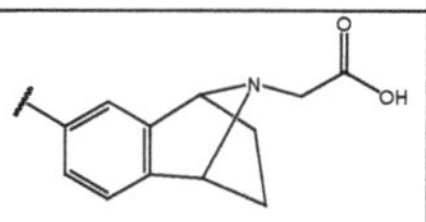	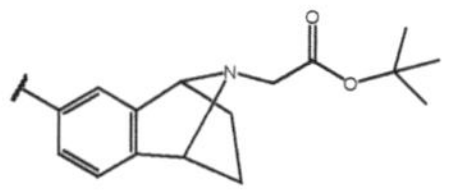	pak4 (2CDZ)	+2.5
Pyrazolopyrimidine	R2	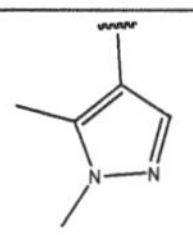	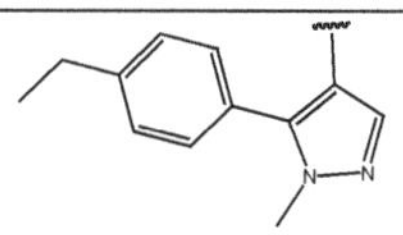	pde2 (in-house)	+1.8
	R3	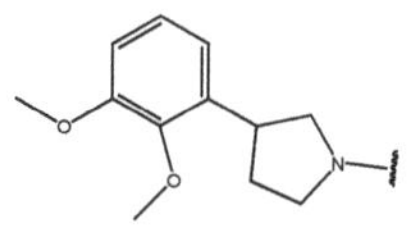	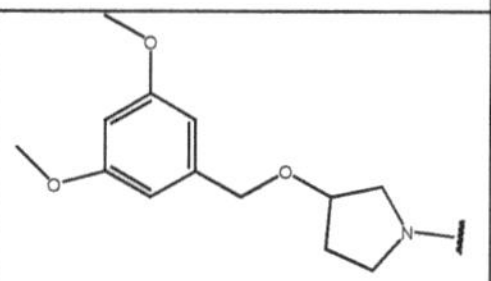	pde2 (in-house)	+1.1

logarithmic units was found using Free–Wilson calculations for estimating the activity contributions of these R-groups. The only structural difference between the two is a methyl at position 5 of the pyridine ring. Although the docking study showed the same binding mode, the methyl moiety in R-group^B is now buried into the protein kinase active site and pointing toward a small lipophilic pocket (F67, V70, K85, V87), explaining the increase in activity predicted by the Free–Wilson model (**Fig. 5.4a**). A different combination of R-groups/protein kinase was examined using the R2 position of Diaminopyrimidine. **Figure 5.4b** shows the resulting poses for R-group^A and R-group^B (**Table 5.2**) when docked into the PAK4 protein kinase-binding site (PDB entry: 2CDZ). Changing from the carboxy- to the *tert*-butyloxy-moiety forces a different binding orientation of the R-groups within the active site. The structure-based rationalization for pIC_{50} difference ($\Delta pIC_{50} = 2.5$) is the R-group^B which undergoes a 45° rotation, around the C–N single bond linking the R-group to the Diaminopyrimidine core, allowing the *tert*-butyloxy tail to orient in the direction of a buried lipophilic pocket made by cavity-flanking residues L448, M585, and R586 (**Fig. 5.4b**).

Similar conclusions can be derived when analyzing the core-docking results for the Pyrazolopyrimidine series. The in-house X-ray structure of PDE2 was used to elucidate the differences in activity ($\Delta pIC_{50} = 1.8$) when moving from R-group^A to R-group^B (**Table 5.2**) at position R2 of the Pyrazolopyrimidine core. **Figure 5.5a** highlights the structural explanation for that, where the presence of the additional phenyl ring at this site is not influencing the R-group binding mode, but is extending the staked hydrophobic interaction toward residue L809. When position R3 of the Pyrazolopyrimidine series was examined, a ΔpIC_{50} of 1.1 units was obtained by substituting two highly similar R-groups in the PDE2 biochemical assay (R-group^A and R-group^B in **Table 5.2**). **Figure 5.5b** shows how the variation in R-group composition determines a different binding mode for the two R-groups, with the 1,3-dimethoxy benzene portion of R-group^B now filling a hydrophobic pocket in the active site made up of a combination of lipophilic (L770, L809, I866, I870) residues, and optimizing stacked hydrophobic interactions with the isopropyl moiety of residue L770 (**Fig. 5.5b**).

3. Notes

1. The Free–Wilson approach has proven to be a successful strategy for the analysis of data sets where large library collections of compounds obtained through combinatorial chemistry have been screened against a panel of related proteins or target families, thus boosting the overall quest for selective inhibitors.
2. A key advantage of the Free–Wilson method over standard descriptors-based QSAR techniques is the estimation of activity contribution for individual R-group structures that are readily interpretable to medicinal chemists.
3. The possibility to expand the original chemical space of a given chemical series into a complete virtual library provided us with the identification of compounds with desirable selectivity profiles.
4. The major disadvantage relies on the use of R-groups as descriptors in model building which gives the models a well-defined boundary of the chemical space that can be predicted. It can only explore the chemical space defined by the R-group combinations present in the training set compounds and cannot be applied, as it is, for predicting the activity of new compounds with R-groups beyond those used in the analysis.

5. Data preparation and quality control is a key step in applying Free–Wilson methodology to model biological data. Care must be taken to make sure the underlying data complies with F–W additive assumption.
6. Compounds with correlated R-groups and outlier compounds whose R-groups did not occur in other compounds were removed from the data set as the activity contribution for these R-groups could not be estimated.
7. In case of sparse structural matrices, these were normally rearranged into independent blocks where R-groups from one block would not cross over with other blocks, and statistical analysis was applied to each block separately to estimate the activity contribution for each R-group. Blocks whose R-group activity contributions could not be estimated due to a lack in R-group crossovers were further eliminated. The block separation and compound removal procedure maximized the total number of R-group activity contributions that could be estimated.
8. LOO cross-validation analysis of F–W QSAR models showed an overall agreement between predicted and experimental pIC_{50} for each individual combination of chemical series and protein target.
9. The construction of R-group selectivity profiles based on in silico R-group contributions allowed us to identify structural determinants for selectivity where a small modification in the R-groups results in significant difference in selective profiles.
10. The R-group selectivity knowledge coupled with the availability of X-ray data for many of the kinase/PDE structures provides substrates for scientists to formulate novel lead transformation ideas for inhibitor compounds with better physicochemical properties.

Acknowledgment

This chapter is adapted in part with permission from Simone Sciabola et al. (2008) *J Chem Info Model* **48**, 1851–1867.

References

1. Manning, G., Whyte, D. B., Martinez, R., Hunter, T., Sudarsanam, S. (2002) The protein kinase complement of the human genome. *Science* **298**, 1912–1934.
2. Kostich, M., English, J., Madison, V., et al. (2002) Human members of the eukaryotic protein kinase family. *Genome Biol* **3**(9), 0043.1–0043.12.

3. Johnson, L. N., Lewis, R. J. (2001) Structural basis for control by phosphorylation. *Chem Rev* **101**, 2209–2242.
4. Nagar, B., Bornmann, W. G., Pellicena, P., et al. (2002) Crystal structures of the kinase domain of c-Abl in complex with the small molecule inhibitors PD173955 and Imatinib (STI-571). *Cancer Res* **62**, 4236–4243.
5. George, S. (2007) Sunitinib, a multitargeted tyrosine kinase inhibitor, in the management of gastrointestinal stromal tumor. *Curr Oncol Rep* **9**(4), 323–327.
6. Yun, C. -H., Boggon, T. J., Li, Y., et al. (2007) Structures of lung cancer-derived EGFR mutants and inhibitor complexes: mechanism of activation and insights into differential inhibitor sensitivity. *Cancer Cell* **11**(3), 217–227.
7. Stamos, J., Sliwkowski, M. X., Eigenbrot, C. (2002) Structure of the epidermal growth factor receptor kinase domain alone and in complex with a 4-anilinoquinazoline inhibitor. *J Biol Chem* **277**(48), 46265–46272.
8. Fabian, M. A., Biggs, W. H., Treiber, D. K., et al. (2005) A small molecule–kinase interaction map for clinical kinase inhibitors. *Nat Biotech* **23**(3), 329–336.
9. Beavo, J. A. (1995) Cyclic nucleotide phosphodiesterases: functional implications of multiple isoforms. *Physiologic Rev* **75**(4), 725–748.
10. Soderling, S. H., Beavo, J. A. (2000) Regulation of cAMP and cGMP signaling: new phosphodiesterases and new functions. *Curr Opin Cell Biol* **12**(2), 174–179.
11. Manallack, D. T., Hughes, R. A., Thompson, P. E. (2005) The next generation of phosphodiesterase inhibitors: structural clues to ligand and substrate selectivity of phosphodiesterases. *J Med Chem* **48**(10), 3449–3462.
12. Conti, M., Jin, S. L. (1999) The molecular biology of cyclic nucleotide phosphodiesterases. *Prog Nucleic Acid Res Mol Biol* **63**, 1–38.
13. Mehats, C., Andersen, C. B., Filopanti, M., Jin, S. L. C., Conti, M. (2002) Cyclic nucleotide phosphodiesterases and their role in endocrine cell signaling. *Trends Endocrinol Metab* **13**(1), 29–35.
14. Perry, M. J., Higgs, G. A. (1998) Chemotherapeutic potential of phosphodiesterase inhibitors. *Curr Opin Chem Biol* **2**(4), 472–481.
15. Torphy, T. (1998) Phosphodiesterase isozymes: molecular targets for novel antiasthma agents. *Am J Respir Crit Care Med* **157**, 351.
16. Rotella, D. P. (2002) Phosphodiesterase 5 inhibitors: current status and potential applications. *Nat Rev Drug Discov* **1**(9), 674–682.
17. Conti, M., Nemoz, G., Sette, C., Vicini, E. (1995) Recent progress in understanding the hormonal regulation of phosphodiesterases. *Endocr Rev* **16**(3), 370–389.
18. Weishaar, R. E., Cain, M. H., Bristol, J. A. (1985) A new generation of phosphodiesterase inhibitors: multiple molecular forms of phosphodiesterase and the potential for drug selectivity. *J Med Chem* **28**(5), 537–545.
19. Appleman, M. M., Thompson, W. J. (1971) Multiple cyclic nucleotide phosphodiesterase activities from rat brain. *Biochemistry* **10**(2), 311–316.
20. Manganiello, V. C., Degerman, E. (1999) Cyclic nucleotide phosphodiesterases (PDEs): diverse regulators of cyclic nucleotide signals and inviting molecular targets for novel therapeutic agents. *Thromb Haemostasis* **82**, 407.
21. Houslay, M. D., Adams, D. R. (2003) PDE4 cAMP phosphodiesterases: modular enzymes that orchestrate signalling cross-talk, desensitization and compartmentalization. *Biochem J* **370**, 1.
22. Corbin, J. D., Francis, S. H. (1999) Cyclic GMP phosphodiesterase-5: target of sildenafil. *J Biol Chem* **274**, 13729–13732. doi:10.1074/jbc.274.20.13729.
23. Francis, S. H. T. I., Corbin, J. D. (2001) Cyclic nucleotide phosphodiesterases: relating structure and function. *Prog Nucleic Acid Res Mol Biol* **65**, 1.
24. Conti, M., Richter, W., Mehats, C., Livera, G., Park, J. Y. (2003) Cyclic AMP-specific PDE4 phosphodiesterases as critical components of cyclic AMP signaling. *J Biol Chem* **278**, 5493.
25. Card, A., Caldwell, C., Min, H., et al. (2009) High-throughput biochemical kinase selectivity assays: panel development and screening applications. *J Biomol Screen* **14**(1), 31–42.
26. Free, S. M., Wilson, J. W. (1964) A mathematical contribution to structure-activity studies. *J Med Chem* **7**(4), 395–399.
27. Craig, P. N. (1972) Structure-activity correlations of antimalarial compounds. 1. Free-Wilson analysis of 2-phenylquinoline-4-carbinols. *J Med Chem* **15**(2), 144–149.
28. Nisato, D., Wagnon, J., Callet, G., et al. (1987) Renin inhibitors. Free-Wilson and correlation analysis of the inhibitory potency of a series of pepstatin analogs on plasma renin. *J Med Chem* **30**(12), 2287–2291.

29. Schaad, L. J., Hess, B. A., Purcell, W. P., Cammarata, A., Franke, R., Kubinyi, H. (1981) Compatibility of the Free-Wilson and Hansch quantitative structure-activity relations. *J Med Chem* **24**(7), 900–901.
30. Tomic, S., Nilsson, L., Wade, R. C. (2000) Nuclear receptor-DNA binding specificity: a COMBINE and Free-Wilson QSAR analysis. *J Med Chem* **43**(9), 1780–1792.
31. Fujita, T., Ban, T. (1971) Structure-activity study of phenethylamines as substrates of biosynthetic enzymes of sympathetic transmitters. *J Med Chem* **14**(2), 148–152.
32. Hernandez-Gallegos, Z., Lehmann, P. A. (1990) A Free-Wilson/Fujita-Ban analysis and prediction of the analgesic potency of some 3-hydroxy- and 3-methoxy-*N*-alkylmorphinan-6-one opioids. *J Med Chem* **33**(10), 2813–2817.
33. Kubinyi, H., Kehrhahn, O. H. (1976) Quantitative structure-activity relationships. 1. The modified Free-Wilson approach. *J Med Chem* **19**(5), 578–586.
34. Kubinyi, H., Kehrhahn, O. H. (1976) Quantitative structure-activity relationships. 3. A comparison of different Free-Wilson models. *J Med Chem* **19**(8), 1040–1049.
35. Ekins, S., Gao, F., Johnson, D. L., Kelly, K. G., Meyer, R. D. inventors (2001) Single point interaction screen to predict IC50 patent EP 1 139 267 A2.26.03.2001.
36. Sciabola, S., Stanton, R. V., Wittkopp, S., et al. (2008) Predicting kinase selectivity profiles using Free-Wilson QSAR analysis. *J Chem Inform Model* **48**(9), 1851–1867.
37. Rogers, D., Brown, R. D., Hahn, M. (2005) Using extended-connectivity fingerprints with Laplacian-modified Bayesian analysis in high-throughput screening follow-up. *J Biomol Screeni* **10**, 682–686. First published on September 16, 2005 doi:10.1177/1087057105281365.
38. Durant, J. L., Leland, B. A., Henry, D. R., Nourse, J. G. (2002) Reoptimization of MDL keys for use in drug discovery. *J Chem Inform Comput Sci* **42**(6), 1273–1280.
39. Wittkopp, S., Penzotti, J. E., Stanton, R. V., Wildman, S. A. (2007) Knowledge-based docking for kinases with minimal bias. In: 234th ACS National Meeting, Boston, MA, United States.

Chapter 6

Application of QSAR and Shape Pharmacophore Modeling Approaches for Targeted Chemical Library Design

Jerry O. Ebalunode, Weifan Zheng, and Alexander Tropsha

Abstract

Optimization of chemical library composition affords more efficient identification of hits from biological screening experiments. The optimization could be achieved through rational selection of reagents used in combinatorial library synthesis. However, with a rapid advent of parallel synthesis methods and availability of millions of compounds synthesized by many vendors, it may be more efficient to design targeted libraries by means of virtual screening of commercial compound collections. This chapter reviews the application of advanced cheminformatics approaches such as quantitative structure–activity relationships (QSAR) and pharmacophore modeling (both ligand and structure based) for virtual screening. Both approaches rely on empirical SAR data to build models; thus, the emphasis is placed on achieving models of the highest rigor and external predictive power. We present several examples of successful applications of both approaches for virtual screening to illustrate their utility. We suggest that the expert use of both QSAR and pharmacophore models, either independently or in combination, enables users to achieve targeted libraries enriched with experimentally confirmed hit compounds.

Key words: QSAR modeling, pharmacophore modeling, model validation, virtual screening.

1. Introduction

There is an increased realization that rationally designed chemical libraries facilitate significantly the process of discovering new drug candidates. The library is described as focused (or targeted) when compounds selected into the library are optimized with respect to at least one target property [the property(-ies) can be specific biological activities and/or various desired parameters of drug likeness, including drug safety, that are generally covered by the optimal ADME/Tox paradigm].Naturally, rational design of

J.Z. Zhou (ed.), *Chemical Library Design*, Methods in Molecular Biology 685,
DOI 10.1007/978-1-60761-931-4_6,

such libraries is only enabled when sufficient amount of experimental data (e.g., results of biological testing for ligands and/or target structural information) relevant to the target property(-ies) is available.

In the early days of combinatorial chemistry, rational design of chemical libraries frequently implied the selection of building blocks (from a large available pool) that would produce a reduced library enriched with potential hit compounds. For instance, in one of our earlier studies we have developed an approach termed FOCUS-2D (1, 2) for designing targeted libraries via rational selection of building blocks. The approach was based on a virtual combinatorial synthesis procedure where the products were assembled by combining reagents (or building blocks) into virtual compounds. The building blocks were sampled using stochastic optimization procedure, and the scoring function optimized in this process was either the similarity of products to a known active compound(-s) or target activity predicted from independently developed quantitative structure–activity relationship (QSAR) models. The virtual library of high scoring (i.e., predicted to be active) compounds was assembled and analyzed in terms of building blocks found with the highest frequency within selected compounds; thus, the ultimate goal of the study was the rational selection of building blocks that would be used to build a complete chemical library (as opposed to "cherry-picking" selected compounds).

Although studies into rational building block selection such as those described above were popular in the early days of computational combinatorial chemistry, the alternative approaches looking into rational selection of compounds from commercial libraries of already synthesized or synthetically feasible compounds have gradually prevailed. In fact, in a popular review Jamois (3) has compared reagent-based vs. product-based strategies for library design and concluded that "several studies have demonstrated the superiority of product-based designs in yielding diverse and representative subsets." Nowadays, large commercial libraries and services that provide integrated links to commercially available compounds are widely available (for instance, ca. 10 M compounds have been compiled in publicly available ZINC database (4)); see a recent review (5) for a partial list of additional chemical databases. Thus, most of the current approaches employ various virtual screening strategies to select specific compound subsets for subsequent experimental exploration.

This chapter discusses the application of popular cheminformatics approaches, such as rigorously built QSAR models and shape pharmacophore models, to the problem of targeted library design. QSAR models offer unique ability to rationalize existing experimental SAR data in the form of robust quantitative models that predict target property directly from structural chemical descriptors; thus, they can be used to screen an external chemical library to select compounds predicted to be active against the target. Conversely, shape pharmacophore models utilize the representative shape of active ligands or the negative image (or pseudo molecule) extracted from the binding site of the target protein to query 3D conformational databases of virtual or real molecular libraries. With enough attention paid to critical issues of model validation and applicability domain definition, both QSAR and shape pharmacophore models could be used successfully (and concurrently) to mine external virtual libraries to identify putative compounds with the desired target properties. The selected compounds could be chosen as candidates for thereby rationally designed compound library.

This chapter will initially discuss current algorithms for developing *externally* predictive QSAR models and present experimentally confirmed examples of identifying novel bioactive compounds by the means of QSAR model-based virtual screening. It will also present a novel shape pharmacophore modeling method and its validation through retrospective analysis of known biologically active compounds. Of course, many approaches, both structure based and ligand based, have been used for virtual screening.We have decided to focus on these specific methodologies, i.e., QSAR and pharmacophore modeling because both approaches are well known to both computational and medicinal chemists as structure *optimization* tools used at later stages of drug discovery after the lead compounds have been identified experimentally. However, in recent years these approaches, among other cheminformatics methods (6), have found new applications as virtual screening tools. The methods and applications discussed in this chapter should be of interest to both computational and synthetic chemists and experimental biologists working in the areas of biological screening of chemical libraries.

2. Predictive QSAR Models as Virtual Screening Tools

QSAR modeling has been traditionally viewed as an evaluative approach, i.e., with the focus on developing retrospective and explanatory models of existing data. Model extrapolation has been considered only in hypothetical sense in terms of potential modifications of known biologically active chemicals that could improve compounds' activity. Nevertheless recent studies suggest that current QSAR methodologies may afford robust and validated models capable of accurate prediction of compound properties for molecules not included in the training sets.

Below, we discuss a data analytical modeling workflow developed in our laboratory that incorporates modules for combinatorial QSAR model development (i.e., using all possible binary combinations of available descriptor sets and statistical data modeling techniques), rigorous model validation, and virtual screening of available chemical databases to identify novel biologically active compounds. Our approach places particular emphasis on model validation as well as the need to define model applicability domains in the chemistry space. We present examples of studies where the application of rigorously validated QSAR models to virtual screening identified computational hits that were confirmed by subsequent experimental investigations. This approach enables to identify subsets of putative active compounds that form a targeted chemical library expected to be enriched with target-specific bioactive compounds.

2.1. Basic QSAR Modeling Concepts

Any QSAR method can be generally defined as an application of mathematical and statistical methods to the problem of finding empirical relationships (QSAR models) of the form $P_i = \hat{k}(D_1, D_2, \ldots, D_n)$, where P_i are biological activities (or other properties of interest) of molecules, D_1, D_2,. . .,D_n are calculated (or, sometimes, experimentally measured) structural properties (molecular descriptors) of compounds, and $\hat{k}$ is some empirically established mathematical transformation that should be applied to descriptors to calculate the property values for all molecules (**Fig. 6.1**). The goal of QSAR modeling is to establish a trend in the descriptor values, which parallels the trend in biological activity. In essence, all QSAR approaches imply, directly or indi-

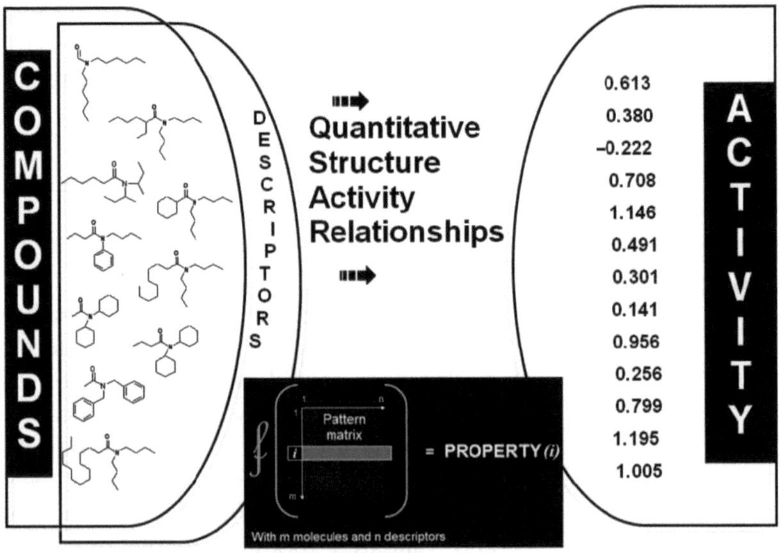

Fig. 6.1. General framework of QSAR modeling.

rectly, a simple similarity principle, which for a long time has provided a foundation for the experimental medicinal chemistry: compounds with similar structures are expected to have similar biological activities. The detailed description of major tenets of QSAR modeling is beyond the scope of this chapter; the overview of popular QSAR modeling techniques could be found in multiple reviews, e.g., (7). Here, we comment on most critical general aspects of model development and, most importantly, validation that are especially important in the context of using QSAR models for virtual screening.

2.1.1. Critical Importance of Model Validation

In our important paper titled "Beware of q^2!" (8), we have demonstrated the insufficiency of the training set statistics for developing externally predictive QSAR models and formulated the main principles of model validation. Despite earlier observations and warnings of several authors (9–11) that high cross-validated correlation coefficient R^2 (q^2) is a necessary but insufficient condition for the model to have high predictive power, many studies continue to consider q^2 as the only parameter characterizing the predictive power of QSAR models. In reference (8) we have shown that the predictive power of QSAR models can be claimed only if the model was successfully applied for prediction of the external test set compounds, which were not used in the model development. We have demonstrated that the majority of the models with high q^2 values have poor predictive power when applied for prediction of compounds in the external test set. In the subsequent publication (12) the importance of rigorous validation was again emphasized as a crucial, integral component of model development. Several examples of published QSAR models with high fitted accuracy for the training sets, which failed rigorous validation tests, have been considered. We presented a set of simple guidelines for developing validated and predictive QSAR models and discussed several validation strategies such as the randomization of the response variable (Y-randomization) and external validation using rational division of a data set into training and test sets. We highlighted the need to establish the domain of model applicability in the chemical space to flag molecules for which predictions may be unreliable, and discussed some algorithms that can be used for this purpose. We advocated the broad use of these guidelines in the development of predictive QSPR models (12–14).

At the 37th Joint Meeting of Chemicals Committee and Working Party on Chemicals, Pesticides and Biotechnology, held in Paris on 17–19 November 2004, the OECD (Organization for Economic Co-operation and Development) member countries adopted the following five principles that valid (Q)SAR models should follow to allow their use in regulatory assessment of chemical safety: (i) a defined endpoint; (ii) an unambiguous algorithm;

(iii) a defined domain of applicability; (iv) appropriate measures of goodness-of-fit, robustness, and predictivity; and (v) a mechanistic interpretation, if possible. Since then, most of the European authors publishing in QSAR area include a statement that their models fully comply with OECD principles (e.g., *see* (15–18)).

Validation of QSAR models is one of the most critical problems of QSAR. Recently, we have extended our requirements for the validation of multiple QSAR models selected by acceptable statistics criteria of prediction for the test set (19). Additional studies on this critical component of QSAR modeling should establish reliable and commonly accepted "good practices" for model development, which should make models increasingly useful for virtual screening.

2.1.2. Applicability Domains and QSAR Model Acceptability Criteria

One of the most important problems in QSAR analysis is establishing the domain of applicability for each model. In the absence of the applicability domain restriction, each model can formally predict the activity of any compound, even with a completely different structure from those included in the training set. Thus, the absence of the model applicability domain as a mandatory component of any QSAR model would lead to the unjustified extrapolation of the model in the chemistry space and, as a result, a high likelihood of inaccurate predictions. In our research we have always paid particular attention to this issue (12, 20–27). A good overview of commonly used applicability domain definitions can be found in reference (28).

In our earlier publications (8, 12) we have recommended a set of statistical criteria which must be satisfied by a predictive model. For continuous QSAR, criteria that we will follow in developing activity/property predictors are as follows: (i) correlation coefficient R between the predicted and the observed activities; (ii) coefficients of determination (29) (predicted versus observed activities R_0^2 and observed versus predicted activities $R_0'^2$ for regressions through the origin); (iii) slopes k and k' of regression lines through the origin. We consider a QSAR model *predictive* if the following conditions are satisfied (i) $q^2>0.5$; (ii) $R^2>0.6$; (iii) $\frac{(R^2-R_0^2)}{R^2}<0.1$ and $0.85 \le k \le 1.15$ or $\frac{(R^2-R_0'^2)}{R^2}<0.1$ and $0.85 \le k' \le 1.15$; (iv) $|R_0^2 - R_0'^2| < 0.3$ where q^2 is the cross-validated correlation coefficient calculated for the training set, but all other criteria are calculated for the test set (for additional discussion, *see* (30)).

2.1.3. Predictive QSAR Modeling Workflow

Our experience in QSAR model development and validation has led us to establishing a complex strategy that is summarized in **Fig. 6.2**. It describes the predictive QSAR modeling workflow, which focuses on delivering validated models and ultimately, computational hits confirmed by the experimental validation. We

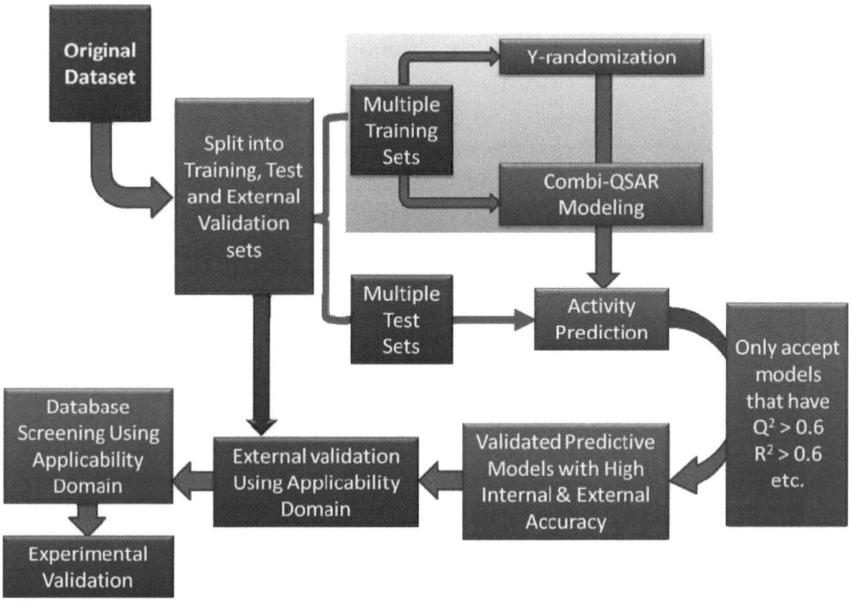

Fig. 6.2. General workflow for predictive QSAR modeling.

start by randomly selecting a fraction of compounds (typically, 10–15%) as an external validation set. The remaining compounds are then divided rationally (e.g., using the Sphere Exclusion protocol implemented in our laboratory (14)) into multiple training and test sets that are used for model development and validation, respectively using criteria discussed in more detail below. We employ multiple QSAR techniques based on the combinatorial exploration of all possible pairs of descriptor sets coupled with various statistical data mining techniques (termed combi-QSAR) and select models characterized by high accuracy in predicting both training and test sets data. Validated models are finally tested using the evaluation set. The critical step of the external validation is the use of applicability domains. If external validation demonstrates the significant predictive power of the models we use all such models for virtual screening of available chemical databases (e.g., ZINC (4)) to identify putative active compounds and work with collaborators who could validate such hits experimentally. The entire approach is described in detail in several recent papers and reviews (e.g., (7, 12, 30, 31))

2.2. Application of QSAR Models to Virtual Screening

In our recent studies we were fortunate to recruit experimental collaborators who have validated computational hits identified by virtual screening of commercially available compound libraries using rigorously validated QSAR models. Examples include anticonvulsants (25), HIV-1 reverse transcriptase inhibitors (32), D1 antagonists (33), antitumor compounds (34), beta-lactamase inhibitors (35), human histone deacetylase (HDAC) inhibitors

(36), and geranylgeranyltransferase-I inhibitors (37). Thus, models resulting from predictive QSAR workflow could be used to prioritize the selection of chemicals for the experimental validation. To illustrate the power of validated QSAR models as virtual screening tools, we shall discuss the examples of studies that resulted in experimentally confirmed hits. We note that such studies could only be done if there are sufficient data available for a series of tested compounds such that robust validated models could be developed using the workflow described in **Fig. 6.2**. The following examples illustrate the use of QSAR models developed with predictive QSAR modeling and validation workflow (**Fig. 6.2**) for virtual screening of commercial libraries to identify experimentally confirmed hits.

2.2.1. Discovery of Novel Anticancer Agents

A combined approach of validated QSAR modeling and virtual screening was successfully applied to the discovery of novel tylophorine derivatives as anticancer agents (34). QSAR models have been initially developed for 52 chemically diverse phenanthrine-based tylophorine derivatives (PBTs) with known experimental EC_{50} using chemical topological descriptors (calculated with the MolConnZ program) and variable selection *k*nearest neighbor (*k*NN) method. Several validation protocols have been applied to achieve robust QSAR models. The original data set was divided into multiple training and test sets, and the models were considered acceptable only if the leave-one-out cross-validated R^2 (q^2) values were greater than 0.5 for the training sets and the correlation coefficient R^2 values were greater than 0.6 for the test sets. Furthermore, the q^2 values for the actual data set were shown to be significantly higher than those obtained for the same data set with randomized target properties (*Y*-randomization test), indicating that models were statistically significant. Ten best models were then employed to mine a commercially available ChemDiv Database (ca. 500K compounds) resulting in 34 consensus hits with moderate to high predicted activities. Ten structurally diverse hits were experimentally tested and eight were confirmed active with the highest experimental EC_{50} of 1.8 μM implying an exceptionally high hit rate (80%). The same 10 models were further applied to predict EC_{50} for four new PBTs, and the correlation coefficient (R^2)between the experimental and the predicted EC_{50} for these compounds plus eight active consensus hits was shown to be as high as 0.57.

2.2.2. Discovery of Novel Histone Deacetylase (HDAC) Inhibitors

Histone deacetylases (HDACs) play a critical role in transcription regulation. Small molecule HDAC inhibitors have become an emerging target for the treatment of cancer and other cell proliferation diseases. We have employed variable selection *k* nearest neighbor approach (*k*NN)and support vector machines (SVM) approach to generate QSAR models for 59 chemically diverse compounds with inhibition activity on class I HDAC. MOE (38)- and MolConnZ (39)-based 2D descriptors were combined with *k*nearest neighbor (*k*NN) and support vector machines (SVM) approaches independently to improve the predictive power of models. Rigorous model validation approaches were employed including randomization of target activity (*Y*-randomization test) and assessment of model predictability by consensus prediction on two external data sets. Highly predictive QSAR models were generated with leave-one-out cross-validation R^2 (q^2) values for the training set and R^2 values for the test set as high as 0.81 and 0.80, respectively, with MolconnZ/*k*NN approach and 0.94 and 0.81, respectiveley, with MolconnZ/SVM approach. Validated QSAR models were then used to mine four chemical databases: National Cancer Institute (NCI) database, Maybridge database, ChemDiv database, and one ZINC database, including a total of over 3 million compounds. The searches resulted in 48 consensus hits, including two reported HDAC inhibitors that were not included in the original data set. Four hits with novel structural features were purchased and tested using the same biological assay that was employed to assess the inhibition activity of the training set compounds. Three of these four compounds were confirmed active with the best inhibitory activity (IC_{50}) of 1 μM. The overall workflow for model development, validation, and virtual screening is illustrated in **Fig. 6.3**.

2.2.3. Discovery of Novel Histone Deacetylase (HDAC) Inhibitors

In another recent study (37), we employed our standard QSAR modeling workflow (**Fig. 6.2**) to discover novel geranylgeranyltransferase type I (GGTase-I) inhibitors. Geranylgeranylation is critical to the function of several proteins including Rho, Rap1, Rac, Cdc42, and G protein gamma subunits. GGTase-I inhibitors

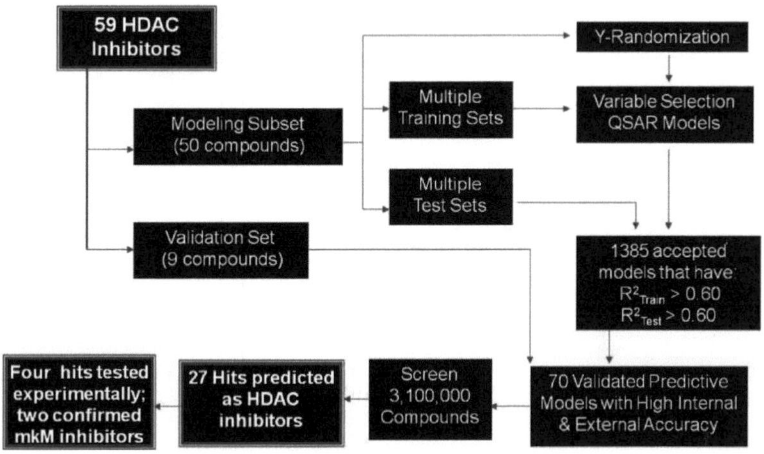

Fig. 6.3. Application of predictive QSAR workflow including virtual screening to discover novel HDAC inhibitors.

(GGTIs) have therapeutic potential to treat inflammation, multiple sclerosis, atherosclerosis, and many other diseases. Following our standard QSAR modeling workflow, we have developed and rigorously validated models for 48 GGTIs using variable selection*k* nearest neighbor (40) and automated lazy learning (26) and genetic algorithm-partial least square (41) QSAR methods. The QSAR models were employed for virtual screening of 9.5 million commercially available chemicals yielding 47 diverse computational hits. Seven of these compounds with novel scaffolds and high predicted GGTase-I inhibitory activities were tested in vitro, and all were found to be bona fide and selective micromolar inhibitors.

Figure 6.4 shows the structures of both representative training set compounds and confirmed computational hits. We should emphasize that QSAR models have been traditionally viewed as lead optimization tools capable of predicting compounds with chemical structure similar to the structure of molecules used for the training set. However, this study clearly indicates (**Fig. 6.4**) that with enough attention given to the model development process and using chemical descriptors characterizing whole molecules (as opposed to, e.g., chemical fragments), it is indeed possible to discover compounds with novel chemical scaffolds. Furthermore, in our study we have additionally demonstrated that these novel hits could not be identified using tradi-

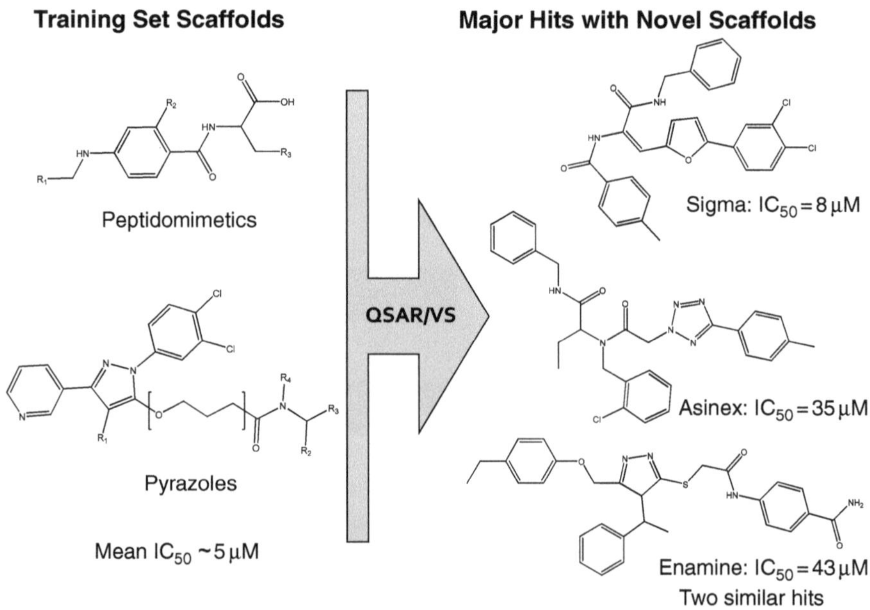

Fig. 6.4. Discovery of GGTase-I inhibitors with novel chemical scaffolds using a combination of QSAR modeling and virtual screening.

tional chemical similarity search (37), which highlights the power of robust QSAR models as the drug discovery tool.

In summary, our studies have established that QSAR models could be used successfully as virtual screening tools to discover compounds with the desired biological activity in chemical databases or virtual libraries (25, 31, 33, 34, 42). It should be stressed that the total number of compounds selected for virtual screening based on QSAR model predictions is typically relatively small, only a few dozen. Obviously, the total number of computational hits is controlled by the value of applicability domain. In most published cases, because we were limited in both time and resources, we chose a very conservative applicability domain leading to the selection of a small library of computational hits with an expectation that a large fraction of these would be confirmed as active compounds. In the industrial size projects it may be more reasonable to loosen the applicability domain requirement and increase the size of virtual hit library. One may expect that the increase in the library size will result in lower relative accuracy of prediction but the absolute number of confirmed hits may actually increase. Thus, scientists using QSAR models that incorporate the applicability domain should always be aware of the interplay between the size of the domain, the coverage of the virtual screening library, and the prediction accuracy so they should use the applicability domain as a tunable parameter to control this interplay. The discovery of novel bioactive chemical entities is the primary goal of computational drug discovery, and the development of validated and predictive QSAR models is critical to achieve this goal.

3. Shape Pharmacophore Modeling as Virtual Screening Tool

Shape complementarity plays an important role in the process of molecular recognition (43). In a typical 3D structure of ligand–receptor complex, one can observe tight van der Waals contacts between the ligand atoms and the receptor atoms of the binding pocket. Grant et al. (44) pointed out the fundamental reasons for such shape complementarity. They argued that the intermolecular interactions that stabilize the receptor–ligand complex are enthalpically weak, and they become effective only when the chemical groups involved can approach each other closely, which is favored by the shape complementarity. They further argued that the entropic contributions advantageous to binding, which involve the loss of bound water of both the host and the guest, are also favored by shape complementarity. Thus, the concept of shape complementarity is widely adopted by medicinal chemists in structure-based drug design. When the constraints of critical functional groups and their spatial orientation are taken into account, together with shape complementarity, one can create a shape pharmacophore model. This latter model has proved to be more effective in virtual screening experiments. In the following sections, we first describe the basic concept of shape and shape pharmacophore modeling and then present some recent literature examples.

3.1. Basic Concept of Molecular Shape Analysis

Molecular shape analysis tools can be broadly categorized into two groups. In terms of the methodology employed, they are either superposition based or superposition free. The former calculates a shape-matching measure only after an optimal superposition of the two objects has been obtained. The second category of methods calculates shape similarity score based on rotation-and-translation independent descriptors that are computed from different representations of molecular objects, and thus, it does not depend on the orientation or alignment of the two molecular objects. Zauhar's shape signatures (45) and the more recent USR method (42, 43) belong to this category.

The following two categories of methods can be identified, in terms of the input information for shape analysis tools: (1) ligand-based analysis, where receptor's structural information is not included in the analysis and (2) receptor-based methods, where the structural information of the receptor is an integral part of the analysis process and is essential in formulating the models.

3.1.1. Alignment-Based and Alignment-Free Methods

3.1.1.1. Alignment-Based Algorithms

In alignment-based algorithms, the shape similarity calculation is conducted after an optimal superposition of two molecular objects is achieved. One of the earliest methods, studied by Meyer and Richards (46), performed the alignment and then counted common points between the two objects as a way to quantify the similarity between two molecular objects. The optimization process was slow, which limited its use. The shape similarity concept was further developed by Good and Richards, by employing Gaussian functions as the basis for similarity calculation (47). Grant et al. also employed Gaussian functions to calculate shape similarity (44), based on the calculation of volume overlap between two superposed molecular objects. This latter method has further been modified and implemented in the program ROCS (Rapid Overlay of Compound Structures) (48) and the OE Shape toolkit (49).

Gaussian Shape Similarity by Good and Richards. This method introduced the use of Gaussians for molecular shape matching for the first time (47). The shape of each atom was described as a suitable electron density function, and then three Gaussian functions were fitted to each of the atomic electron density functions. An analytical shape similarity index was formulated according to the Carbo index (50, 51). Molecular superposition was achieved via the optimization of the similarity index.

Shape-Matching Method by Grant and Pickup (40, 51). This method defines a Gaussian density for each atom to replace the hard sphere representation of atoms (52). The molecular volume is expressed as a series of integration terms, representing the intersection volumes between the atoms in a molecule. The Gaussian description was used to compare the shapes of two molecules by optimizing their volume overlap using analytical derivatives with respect to rotations and translations. This idea was later implemented in the ROCS program. However, it has been pointed out (53) that ROCS, by default, gives the same radius value to all heavy atoms in the molecule. This approximation led to the conclusion that the volume calculation in ROCS might not be as accurate as expected from the original theory of Grant and Pickup. Nonetheless, ROCS has been shown to be very successful in many validation studies and actual applications.

3.1.1.2. Alignment-Free Algorithms

The basic idea of alignment-free shape matching is that a set of rotation- and translation-free descriptors are calculated for conformers under consideration, and then some similarity measure is devised to quantify the similarity between two molecular objects. Zauhar's shape signatures (45), Breneman's PEST and PESD methods (54–56), the USR (ultrafast shape recognition) method (53), the atom triplets method (57), and Schlosser's recent TrixX BMI approach (58) are a few examples. One advantage of these algorithms is that they offer much faster computational speed and, thus, are suitable for screening large molecular databases and virtual compound libraries.

The Shape Signatures Method. This method was reported by Zauhar et al. for shape description and comparison (45). Solvent-accessible molecular surface is triangulated using the smooth molecular surface triangulator algorithm (59) (SMART). The molecular surface is divided into regular triangular area elements. The volume defined by the molecular surface is explored using ray tracing, which starts each ray from a randomly selected point on the molecular surface and then allows the ray to propagate by the rules of optical reflection. The tracing and reflection of light stop until some preset conditions are met. The result is a collection of line segments that connect two successive reflection points. The simplest shape signature is the distribution of the lengths of these segments, stored as histogram for each molecule. The similarity between molecular shapes is simply the similarity between their histograms.

The PEST and PESD Method These methods were developed by Breneman's group. The PEST (property-encoded surface translator) method is based on the combination of the TAE descriptors (54) and the shape signatures idea by Zauhar (45). It uses the TAE molecular surface representations to define property-encoded boundaries. It first computes the molecular surface property distributions and then collects ray-tracing path information and lastly generates the shape descriptors. The 2D histograms are generated to represent surface shape profile, encoding both shape and surface properties. Similarly, the property-encoded shape distributions (PESD) descriptors have recently been reported and employed to study ligand–protein binding affinities (56). The PESD algorithm is different from PEST in that it is based on a fixed number of randomly sampled point pairs on the molecular surface that does not require ray tracing. Both PEST and PESD descriptors should account for the distribution of both the polar and non-polar regions and electrostatic potential on the molecular surface.

The USR (Ultrafast Shape Recognition) Method. This method was reported by Ballester and Richards (53) for compound database search on the basis of molecular shape similarity. It was reportedly capable of screening billions of compounds for similar shapes on a single computer. The method is based on the notion that the relative position of the atoms in a molecule is completely determined by inter-atomic distances. Instead of using all inter-atomic distances, USR uses a subset of distances, reducing the computational costs. Specifically, the distances between all atoms of a molecule to each of four strategic points are calculated. Each set of distances forms a distribution, and the three moments (mean, variance, and skewness) of the four distributions are calculated. Thus, for each molecule, 12 USR descriptors are calculated. The inverse of the translated and scaled Manhattan distance between two shape descriptors is used to measure the similarity between the two molecules. A value of "1" corresponds to maximum similarity and a value of "0" corresponds to minimum similarity.

3.2. Examples of Application of Shape and Pharmacophore Models for Virtual Screening

3.2.1. Ligand-Based Studies

When a few ligands are known for a particular target, one can use ligand-based shape-matching technology to search for potential ligands via virtual screening. A ligand-based application of shape-matching methods starts with a ligand with known biological activity. Its 3D conformations are often pregenerated by a conformer generator. Also, a multiconformer database of potential drug molecules is pregenerated to be used by the shape-matching program. For alignment-based methods, the conformers of the known ligand (i.e., the query) will be directly aligned with those of the database molecules. Molecules that align well with the query molecule will be selected for further consideration. In the case of alignment-free methods, both the shape descriptors of the query and those of the database molecules are first calculated, and a similarity value is calculated between the query and each of the database molecules. Molecules with better similarity values to the query are selected for further consideration.

In the validation study by Hawkins et al. (60), the shape-matching method ROCS was compared to 7 well-known docking tools, in terms of their abilities to recover known ligands for 21 different protein targets. The comparative study showed that the 3D shape method (ROCS) performed at least the same as, and often better than, the docking tools studied. Their work indicated that shape-based virtual screening method could be both efficient (in terms of the computational speed) and effective (in terms of hit enrichment) in virtual screening projects.

In a comparative study, McGaughey et al. (61) investigated several 2D similarity methods (including Daylight fingerprint similarity (62) and TOPOSIM (63)), 3D shape similarity methods (ROCS and SQ (64)), and several known docking tools (FLOG (65), FRED (66), and Glide (67, 68)). Based on the performance on a benchmark set of 11 protein targets, they observed that, on average, the ligand-based shape method with chemistry constraints outperformed more sophisticated docking tools. Their results also demonstrated that shape matching (including chemistry constraints) could select more diverse active compounds than 2D similarity methods. This indicates that shape-matching tools may offer a better "scaffold hopping" capability than 2D methods.

Moffat et al. (69) also compared three ligand-based shape similarity methods, including CatShape (70), FBSS (71), and ROCS. These methods have been compared on the basis of retrospective virtual screening experiments. All three methods have demonstrated significant enrichment, but ROCS with CFF option (CFF: chemical force field) gave the best performance. They reported that shape matching, coupled with chemistry constraints, afforded better enrichment factors than shape-matching alone, indicating the importance of including chemistry information in the search. This observation is consistent with the recent validation study by Hawkins et al. (60) and by Ebalunode et al. (72). In general, flexible methods gave slightly better performance than the respective rigid search methods; however, the increased performance could not justify the increased computational cost. This observation is again consistent with the finding by a different validation study by Ebalunode et al. (73).

Zauhar et al. reported an interesting application of the shape signatures approach to shape matching and similarity search (45). In a validation study (74), they found significant enrichment of ligands for the serotonin receptor using the shape signatures approach. A set of 825 agonists and 400 antagonists as well as roughly 10,000 randomly chosen compounds from the NCI database were used in that study.

Ballester et al. (75) evaluated a new algorithm (Ultrafast Shape Recognition or USR) in the context of retrospective ligand-based virtual screening. They showed that USR performed better, on average, than a commercially available shape similarity method, while screening conformers at a rate that is >2500 times faster. This feature makes USR an ideal virtual screening tool for searching extremely large molecular databases. However, no atomic property information is encoded in this method.

When ROCS or any other ligand-based 3D shape-matching method is used for virtual screening, the choice of the query conformation can have significant impact on the results of virtual screening. This is especially true when no X-ray structure of a bound ligand is available. In a recent study by Tawa et al. (76), the authors developed a rational conformation selection protocol (named CORAL), which allows the selection of conformation that affords better enrichment than using simply the lowest energy conformation as the query. They have demonstrated that this method can significantly improve the effectiveness of ligand-based method (ROCS) for drug discovery. In a related study, Kirchmair et al. (77) described ways to optimize shape-based virtual screening. They discussed how to choose the right query together with chemical information. They have examined various parameters that may improve the performance and offered guidelines on how to achieve the optimum performance using shape-matching techniques in virtual screening.

3.2.2. Receptor-Based Studies

Various variants of the basic shape-matching algorithms have been reported in the literature (69, 78). The general idea of these tools is to extract the shape and pharmacophore information from the binding site structure and represent such information or constraints as pseudo-molecular shapes. Once the pseudo-molecular shape is created, a regular shape-matching algorithm can be employed to compare binding sites with small molecules. Here, we review a few of the recent developments as follows.

To employ the shape-matching algorithm in a receptor-based fashion, Ebalunode et al. developed a method that can be considered as a structure-based variant of ROCS. The method, SHAPE4 (72), utilizes a computational geometry method (the alpha-shape algorithm) to extract and characterize the binding site of a given target. It then uses a grid to approximate the geometric volume of the binding site, defined by the Delaunay simplices generated from the alpha-shape analysis. The pharmacophore centers are derived from the binding site atomic information, using either the LigandScout program (79) or other equivalent approaches. As a result, the extracted binding site shape and the pharmacophore constraints reflect the nature of the binding site. In theory, this approach can overcome the limit imposed by using the bound ligand per se as the query, in that the query in SHAPE4 can cover more diverse characteristics of the binding site than the bound ligand itself. The effectiveness, in terms of enrichment factors and diversity of the hits, has been demonstrated in the SHAPE4 article (72).

Similar to SHAPE4, Lee et al. developed the SLIM program (80), another variant of the ROCS technology. It derives the binding site shape and pharmacophore information based on the X-ray structure of the target. It is different from SHAPE4 in that a more straightforward method for extracting the binding site is employed by SLIM, where a geometric box is defined based on the knowledge of the binding site. Visualization by human expert is often needed to help define the binding pocket, and it is harder to use in cases where large number of protein targets are being studied. However, as pointed out by the authors, their focus was to test the effect and impact of multiple conformations of the target protein in order to address the conformational flexibility issue. Thus, SLIM worked very well for their purpose.

3.3. Prospective Applications of Pharmacophore Shape Technologies

Markt et al. (81) reported the discovery of PPAR ligands using an integrated screening protocol. Using a combination of pharmacophore, 3D shape similarity, and electrostatic similarity, they discovered 10 virtual screening hits, of which 5 tested positive against the ligand-binding domain (LBD) of human PPAR in transactivation assays and showed affinities for PPAR in a competitive binding assay. Therefore, this represents a successful application of multiple complementary technologies in drug discovery, where the 3D shape technology was part of the workflow.

An application of the ROCS program has been reported recently (82). New scaffolds for small molecule inhibitors of the ZipA-FtsZ protein–protein interaction have been found. The shape comparisons are made relative to the bioactive conformation of a HTS hit, determined by X-ray crystallography. A follow-up X-ray crystallographic analysis also showed that ROCS accurately predicted the binding mode of the inhibitor. This result offers the first experimental evidence that validates the use of ROCS for scaffold hopping purposes.

Another successful application of a shape similarity method was reported by Cramer et al. (83). Over 400 compounds were synthesized and tested for their inhibition of angiotensin II. The

63 compounds that were identified by topomer shape similarity as most similar to one of the four query structures covered all the compounds found to be highly active. None of the remaining 362 structures were highly active. Thus, this report is a nice demonstration of the ability of a shape similarity method for discovering new biologically active compounds. In another study, Cramer et al. (84) reported the application of topomer shape similarity for lead hopping. The hit rate averaged over all assays was 39%. The average 2D fingerprint Tanimoto similarity between a query and the newly found structures was 0.36, similar to the Tanimoto similarity between random drug-like structures. Thus, this is a good indication of the lead hopping ability of the topomer shape method.

A successful application of the shape and electrostatic similarity methods to prospective drug discovery has been reported by Muchmore et al. (85). To identify novel melanin-concentrating hormone receptor 1 (MCHR1) antagonists, a library of virtual molecules was designed. Over 3 million molecules were searched using 3D shape similarity methods (in conjunction with an electrostatic similarity-matching algorithm). One of the top scoring hits was made and tested for MCHR1 activity. A threefold improvement in binding affinity and cellular potency has been achieved compared to the parent ligand. This example demonstrated the power of the ligand-based shape method for the discovery of new compounds from a large virtual library for targets without crystallographic information.

In a study that combined a variety of ligand-based and structure-based methods, Perez-Nueno et al. reported the success of a prospective virtual screening project (86). They first established a screening protocol based on a retrospective virtual screening, using a database of CXCR4 inhibitors and inactive compounds compiled from the literature. A large virtual combinatorial library of molecules was designed. The virtual screening protocol has been employed to select five compounds for synthesis and testing. Experimental binding assays of those compounds confirmed that their mode of action was blocking the CXCR4 receptor. This represents another successful example of using a shape similarity method for the discovery of new compounds via virtual screening.

In a more recent virtual screening study, Ballester et al. reported the successful identification of novel inhibitors of arylamine *N*-acetyltransferases using the USR algorithm (87). A computational screening of 700 million molecular conformers was conducted very efficiently. A small number of the predicted hits were purchased and experimentally tested. An impressive hit rate of 40% has been achieved. The authors also showed the ability of USR to find biologically active compounds with different chemical structures (i.e., scaffold hopping), evidenced by low Tanimoto coefficients between the found hits and the query molecule. Visual inspection also confirmed that none of the nine actives found shared a common scaffold with the template. Thus, this example demonstrated the power of a pure shape similarity method for scaffold hopping projects.

Finally, Ebalunode et al. reported a structure-based shape pharmacophore modeling for the discovery of novel anesthetic compounds (88). The 3D structure of apoferritin, a surrogate target for $GABA_A$, was used as the basis for the development of several shape pharmacophore models. They demonstrated that (1) the method effectively recovered known anesthetic agents from a diverse database of compounds; (2) the shape pharmacophore scores had a significant linear correlation with the measured binding data of several anesthetic compounds, without prior calibration and fitting; and (3) the computed scores also correctly predicted the trend of the EC_{50} values of a set of anesthetics.

4. Summary and Conclusions

We have discussed the application of cheminformatics approaches such as QSAR and shape pharmacophore modeling to the problem of targeted library design by means of virtual screening. Both approaches offer unique abilities to rationalize existing experimental SAR data in the form of models that could identify novel compounds predicted to interact with the specific target. Pharmacophore models achieve this task by establishing that a compound contains specific chemical features characteristic of known bioactive compounds, whereas QSAR models have the ability to predict the target activity quantitatively from structural chemical descriptors of compounds. As with any computational molecular modeling approach, it is imperative that both QSAR and pharmacophore modeling approaches are used expertly. Therefore, this chapter has focused on the discussion of critical components of both approaches that should be studied and executed rigorously to enable their successful application. We have shown that with enough attention paid to critical issues of model validation and (in the case of QSAR modeling) applicability domain definition, the models could be indeed used successfully to mine external virtual libraries, especially of commercially available chemicals, to create targeted compound libraries with desired properties. The methods and applications discussed in this chapter should be of help to both computational and synthetic chemists and experimental biologists working in the areas of biological screening of chemical libraries.

Acknowledgments

AT acknowledges the support from NIH (grant R01GM066940). J.E. and W.Z. acknowledge the financial support by the Golden Leaf Foundation via the BRITE Institute, North Carolina Central University. W.Z. also acknowledges funding from NIH (grant SC3GM086265).

References

1. Zheng, W., Cho, S. J., Tropsha, A. (1998) Rational combinatorial library design. 1. Focus-2D: a new approach to the design of targeted combinatorial chemical libraries. *J Chem Inf Comput Sci* **38**, 251–258.
2. Cho, S. J., Zheng, W., Tropsha, A. (1998) Focus-2D: a new approach to the design of targeted combinatorial chemical libraries, in (Altman, R. B., Dunker, A. K., Hunter, L., Klein, T. E. eds.) *Pacific Symposium on Biocomputing '98*, Hawaii, Jan 4–9, 1998. World Scientific, Singapore, pp. 305–316.
3. Jamois, E. A. (2003) Reagent-based and product-based computational approaches in library design. *Curr Opin Chem Biol* 7, 326–330.
4. Irwin, J. J., Shoichet, B. K. (2005) ZINC–a free database of commercially available compounds for virtual screening. *J Chem Inf Model* **45**, 177–182.
5. Oprea, T., Tropsha, A. (2006) Target, chemical and bioactivity databases – integration is key. *Drug Discov Today* **3**, 357–365.
6. Varnek, A., Tropsha, A. (2008) *Cheminformatics Approaches to Virtual Screening*, RSC, London.
7. Tropsha, A. (2006) *I* in (Martin, Y. C., ed.) *Comprehensive Medicinal Chemistry I.* pp. 113–126, Elsevier, Oxford.
8. Golbraikh, A. and Tropsha, A. (2002)Beware of q^2! *J Mol Graph Model 20*, 269–276.
9. Novellino, E., Fattorusso, C., Greco, G. (1995) Use of comparative molecular field analysis and cluster analysis in series design. *Pharm Acta Helv* **70**, 149–154.
10. Norinder, U. (1996) Single and domain made variable selection in 3D QSAR applications. *J Chemomet* **10**, 95–105.
11. Tropsha, A., Cho, S. J. (1998) in (Kubinyi, H., Folkers, G., and Martin, Y. C., eds.) *3D QSAR in Drug Design*. Kluwer, Dordrecht, The Netherlands, pp. 57–69.
12. Tropsha, A., Gramatica, P., Gombar, V. K. (2003) The importance of being earnest: validation is the absolute essential for successful application and interpretation of QSPR models. *Quant Struct Act Relat Comb Sci* **22**, 69–77.
13. Golbraikh, A., Tropsha, A. (2002) Predictive QSAR modeling based on diversity sampling of experimental datasets for the training and test set selection. *J Comput Aided Mol Des* **16**, 357–369.
14. Golbraikh, A., Shen, M., Xiao, Z., Xiao, Y. D., Lee, K. H., Tropsha, A. (2003) Rational selection of training and test sets for the development of validated QSAR models. *J Comput Aided Mol Des* **17**, 241–253.
15. Pavan, M., Netzeva, T. I., Worth, A. P. (2006) Validation of a QSAR model for acute toxicity. *SAR QSAR Environ Res* **17**, 147–171.
16. Vracko, M., Bandelj, V., Barbieri, P., Benfenati, E., Chaudhry, Q., Cronin, M., Devillers, J., Gallegos, A., Gini, G., Gramatica, P., Helma, C., Mazzatorta, P., Neagu, D., Netzeva, T., Pavan, M., Patlewicz, G., Randic, M., Tsakovska, I., Worth, A. (2006) Validation of counter propagation neural network models for predictive toxicology according to the OECD principles: a case study. *SAR QSAR Environ Res* **17**, 265–284.
17. Saliner, A. G., Netzeva, T. I., Worth, A. P. (2006) Prediction of estrogenicity: validation of a classification model. *SAR QSAR Environ Res* **17**, 195–223.
18. Roberts, D. W., Aptula, A. O., Patlewicz, G. (2006) Mechanistic applicability domains for non-animal based prediction of toxicological endpoints. QSAR analysis of the Schiff base applicability domain for skin sensitization. *Chem Res Toxicol* **19**, 1228–1233.
19. Zhang, S., Golbraikh, A., Tropsha, A. (2006) Development of quantitative structure-binding affinity relationship models based on novel geometrical chemical descriptors of the protein-ligand interfaces. *J Med Chem 49*, 2713–2724.
20. Golbraikh, A., Bonchev, D., Tropsha, A. (2001) Novel chirality descriptors derived

from molecular topology. *J Chem Inf Comput Sci* **41**, 147–158.
21. Kovatcheva, A., Buchbauer, G., Golbraikh, A., Wolschann, P. (2003) QSAR modeling of alpha-campholenic derivatives with sandalwood odor. *J Chem Inf Comput Sci* **43**, 259–266.
22. Kovatcheva, A., Golbraikh, A., Oloff, S., Xiao, Y. D., Zheng, W., Wolschann, P., Buchbauer, G., Tropsha, A. (2004) Combinatorial QSAR of ambergris fragrance compounds. *J Chem Inf Comput Sci* **44**, 582–595.
23. Shen, M., Xiao, Y., Golbraikh, A., Gombar, V. K., Tropsha, A. (2003) Development and validation of k-nearest-neighbor QSPR models of metabolic stability of drug candidates. *J Med Chem* **46**, 3013–3020.
24. Shen, M., LeTiran, A., Xiao, Y., Golbraikh, A., Kohn, H., Tropsha, A. (2002) Quantitative structure-activity relationship analysis of functionalized amino acid anticonvulsant agents using k nearest neighbor and simulated annealing PLS methods. *J Med Chem* **45**, 2811–2823.
25. Shen, M., Beguin, C., Golbraikh, A., Stables, J. P., Kohn, H., Tropsha, A. (2004) Application of predictive QSAR models to database mining: identification and experimental validation of novel anticonvulsant compounds. *J Med Chem* **47**, 2356–2364.
26. Zhang, S., Golbraikh, A., Oloff, S., Kohn, H., Tropsha, A. (2006) A Novel Automated Lazy Learning QSAR (ALL-QSAR) approach: method development, applications, and virtual screening of chemical databases using validated ALL-QSAR models. *J Chem Inf Model* **46**, 1984–1995.
27. Golbraikh, A., Shen, M., Xiao, Z., Xiao, Y. D., Lee, K. H., Tropsha, A. (2003) Rational selection of training and test sets for the development of validated QSAR models. *J Comput Aided Mol Des* **17**, 241–253.
28. Eriksson, L., Jaworska, J., Worth, A. P., Cronin, M. T., McDowell, R. M., Gramatica, P. (2003) Methods for reliability and uncertainty assessment and for applicability evaluations of classification- and regression-based QSARs. *Environ Health Perspect* **111**, 1361–1375.
29. Sachs, L. (1984) *Handbook of Statistics.* Springer, Heidelberg.
30. Tropsha, A., Golbraikh, A. (2007) Predictive QSAR modeling workflow, model applicability domains, and virtual screening. *Curr Pharm Des* **13**, 3494–3504.
31. Tropsha, A. (2005) in (Oprea, T., ed.) *Cheminformatics in Drug Discovery.*, Wiley-VCH, New York, pp. 437–455.
32. Medina-Franco, J. L., Golbraikh, A., Oloff, S., Castillo, R., Tropsha, A. (2005) Quantitative structure-activity relationship analysis of pyridinone HIV-1 reverse transcriptase inhibitors using the k nearest neighbor method and QSAR-based database mining. *J Comput Aided Mol Des* **19**, 229–242.
33. Oloff, S., Mailman, R. B., Tropsha, A. (2005) Application of validated QSAR models of D1 dopaminergic antagonists for database mining. *J Med Chem* **48**, 7322–7332.
34. Zhang, S., Wei, L., Bastow, K., Zheng, W., Brossi, A., Lee, K. H., Tropsha, A. (2007) Antitumor Agents 252. Application of validated QSAR models to database mining: discovery of novel tylophorine derivatives as potential anticancer agents. *J Comput Aided Mol Des* **21**, 97–112.
35. Hsieh, J. H., Wang, X. S., Teotico, D., Golbraikh, A., Tropsha, A. (2008) Differentiation of AmpC beta-lactamase binders vs. decoys using classification kNN QSAR modeling and application of the QSAR classifier to virtual screening. *J Comput Aided Mol Des* **22**, 593–609.
36. Tang, H., Wang, X. S., Huang, X. P., Roth, B. L., Butler, K. V., Kozikowski, A. P., Jung, M., Tropsha, A. (2009) Novel inhibitors of human histone deacetylase (HDAC) identified by QSAR modeling of known inhibitors, virtual screening, and experimental validation. *J Chem Inf Model* **49**, 461–476.
37. Peterson, Y. K., Wang, X. S., Casey, P. J., Tropsha, A. (2009) Discovery of geranylgeranyltransferase-I inhibitors with novel scaffolds by the means of quantitative structure-activity relationship modeling, virtual screening, and experimental validation. *J Med Chem* **52**, 4210–4220.
38. CCG. *Molecular Operation Environment.* 2008.
39. MolconnZ. http://www.edusoft-lc.com/molconn/ . 2010.
40. Zheng, W., Tropsha, A. (2000) Novel variable selection quantitative structure–property relationship approach based on the k-nearest-neighbor principle. *J Chem Inf Comput Sci* **40**, 185–194.
41. Cho, S. J., Zheng, W., Tropsha, A. (1998) Rational combinatorial library design. 2. Rational design of targeted combinatorial peptide libraries using chemical similarity probe and the inverse QSAR approaches. *J Chem Inf Comput Sci* **38**, 259–268.
42. Tropsha, A., Zheng, W. (2001) Identification of the descriptor pharmacophores using variable selection QSAR: applications to database mining. *Curr Pharm Des* 7, 599–612.

43. DesJarlais, R. L., Sheridan, R. P., Seibel, G. L., Dixon, J. S., Kuntz, I. D., Venkataraghavan, R. (1988) Using shape complementarity as an initial screen in designing ligands for a receptor binding site of known three-dimensional structure. *J Med Chem* **31**, 722–729.
44. Grant, J. A., Gallardo, M. A., Pickup, B. T. (1996) A fast method of molecular shape comparison: a simple application of a Gaussian description of molecular shape. *J Comput Chem* **17**, 1653–1666.
45. Zauhar, R. J., Moyna, G., Tian, L., Li, Z., Welsh, W. J. (2003) Shape signatures: a new approach to computer-aided ligand- and receptor-based drug design. *J Med Chem* **46**, 5674–5690.
46. Meyer, A. Y., Richards, W. G. (1991) Similarity of molecular shape. *J Comput Aided Mol Des* **5**, 427–439.
47. Good, A. C., Richards, W. G. (1993) Rapid evaluation of shape similarity using Gaussian functions. *J Chem Inf Comput Sci* **33**, 112–116.
48. ROCS. version 3.0.0. 2009. Santa Fe, NM, USA, OpenEye Scientific Software.
49. OEShape Toolkit. **version 1.7.2**. 2009. Santa Fe, NM, USA, OpenEye Scientific Software.
50. Carbo, R., Domingo, L. (1987) Lcao-Mo similarity measures and taxonomy. *Int J Quantum Chem* **32**, 517–545.
51. Carbo, R., Leyda, L., Arnau, M. (1980) An electron density measure of the similarity between two compounds. *Int J Quantum Chem* **17**, 1185–1189.
52. Masek, B. B., Merchant, A., Matthew, J. B. (1993) Molecular shape comparison of angiotensin II receptor antagonists. *J Med Chem* **36**, 1230–1238.
53. Ballester, P. J., Richards, W. G. (2007) Ultrafast shape recognition to search compound databases for similar molecular shapes. *J Comput Chem* **28**, 1711–1723.
54. Breneman, C. M., Thompson, T. R., Rhem, M., Dung, M. (1995) Electron-density modeling of large systems using the transferable atom equivalent method. *Comput Chem* **19**, 161.
55. Breneman, C. M., Sundling, C. M., Sukumar, N., Shen, L., Katt, W. P., Embrechts, M. J. (2003) New developments in PEST shape/property hybrid descriptors. *J Comput Aided Mol Des* **17**, 231–240.
56. Das, S., Kokardekar, A., Breneman, C. M. (2009) Rapid comparison of protein binding site surfaces with property encoded shape distributions. *J Chem Inf Model* **49**, 2863–2872.
57. Nilakantan, R., Bauman, N., Venkataraghavan, R. (1993) New method for rapid characterization of molecular shapes: applications in drug design. *J Chem Inf Comput Sci* **33**, 79–85.
58. Schlosser, J., Rarey, M. (2009) Beyond the virtual screening paradigm: structure-based searching for new lead compounds. *J Chem Inf Model* **49**, 800–809.
59. Zauhar, R. J. (1995) SMART: a solvent-accessible triangulated surface generator for molecular graphics and boundary element applications. *J Comput Aided Mol Des* **9**, 149–159.
60. Hawkins, P. C., Skillman, A. G., Nicholls, A. (2007) Comparison of shape-matching and docking as virtual screening tools. *J Med Chem* **50**, 74–82.
61. McGaughey, G. B., Sheridan, R. P., Bayly, C. I., Culberson, J. C., Kreatsoulas, C., Lindsley, S., Maiorov, V., Truchon, J. F., Cornell, W. D. (2007) Comparison of topological, shape, and docking methods in virtual screening. *J Chem Inf Model* **47**, 1504–1519.
62. Daylight. **version 4.82**. 2003. Aliso Viejo, CA, USA, Daylight Chemical Information Systems Inc.
63. Kearsley, S. K., Sallamack, S., Fluder, E. M., Andose, J. D., Mosley, R. T., Sheridan, R. P. (1996) Chemical similarity using physiochemical property descriptors. *J Chem Inf Comput Sci* **36**, 118–127.
64. Miller, M. D., Sheridan, R. P., Kearsley, S. K. (1999) SQ: a program for rapidly producing pharmacophorically relevant molecular superpositions. *J Med Chem* **42**, 1505–1514.
65. Miller, M. D., Kearsley, S. K., Underwood, D. J., Sheridan, R. P. (1994) FLOG: a system to select 'quasi-flexible' ligands complementary to a receptor of known three-dimensional structure. *J Comput Aided Mol Des* **8**, 153–174.
66. McGann, M. R., Almond, H. R., Nicholls, A., Grant, J. A., aBrown, F. K. (2003) Gaussian docking functions *Biopolymers* **68**, 76–90.
67. Friesner, R. A., Banks, J. L., Murphy, R. B., Halgren, T. A., Klicic, J. J., Mainz, D. T., Repasky, M. P., Knoll, E. H., Shelley, M., Perry, J. K., Shaw, D. E., Francis, P., Shenkin, P. S. (2004) Glide: a new approach for rapid, accurate docking and scoring. 1. Method and assessment of docking accuracy. *J Med Chem* **47**, 1739–1749.
68. Halgren, T. A., Murphy, R. B., Friesner, R. A., Beard, H. S., Frye, L. L., Pollard, W. T., Banks, J. L. (2004) Glide: a new approach

for rapid, accurate docking and scoring. 2. Enrichment factors in database screening. *J Med Chem* **47**, 1750–1759.
69. Moffat, K., Gillet, V. J., Whittle, M., Bravi, G., Leach, A. R. (2008) A comparison of field-based similarity searching methods: CatShape, FBSS, and ROCS. *J Chem Inf Model* **48**, 719–729.
70. Hahn, M. (1997) Three-dimensional shape-based searching of conformationally flexible compounds. *J Chem Inf Comput Sci* **37**, 80–86.
71. Wild, D. J., Willett, P. (1996) Similarity searching in files of three-dimensional chemical structures. Alignment of molecular electrostatic potential fields with a genetic algorithm. *J Chem Inf Comput Sci* **36**, 159–167.
72. Ebalunode, J. O., Ouyang, Z., Liang, J., Zheng, W. (2008) Novel approach to structure-based pharmacophore search using computational geometry and shape matching techniques. *J Chem Inf Model* **48**, 889–901.
73. Ebalunode, J. O., Zheng, W. (2009) Unconventional 2D shape similarity method affords comparable enrichment as a 3D shape method in virtual screening experiments. *J Chem Inf Model* **49**, 1313–1320.
74. Nagarajan, K., Zauhar, R., Welsh, W. J. (2005) Enrichment of ligands for the serotonin receptor using the shape signatures approach. *J Chem Inf Model* **45**, 49–57.
75. Ballester, P. J., Finn, P. W., Richards, W. G. (2009) Ultrafast shape recognition: evaluating a new ligand-based virtual screening technology. *J Mol Graph Model* **27**, 836–845.
76. Tawa, G. J., Baber, J. C., Humblet, C. (2009) Computation of 3D queries for ROCS based virtual screens. *J Comput Aided Mol Des* **23**, 853–868.
77. Kirchmair, J., Distinto, S., Markt, P., Schuster, D., Spitzer, G. M., Liedl, K. R., Wolber, G. (2009) How to optimize shape-based virtual screening: choosing the right query and including chemical information. *J Chem Inf Model* **49**, 678–692.
78. Nagarajan, K., Zauhar, R., Welsh, W. J. (2005) Enrichment of ligands for the serotonin receptor using the shape signatures approach. *J Chem Inf Model* **45**, 49–57.
79. Wolber, G., Langer, T. (2005) LigandScout: 3-d pharmacophores derived from protein-bound Ligands and their use as virtual screening filters. *J Chem Inf Model* **45**, 160–169.
80. Lee, H. S., Lee, C. S., Kim, J. S., Kim, D. H., Choe, H. (2009) Improving virtual screening performance against conformational variations of receptors by shape matching with ligand binding pocket. *J Chem Inf Model* **49**, 2419–2428.
81. Markt, P., Petersen, R. K., Flindt, E. N., Kristiansen, K., Kirchmair, J., Spitzer, G., Distinto, S., Schuster, D., Wolber, G., Laggner, C., Langer, T. (2008) Discovery of novel PPAR ligands by a virtual screening approach based on pharmacophore modeling, 3D Shape, and electrostatic similarity screening. *J Med Chem* **51**, 6303–6317.
82. Rush, T. S., Grant, J. A., Mosyak, L., Nicholls, A. (2005) A shape-based 3-D scaffold hopping method and its application to a bacterial protein protein interaction. *J Med Chem* **48**, 1489–1495.
83. Cramer, R. D., Poss, M. A., ermsmeier, M. A., Caulfield, T. J., Kowala, M. C., Valentine, M. T. (1999) Prospective identification of biologically active structures by topomer shape similarity searching. *J Med Chem* **42**, 3919–3933.
84. Cramer, R. D., Jilek, R. J., Guessregen, S., Clark, S. J., Wendt, B., Clark, R. D. (2004) "Lead Hopping." Validation of topomer similarity as a superior predictor of similar biological activities. *J Med Chem* **47**, 6777–6791.
85. Muchmore, S. W., Souers, A. J., Akritopoulou-Zanze, I. (2006) The use of three-dimensional shape and electrostatic similarity searching in the identification of a melanin-concentrating hormone receptor 1 antagonist. *Chem Biol Drug Des* **67**, 174–176.
86. Perez-Nueno, V. I., Ritchie, D. W., Rabal, O., Pascual, R., Borrell, J. I., Teixido, J. (2008) Comparison of ligand-based and receptor-based virtual screening of HIV entry inhibitors for the CXCR4 and CCR5 receptors using 3D ligand shape matching and ligand-receptor docking. *J Chem Inf Model* **48**, 509–533.
87. Ballester, P. J., Westwood, I., Laurieri, N., Sim, E., Richards, W. G. (2010) Prospective virtual screening with Ultrafast shape recognition: the identification of novel inhibitors of arylamine N-acetyltransferases. *J R Soc Interface* 7, 335–342.
88. Ebalunode, J. O., Dong, X., Ouyang, Z., Liang, J., Eckenhoff, R. G., Zheng, W. (2009) Structure-based shape pharmacophore modeling for the discovery of novel anesthetic compounds. *Bioorg Med Chem* **17**, 5133–5138.

Chapter 7

Combinatorial Library Design from Reagent Pharmacophore Fingerprints

Hongming Chen, Ola Engkvist, and Niklas Blomberg

Abstract

Combinatorial and parallel chemical synthesis technologies are powerful tools in early drug discovery projects. Over the past couple of years an increased emphasis on targeted lead generation libraries and focussed screening libraries in the pharmaceutical industry has driven a surge in computational methods to explore molecular frameworks to establish new chemical equity. In this chapter we describe a complementary technique in the library design process, termed ProSAR, to effectively cover the accessible pharmacophore space around a given scaffold. With this method reagents are selected such that each R-group on the scaffold has an optimal coverage of pharmacophoric features. This is achieved by optimising the Shannon entropy, i.e. the information content, of the topological pharmacophore distribution for the reagents. As this method enumerates compounds with a systematic variation of user-defined pharmacophores to the attachment point on the scaffold, the enumerated compounds may serve as a good starting point for deriving a structure–activity relationship (SAR).

Key words: ProSAR, combinatorial library design, topological pharmacophore, pharmacophore fingerprint, genetic algorithm, Shannon entropy, multi-objective optimisation.

1. Introduction

Effective structure–activity relationship (SAR) generation is at the centre of any medicinal chemistry campaign. Much work has been done to devise effective methods to explain and explore SAR data for medicinal chemistry teams to drive the design cycles within drug discovery projects (1). Recent work on SAR generation highlights the commonly observed discontinuity of SAR and bioactivity data, the so-called activity cliffs (2). This also emphasises the need to empirically determine SAR for each lead

J.Z. Zhou (ed.), *Chemical Library Design*, Methods in Molecular Biology 685,
DOI 10.1007/978-1-60761-931-4_7,

series; indeed, it is often difficult to rationalise existing SAR data even with access to high-resolution X-ray crystal structures of the target-compound complex (3, 4). Another common challenge for the medicinal chemistry teams is that many pharmacokinetic properties are often inherent to the scaffold and breaking out of this property space can be very difficult. Thus, the team needs to quickly explore the chemical space around a novel scaffold to establish SAR and make decisions on the medicinal chemistry strategy.

The art and science of computational library design has been reviewed extensively elsewhere (5–7), but it is interesting and instructive to note the developments in library design over the past 10 years showing the continued importance of the subject for the industry. Today, there is less emphasis on the analysis of molecular properties and diversity as the objectives of library design have shifted towards focussed lead generation. Design of target-directed libraries and the need to establish novel chemical equity have driven the concept of scaffold diversity with a significant effort to identify novel methods for scaffold hopping. A key enabler for this work is that direct structural descriptions of molecules and common framework/substructure analysis have been more computationally accessible (8–10).

Pharmacophore-based approaches are widely used in library design. A pharmacophore refers to the topological (2D) or 3D arrangement of functional groups that capture the key interactions of a ligand with its enzyme/receptor. The major attractiveness of pharmacophore-based methods is that they do not rely on the 3D structural information of protein target and thus are applicable for all target classes and therefore for all drug discovery projects. The concept of pharmacophore fingerprint (11–12) was introduced to describe the pharmacophore patterns present in a molecule in a manner analogous to substructural fingerprints (13). A pharmacophore fingerprint is normally encoded as a binary bit string where each bit refers to a pharmacophore pattern, i.e. a set of pharmacophore points separated by a given distance or distance range. The pharmacophore pattern can be atom/pharmacophore pair, pharmacophore triplet (11, 14, 15) or quartet (12, 16). The distance between a pair of pharmacophore points is usually binned to capture variations in conformation (3D) or bond distances (2D). The pharmacophore types normally comprise hydrogen bond acceptor and donor, positive and negative charge centre and hydrophobic group, but most softwares allow for user-defined types to capture target-specific features such as metal-chelating groups. Pharmacophore fingerprints are often used in diverse library design to cover a broad pharmacophore space (12, 14–16). Chem-Diverse (17) was the first commercial software to exploit 3-point and 4-point pharmacophore in diversity analysis. Since then many efforts

(18–26) have been made in applying pharmacophore fingerprints to combinatorial library design. For example, Good et al. (18) reported their HARPick program which makes combinatorial library design in reagent space. A Monte Carlo simulated annealing optimisation method was used to optimise the reagent selection to achieve maximal diversity fitness score. Chemical diversity (26–28) is often used as an optimisation function for combinatorial library design, either on the reagent side (29, 30) or on the product side (31, 32). Such library design strategies are often very efficient at selecting diverse compounds. However, they may lead to libraries where it is hard to derive a clear structure–activity relationship (SAR) from the experimental data as the selected building blocks might have little or no relationship to one another.

Recently, we have reported (33) a reagent-based library design strategy 'ProSAR' to tackle these issues. The ProSAR method relies on topological 2-point pharmacophores to enumerate and optimise a selection of reagents to systematically explore novel scaffolds. Thus, the ProSAR method is complementary to scaffold analysis and computational scaffold hopping tools and addresses a separate step in the library design workflow.

In this chapter we will exemplify this method with selected library design problems and also demonstrate how to apply ProSAR designs with concurrent optimisation of product property profile to design libraries that will not only help to derive a SAR, but also have an attractive property profile.

2. Materials

The 2-point pharmacophores were created by an in-house tool TRUST (34) and a shell script was written to create reagent pharmacophore fingerprint based on TRUST output. The 'greedy' search algorithm (35) was implemented in Python (36) to read in reagent pharmacophore fingerprint and optimise pharmacophore entropy. An in-house genetic algorithm-driven library design tool GALOP was used to optimise library under user-supplied multiple constraints. Library product properties were calculated by various in-house prediction tools. Tanimoto similarity for reagents was calculated using FOYFI fingerprint which is an in-house developed fingerprint (37) and is similar in spirit to standard Daylight fingerprint (38). Database similarity searches were done by using an in-house 2D similarity search tool (34) with FOYFI fingerprint. An in-house program FLUSH (37) was used for the structure clustering.

Three sets of commercially available reagents are used in this study: a set of 493 aliphatic primary amines for selecting 20 reagents subset, 2,518 aldehydes and 634 amino acids as reagent pool for making various 20×20 2D libraries and 112 aliphatic bromides together with 127 aliphatic amines as reagent collection for designing concurrent pharmacophore entropy and library property profile optimised 2D libraries. All the reagents are from ACD (39). In the second example, 139 known active compounds which share the same scaffold and are used as validation set are taken from GVKBio Medchem database (40).

3. Method

3.1. Methodology

3.1.1. Definition of the Pharmacophore Fingerprint

Three-point and 4-point pharmacophores (12, 14–16) have been widely used to represent chemical information of library products. In ProSAR, we extend this concept to a 2-point pharmacophore to encode the chemical information of a reagent. The ProSAR reagent pharmacophore is composed of a single pharmacophore point plus the attachment point of the reagent. Here, we use the five common pharmacophore types: hydrogen bond donor (HD), hydrogen bond acceptor (HA), positive charge centre (POS), negative charge centre (NEG) and lipophilic groups (LIP). In our in-house implementation, these are encoded as SMARTS strings (38).

For each reagent the information of the pharmacophores and their respective distance to the attachment point are incorporated into a fingerprint (as shown in **Fig. 7.1**). Note that even rather simple reagents will have multiple pharmacophores. As we normally would want to select low-complexity reagents and avoid adding long side chains to the scaffold, the maximal topological distance (bond distance) between the pharmacophore element and the attachment point is restricted to six bonds, and the sum of donor, acceptor, positive and negative charge groups in a reagent should be less than or equal to two. Thus, the total number of unique 2-point pharmacophores in a reagent is 30 (5×6) and the

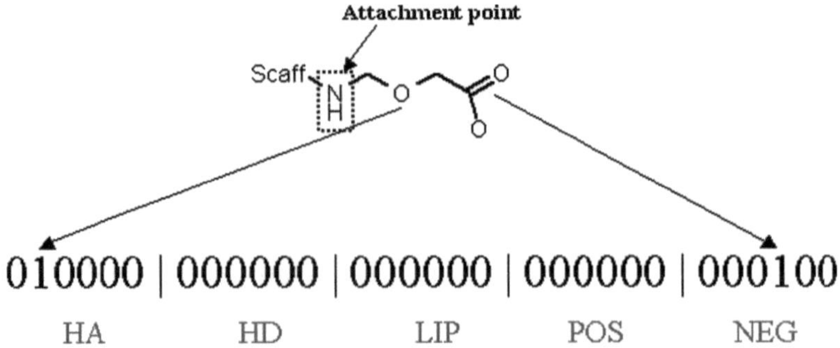

Fig. 7.1 Reagent pharmacophore fingerprint encoding (adapted from (33)).

reagent information is represented by a 30-bin pharmacophore fingerprint. Each bin of the fingerprint refers to a specific 2-point pharmacophore and the count of the specific pharmacophore in the reagent is recorded into this bin. A pharmacophore fingerprint for an amine reagent is exemplified in **Fig. 7.1**. Compared with other pharmacophore fingerprints, this method explicitly captures reagent pharmacophores where one endpoint for the fingerprint is always the attachment point on the scaffold. The advantage of such a fingerprint is that the pharmacophore variability in the fingerprint is always relative to the same position and thus provides a common framework to compare pharmacophore variations for different reagents to further derive SAR information.

3.1.2. Reagent Selection Based on Optimisation in Pharmacophore Space

Once reagent pharmacophore fingerprints are created for all reagents, the next step in the ProSAR strategy is to do reagent selection to optimally cover the 'pharmacophore fingerprint space' and keep the pharmacophore distribution as even as possible. Shannon entropy (SE) (41) has been shown to be an efficient way to characterise the variation of molecular descriptors in compound databases (42). Groothhuis et al. (35, 43) and Miller et al. (44) used SE to measure the chemical diversity of libraries; here, we use SE to represent the pharmacophore distribution of a selected reagent set. SE is defined as follows:

$$SE = -\sum_i p_i \log_2 p_i \quad [1]$$

where p_i is the probability of having a certain pharmacophore in the whole reagent set. p_i is calculated as follows:

$$p_i = c_i / \sum_i c_i \quad [2]$$

where c_i is the population of pharmacophore i in the whole reagent set. A larger SE corresponds to a greater information content, i.e. a more even distribution of reagent pharmacophores. Hence the optimal selection from the set of available reagents will maximise the SE value after library optimisation. A Python (36) program, in which a greedy search algorithm (35) is used as the optimisation search engine, has been developed to make the ProSAR reagent selection.

The general procedure for doing ProSAR library design is as follows: first, all the reagents are collected in smiles file format, and second, a shell script is run to do prefiltering on reagents (remove too complex reagents as described in **Section 3.1.1**) and create the 2-point pharmacophore fingerprints for the remaining reagents; a greedy search optimisation is done by running the python script with the generated pharmacophore fingerprint to select the desired number of reagents.

3.1.3. Concurrent Pharmacophore Entropy and Library Property Profile Optimisation in ProSAR Library Design

Physico-chemical properties and evaluation of potential safety liabilities are important aspects of the library design process. Predicted properties like hERG liability (45), compound aqueous solubility, etc. (46–48) have been extensively studied and included in various library design strategies (49, 50) as a part of multiple constraints optimisation. We have therefore further extended the ProSAR concept to take the library property profile into account in the design process. Several in-house calculated properties are considered; these include a compound novelty check (that checks in in-house and external compound databases to see if the compound is novel), predicted aqueous solubility (51), predicted hERG liability (52) and an in-house developed lead profile score (53, 54).

An in-house library design tool GALOP (33) was extended to include ProSAR designs and it is used in the extended ProSAR library design procedure to replace the greedy search optimisation. GALOP uses a genetic algorithm (GA) (55) to optimise the reagent pharmacophore SE and the product properties simultaneously. In the GA algorithm, each chromosome corresponds to a selected library and it consists of an array of binary bins. Each bin refers to the presence of a reagent. The GA fitness function is a linear combination of the reagent pharmacophore SE term and the product property profile term (as shown in equation [3]):

$$Score = w_p F + w_e \sum_j SE_j \quad [3]$$

Here, F refers to the property profile term and is measured by the fraction of 'good' compounds in the designed library and SE_j refers to the SE for the reagent set which is used for side chain j. A compound is regarded as 'good' only if it meets all the specified property criteria. w_p and w_e are weighting factors for the properties and SE, respectively. In our experience, a weight ratio (w_e/w_p) of 2 works well and is used throughout the libraries designed in this study.

3.2. Application Examples

3.2.1. Selection of Primary Amine Reagents

As the first test case, we selected 20 aliphatic primary amines from a set of 493 commercially available ones by three different methods: random reagent selection, entropy-optimised ProSAR selection and an occupancy-optimised method which purely maximises the occupancy of pharmacophore bins (56), i.e. ensures that as many bins as possible are covered by the reagent selection regardless of the pharmacophore distribution. As the greedy algorithm is a deterministic method in nature, ProSAR and occupancy optimisation give the optimal reagent selection (within the given constraints), while random selections was repeated ten times to get ten different reagent sets.

The distributions of reagent pharmacophore bins for the ProSAR reagent set, one random reagent set and the occupancy-optimised reagent set are compared in **Fig. 7.2**. The total SE of the ProSAR selection is 4.15, average SE of ten random selections is 3.3 and the SE for occupancy-based selection is 3.4. Entropy-driven ProSAR selections have the same pharmacophore coverage as the occupancy-optimised set and both optimisation techniques achieve better coverage than random selections. Additionally, entropy-optimised ProSAR has the most even bin distribution of the selections. As an example, for the lipophilic bins from no. 14 to 17, the reagent count for random selection is 39, the count for occupancy selection is 33, while the count of these bins has been reduced to 15 in the ProSAR library. Entropy-based optimisation achieves the same level of pharmacophore coverage as occupancy optimisation but has a more even distribution of pharmacophores in the reagent set and does not bias the selection towards reagents with lipophilic pharmacophores.

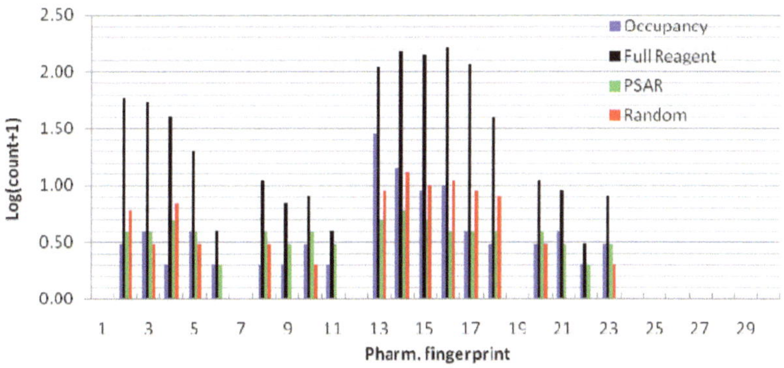

Fig. 7.2 Pharmacophore fingerprint distribution for 20 primary amines selected by using the 'ProSAR' strategy, random selection and occupancy optimisation of fingerprint bins, respectively.

3.2.2. Affymax Library Example

A pending question for ProSAR library design is how well the design covers the pharmacophores from real active compounds. Therefore, we compared the pharmacophores of a ProSAR library with those of active compounds for a specific scaffold. A library example from Affymax (57–59) is selected as the test case here (shown in **Fig. 7.3**). The library diversity is generated from aldehydes (R1) and amino acids (R2) and active compounds for several targets were identified by screening the library. A total of 139 known active compounds with this scaffold were retrieved from the GVKBio MedChem database (40) and are used as validation set.

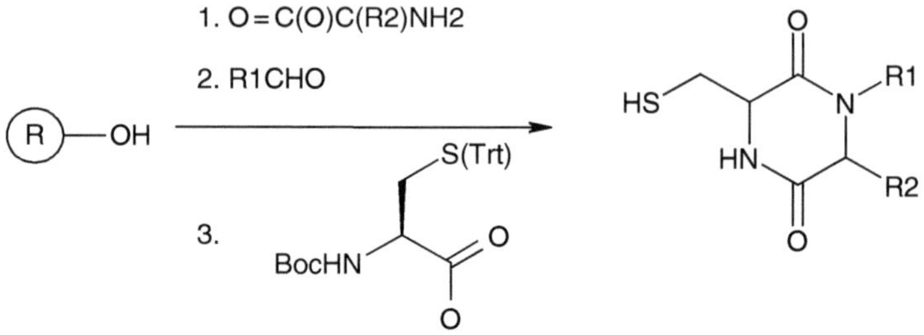

Fig. 7.3 Combinatorial library example from Affymax (57).

In this study, 2,518 aldehydes and 634 amino acids were selected from ACD (39) and used as reagent pool for the libraries (20×20). A ProSAR library was built using the greedy algorithm with ten conventional diversity libraries and ten random reagent selections as a comparison. The diversity libraries were built by using GALOP with the average Tanimoto dissimilarity for the reagent ensemble (based on the in-house FOYFI (37) structural fingerprint) used as the GA fitness function. The pharmacophore distributions of R1 and R2 for the different reagent collections are compared in **Fig. 7.4** and the results for the libraries from different design strategies are summarised in **Table 7.1**. It can be seen that the ProSAR reagent sets cover almost all of the pharmacophore bins (27 bins covered in both R1 and R2) while having an even reagent distribution in the covered bins (SE for R1 and R2 reagents are 4.61 and 4.65, respectively). Random and diversity libraries have marked lower pharmacophore bin coverage (**Fig. 7.4**).

Comparing pharmacophores from known active compounds, all the R1 pharmacophore bins in active set are covered by the ProSAR library while two bins (no. 8 and no. 20) are missing in the random and diversity libraries. For the R2 reagents, one pharmacophore bin from the active molecules (no. 12) is not found in the ProSAR reagents. For the random and the diversity libraries there are ten and six bins not covered, respectively. In this example, SE-driven optimisation of ProSAR pharmacophores has a marked better coverage of potentially important pharmacophore elements present in the known active compounds set. In addition to the pharmacophores present in the active molecules, the ProSAR library also covers many more additional pharmacophores compared to the structural fingerprint diversity library and random selections (**Table 7.1**).

To further estimate the likelihood of obtaining active molecules from the compounds in the designed libraries, compounds in the designed libraries were used as queries and similarity searches against the GVKBio database with a high similarity cut-off were performed to investigate how many active compounds could be retrieved. Taking the observation that similar

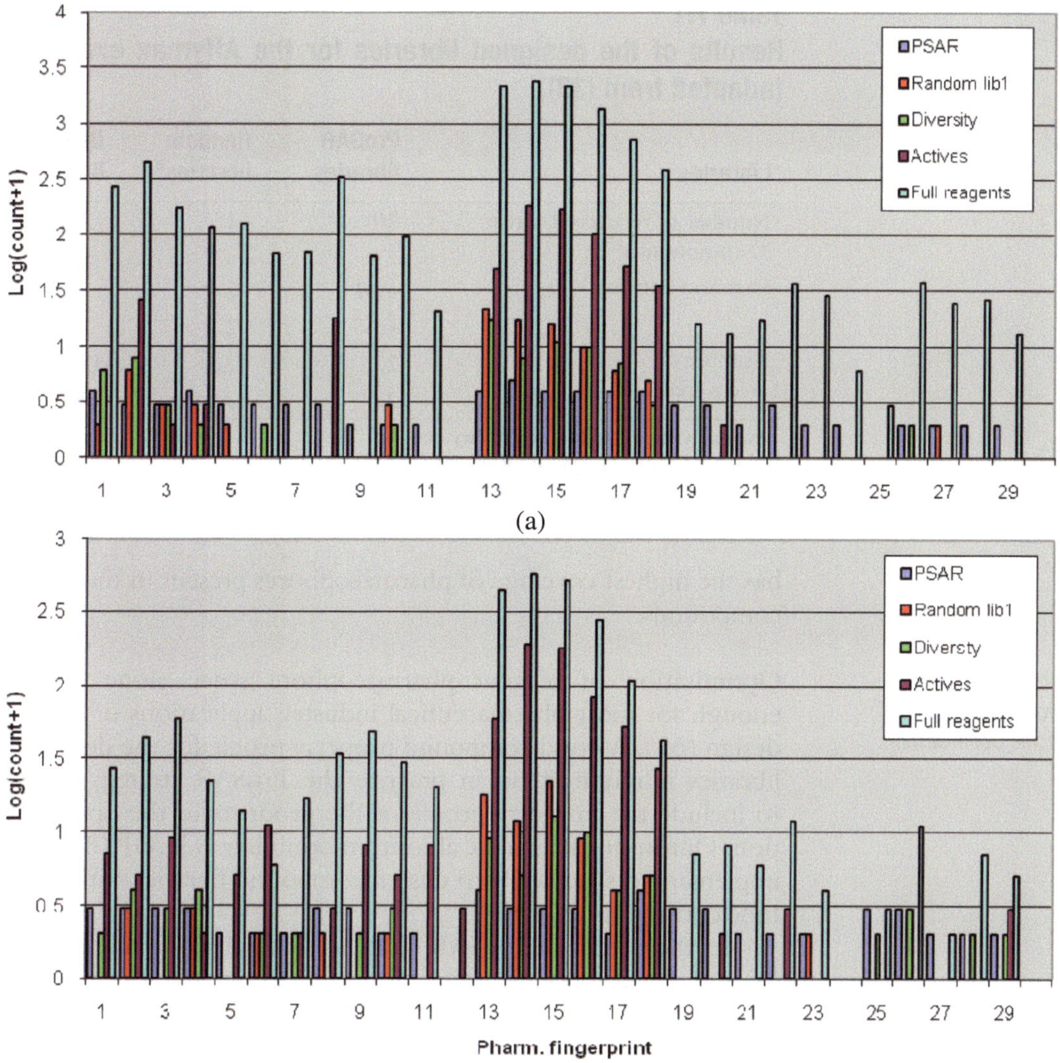

(b)

Fig. 7.4 (**a**) Pharmacophore fingerprint distribution for the R1 reagents. (**b**) Pharmacophore fingerprint distribution for the R2 reagents (adapted from (33)).

compounds tend to have similar bioactivity (60) as an axiom, a high retrieval rate from the GVKBio database is taken as an indication that potentially active molecules are present in the library. Library products are therefore used as query structures to search against the GVKBio database to retrieve active compounds with the conservative similarity cut-off (based on FOYFI fingerprint) of 0.85. From these searches (**Table 7.1**) the ProSAR library retrieves 20 compounds, while the random and diversity libraries retrieve on average 11.7 and 1.1 compounds, respectively. The ProSAR library clearly has the best retrieval rates for active compounds among all the designed libraries, and at the same time

Table 7.1
Results of the designed libraries for the Affymax example (adapted from (33))

Libraries		ProSAR libraries	Random libraries[a]	Diversity libraries[a]
Number of recovered active compounds[b]		20	11.7	1.1
Shannon entropy	R1	4.61	3.1	3.2
	R2	4.65	2.9	3.6
Number of covered bins	R1	27	13.8	12.1
	R2	27	13.3	17.2

[a]Average values based on ten library designs.
[b]Retrieved active compound from the GVKBio database in the similarity search with a Tanimoto similarity cut-off of 0.85.

has the highest coverage of pharmacophores present in the active compounds.

3.2.3. Concurrent ProSAR and Property Profile Optimisation

Optimisation of reagent pharmacophore space alone is not enough for most pharmaceutical industry applications of library design (61). A good compound property profile for the designed libraries is required, so in practice the ProSAR strategy needs to include the property profile of the products in the optimisation. Our in-house genetic algorithm optimiser GALOP (33) was implemented specifically to design compound libraries with multiple constraints (62, 63).

In the extended ProSAR strategy, both the pharmacophore SE and the compound property profile are included in the GA fitness function as shown in equation [3]. Compound properties considered in the algorithm implementation include (1) novelty check, (2) in silico predicted aqueous solubility (51), (3) in silico predicted hERG liability (52) and (4) in-house lead-like criteria (53, 54). A 'good' compound has to pass all the four criteria. One library example (**Fig. 7.5**) is used to demonstrate this extended ProSAR strategy. A set of 112 aliphatic bromides (R1 reagent) and a set of 127 aliphatic amines (R2 reagent) are used as the reagent pool. Ten ProSAR libraries, ten diversity combined with property-optimised libraries and ten libraries only optimised by property were created using GALOP with different fitness functions. As a reference, ten libraries are created with random reagent selections. Each library was clustered using FOYFI structural fingerprints such that we can use a number of clusters as a simple estimate of the structural diversity.

Property-optimised ProSAR libraries have the best pharmacophore Shannon entropy of all the libraries and 99.7% of the compounds have 'good' properties (**Table 7.2**). In terms of phar-

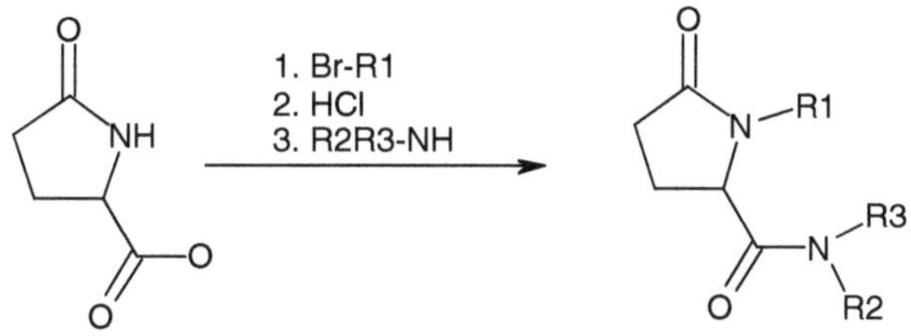

Fig. 7.5 Library example for concurrent reagent pharmacophore entropy and library property profile optimisation.

macophore coverage, the ProSAR libraries cover on average 10.7 bins in R1, slightly lower than the coverage of random libraries. This could be due to the limited variation in R1 for compounds with a good property profile. In the R2 reagents, ProSAR libraries cover on average 15.4 bins, markedly better than any other design strategies. Diversity/property optimisation produces most diverse libraries; this can be seen from its highest average FOYFI Tanimoto dissimilarity value and number of clusters. These libraries have 99.7% good compounds. As expected, property-optimised libraries have a perfect profile (100% good compounds) but low SE and diversity (**Table 7.2**). The random libraries have the worst property profile with medium entropy and diversity values.

Table 7.2
Results for the GA-optimised libraries[a] (adapted from (33))

Libraries		ProSAR+ property[b]	Diversity+ property[c]	Property[d]	Random library[e]	Full library
% of good compounds		99.7	99.7	100	62.2	62
Number of clusters		21	46.1	14.1	23	NC
Shannon entropy	R1	3.03	2.86	2.38	2.71	2.83
	R2	3.52	2.62	2.32	2.81	2.94
Dissimilarity index	R1	0.74	0.80	0.64	0.72	0.74
	R2	0.69	0.80	0.65	0.71	0.73
Number of covered bins	R1	10.5	10.3	7	10.7	21
	R2	15.4	10.2	10.5	12	20

[a]The values listed in the table are averaged over ten library designs, except for the full library.
[b]Libraries obtained by optimising both the pharmacophore entropy and the property profile simultaneously.
[c]Libraries obtained by optimising both the diversity and the property profile simultaneously.
[d]Libraries obtained by only optimising the property profile.
[e]Libraries obtained by randomly selecting reagents.

As an illustration, one ProSAR library and one diversity library were selected for a closer investigation. The R1 and R2 pharmacophore distributions are shown in **Fig. 7.6** with the structures of the selected R1 and R2 reagents shown in **Figures 7.7**, **7.8**, **7.9** and **7.10**. For the R1 reagents the diversity library lacks bin no. 5 (acceptor five bonds distant to the attachment point) and 11 (donor five bonds distant to the attachment point) while both of these pharmacophores are present in the ProSAR library. For the R2 reagent set, bin no. 9, 10, 21, 22 and 27 are missing in the diversity library while being present in the ProSAR library. Again in this example the ProSAR library has a more balanced reagent set in terms of pharmacophoric features and pharmacophore variations than the diversity library. On examination of the R1 and R2 reagents for the two libraries, one sees that the ProSAR reagent set has more structurally related compounds. For example, reagents **1**, **2** and **3** of ProSAR R2 reagent set (**Fig. 7.9**) are similar structures with variations on the alcohol functionality and lipophilic bulk; this could potentially help to derive a SAR around the HD functionality on the side chain. Similarly, structures **12** and **13** may provide SAR around the positive charge

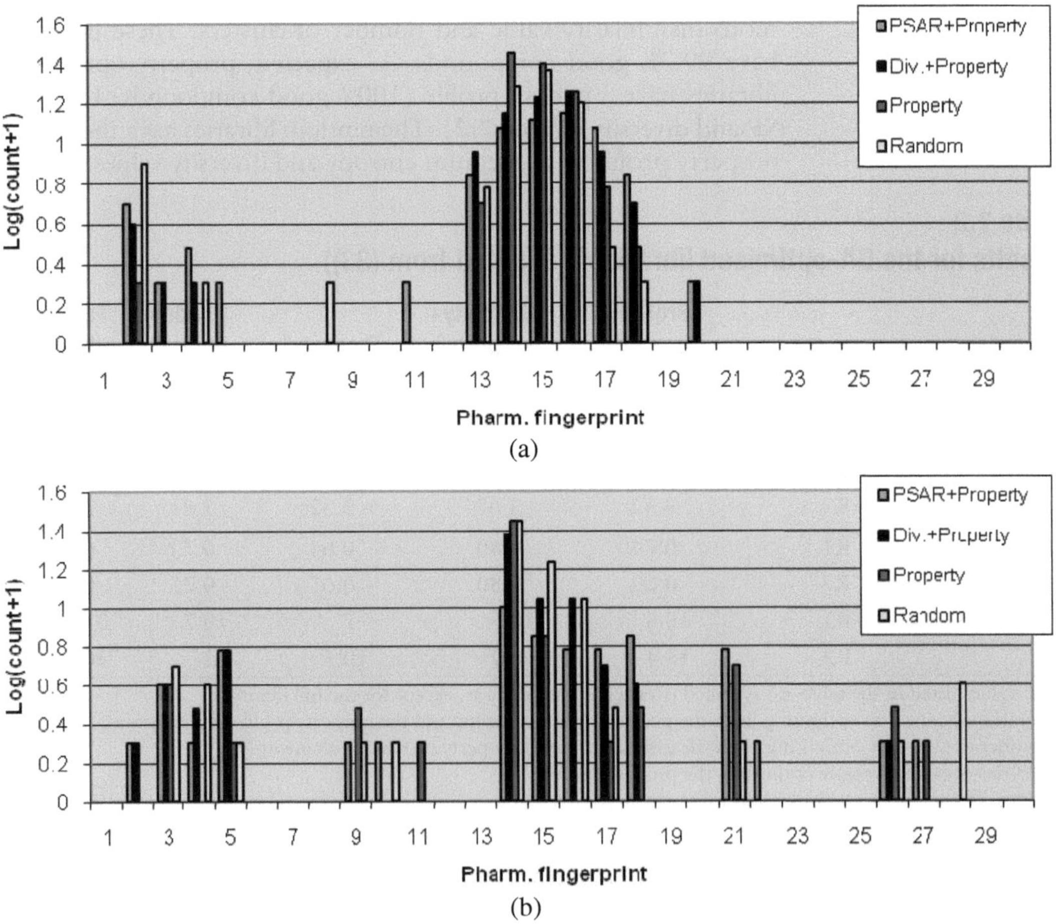

Fig. 7.6 Comparison of pharmacophore fingerprint distribution for libraries with different design strategies. (**a**) Pharmacophore fingerprint distribution for R1 reagents. (**b**) Pharmacophore fingerprint distribution for R2 reagents (adapted from (33)).

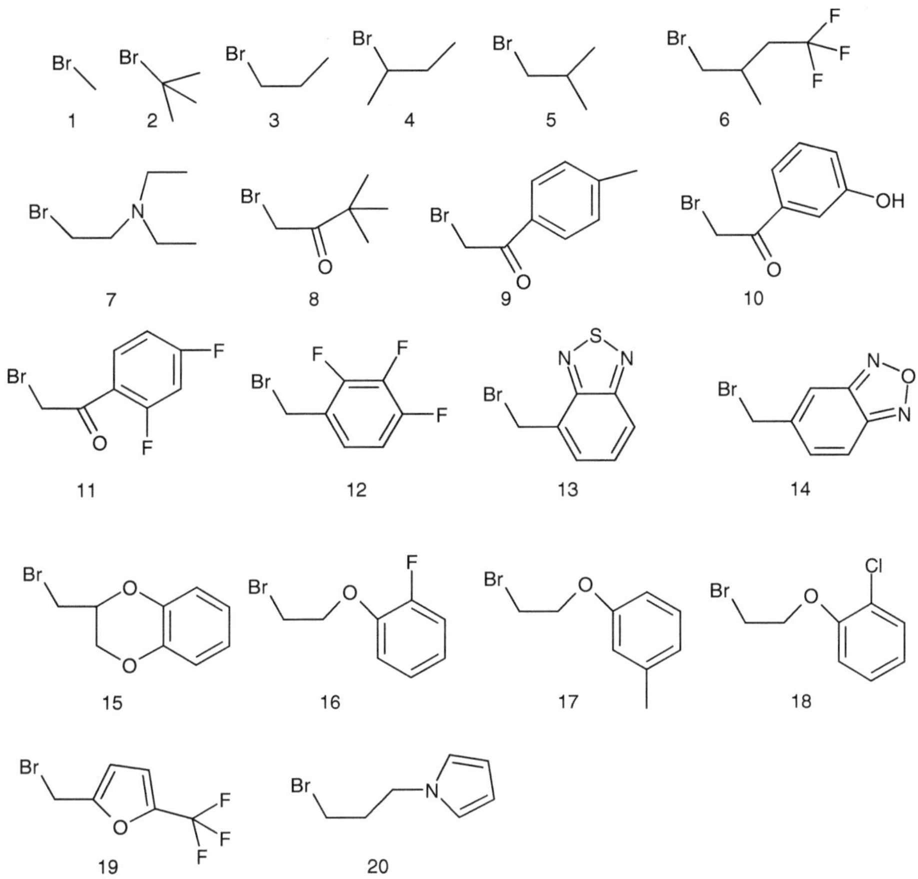

Fig. 7.7 Selected R1 reagents for the ProSAR library (adapted from (33)).

functionality and structures **4–11** may show some SAR around the piperazine ring. These structurally related reagents will have less chance to be selected in the diversity-based design strategy due to the low Tanimoto dissimilarity value (*see* **Section 4**). In summary, the ProSAR libraries tend to include structurally related reagents with systematic variation of side chain pharmacophore elements. These designs are helpful to chemists attempting to derive SAR.

4. Conclusion

The ProSAR strategy for library design selects reagents by optimising the reagent pharmacophore space to achieve a systematic variation of the pharmacophores relative to a scaffold attachment. We show that optimising the Shannon entropy of the reagent

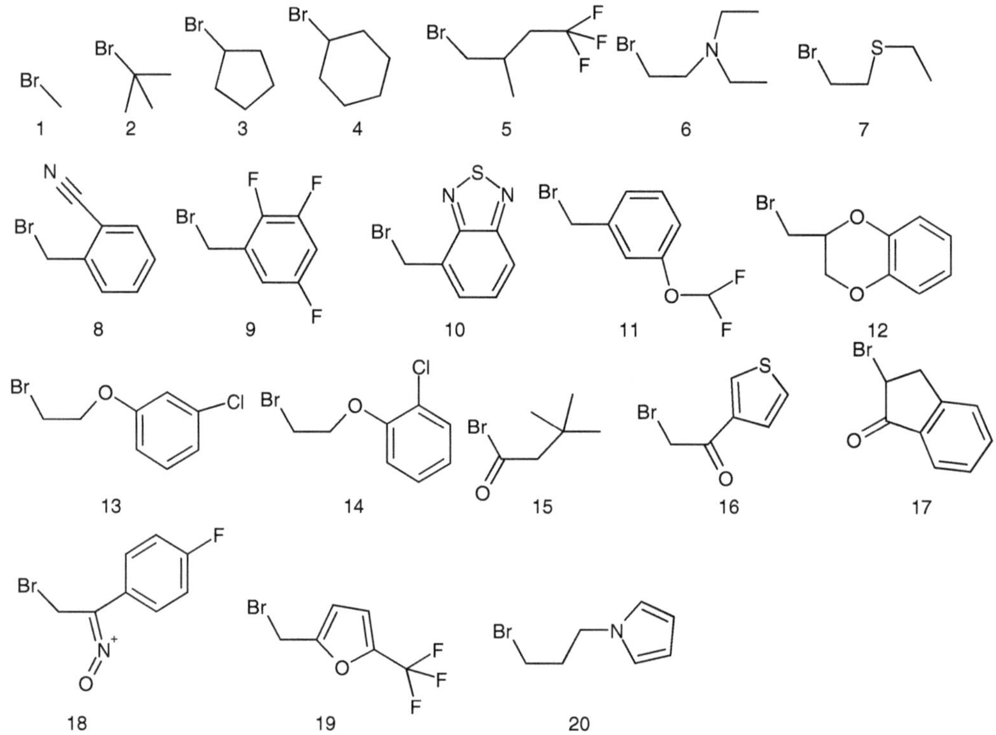

Fig. 7.8 Selected R1 reagents for the diversity library (adapted from (33)).

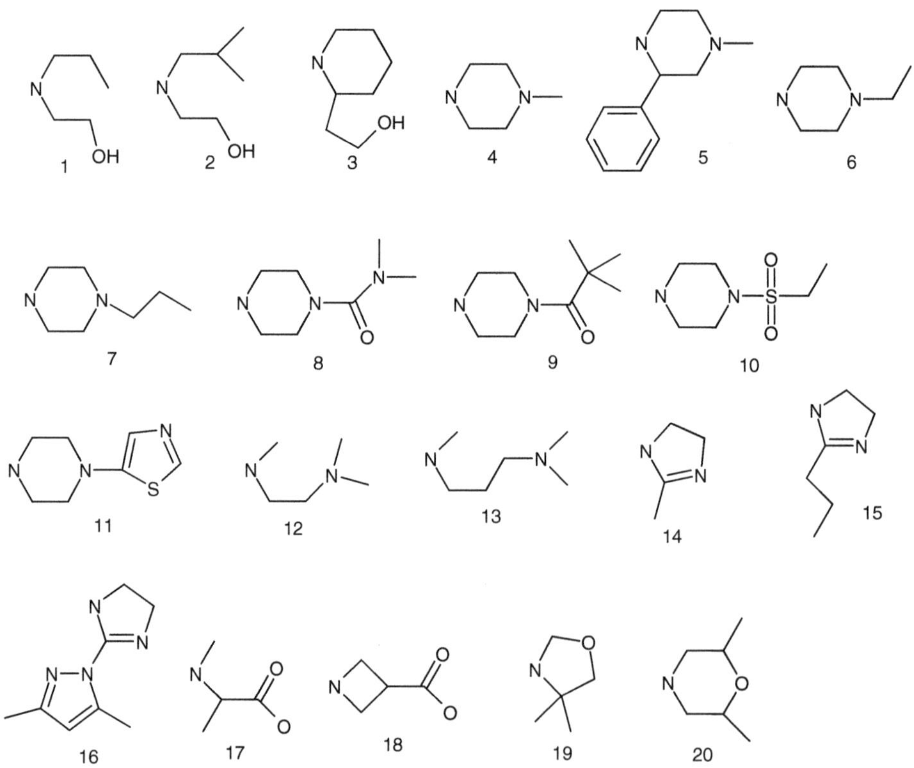

Fig. 7.9 Selected R2 reagents for the ProSAR library (adapted from (33)).

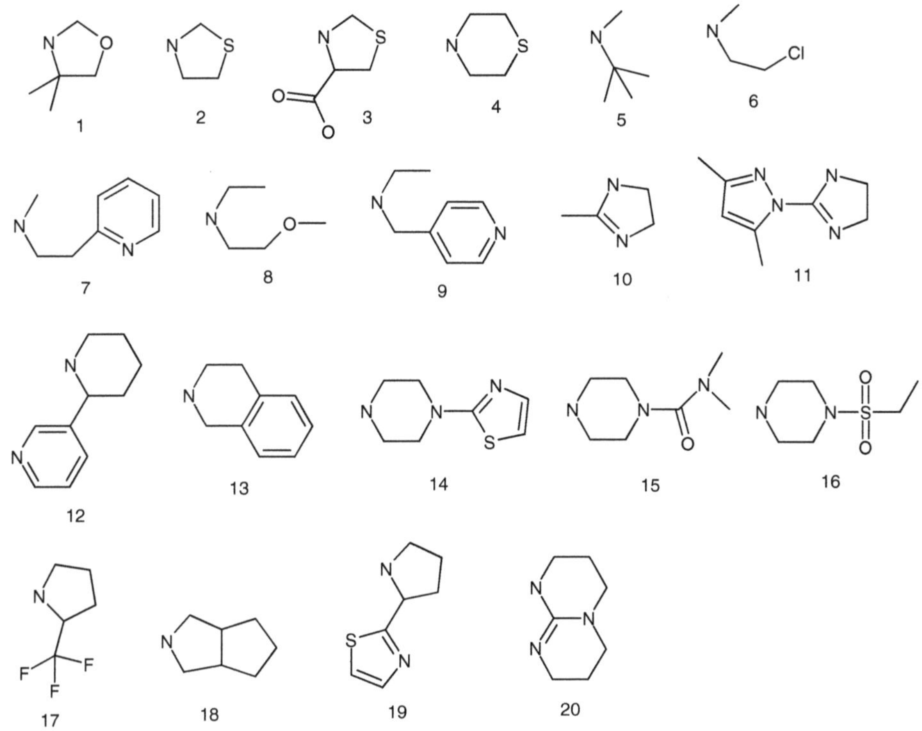

Fig. 7.10 Selected R2 reagents for the diversity library (adapted from (33)).

pharmacophores effectively covers the available pharmacophores among the reagents. It also reduces bias of over-represented pharmacophores and evens the distribution among the reagents, thus potentially making it easier for medicinal chemists to derive SAR. A ProSAR-derived library can also retrieve more bioactive compounds from a database than other design strategies evaluated. In practice, the full ProSAR strategy includes compound properties to obtain libraries which possess not only a wide pharmacophore coverage from the reagents but also satisfactory physico-chemical properties.

It should be borne in mind that diversity in pharmacophore space is not equivalent to the structural diversity. As we can see from the third application example, optimising the average Tanimoto dissimilarity will create a more structurally diverse compound set with little relationship among the compounds, while ProSAR-optimised reagent set tends to include several clusters of structure-related compounds which have systematic variation on reagent pharmacophore. However, ultimately the choice of library design strategy depends on the design objective.

Acknowledgements

The authors are grateful to the following colleagues at AstraZeneca: Dr. David Cosgrove for providing the FOYFI fingerprint programs, Dr. Jens Sadowski for providing the tool to extract the R-groups for the library compounds and Dr. Ulf Börjesson for developing the GALOP program.

References

1. Bajorath, J., Peltason, L., Wawer, M., Guha, R., Lajiness, M. S., van Drie, J. H. (2009) Navigating structure activity landscapes. *Drug Discovery Today* **14**, 698–705.
2. Maggiora, G. M. (2006) On outliers and activity cliffs – why QSAR often disappoints. *J Chem Inf Model* **46**, 1535.
3. Sisay, M. H., Peltason, L., Bajorath, J. (2009) Structural interpretation of activity cliffs revealed by systematic analysis of structure–activity relationships in analog series. *J Chem Inf Model* **49**, 2179–2189.
4. Boström, J., Hogner, A., Schmitt, S. (2006) Do structurally similar ligands bind in a similar fashion? *J Med Chem* **49**, 6716–6725.
5. Spellmeyer, D. C., Grootenhuis, P. D. J. (1999) Recent developments in molecular diversity: computational approaches to combinatorial chemistry. *Annu Rep Med Chem Rev* **34**, 287–296.
6. Beno, B. R., Mason, J. S. (2001) The design of combinatorial libraries using properties and 3D pharmacophore fingerprints. *Drug Discovery Today* **6**, 251–258.
7. Willett, P. (2000) Chemoinformatics – similarity and diversity in chemical libraries. *Curr Opin Biotechnol* **11**, 85–88.
8. Bemis, G. W., Murcko, M. A. (1996) The properties of known drugs. 1. Molecular frameworks. *J Med Chem* **39**, 2887–2893.
9. Xu, Y. J., Johnson, M. (2002) Using molecular equivalence numbers to visually explore structural features that distinguish chemical libraries. *J Chem Inf Comp Sci* **42**, 912–926.
10. Pitt, W., Parry, D. M., Perry, B. G., Groom, C. R. (2009) Heteroaromatic rings of the future. *J Med Chem* **52**, 2952–2963.
11. Good, A. C., Kuntz, I. D. (1995) Investigating the extension of pairwise distance pharmacophore measures to triplet-based descriptors. *J Mol Comput Aided Mol Des* **9**, 373–379.
12. Mason, J. S., Morize, I., Menard, P. R., Cheney, D. L., Hulme, C., Labaudiniere, R. F. (1999) New 4-point pharmaophore method for molecular similarity and diversity applications: overview of the method and applications, including a novel approach to the design of combinatorial libraries containing privileged substructures. *J Med Chem* **42**, 3251–3264.
13. Symyx, 2.5; Symyx Technologies Inc., Santa Clara, CA 95051, USA.
14. McGregor, M. J., Muskal, S. M. (1999) Pharmacophore fingerprinting. 1. Application to QSAR and focused library design. *J Chem Inf Comput Sci* **39**, 569–574.
15. Pickett, S. D., Mason, J. S., Mclay, I. M. (1996) Diversity profiling and design using 3D pharmacophore: pharmacophore-derived queries (PDQ). *J Chem Inf Comput Sci* **36**, 1214–1223.
16. Mason, J. S., Beno, B. R. (2000) Library design using BCUT chemistry-space descriptors and multiple four-point pharmacophore fingerprints: simultaneous optimization and structure-based diversity. *J Mol Graph Mod* **18**, 438–451.
17. Cato, S. J. (2000) Exploring pharmacophores with Chem-X, in (Güner, O., ed.) *Pharmacophore Perception, Development, and Use in Drug Designer.* International University Line, La Jolla, CA, pp. 107–125.
18. Good, A. C., Lewis, R. A. (1997) New methodology for profiling combinatorial libraries and screening sets: cleaning up the design process with HARPick. *J. Med Chem* **40**, 3926–3936.
19. Chen. X., Rusinko, A., III, Young, S. S. (1998) Recursive partitioning analysis of a large structure-activity data set using three-dimensional descriptors. *J Chem Inf Comput Sci* **38**, 1054–1062.
20. Matter, H., Pötter, T. (1999) Comparing 3D pharmacophore triplets and 2D fingerprints for selecting diverse compound subsets. *J Chem Inf Comput Sci* **39**, 1211–1225.
21. Eksterowicz, J. E., Evensen, E., Lemmen, C., Brady, G. P., Lanctot, J. K., Bradley, E. K., Saiah, E., Robinson, L. A.,

Grootenhuis, P. D. J., Blaney, J. M. (2002) Coupling structure-based design with combinatorial chemistry: application of active site derived pharmaophores with informative library design. *J Mol Graph Model* **20**, 469–477.
22. Good. A. C., Masson, J. S., Green, D. V. S., Leach, A. R. (2001) Pharmacophore-based approaches to combinatorial library design„ in (Ghose, A. K., Viswanadhan, V. N., eds.) *Combinatorial Library Design and Evaluation*. Marcel Dekker, New York, pp. 399–428.
23. McGregor, M. J., Muskal, S. M. (2000) Pharmacophore fingerprinting. 2. Application to primary library design. *J Chem Inf Comput Sci* **40**, 117–125.
24. SYBYL Pharmacophore triplet is distributed by Tripos, Inc., 1699 S. Hanley Rd., St. Louis, MO 63144, USA.
25. Schneider, G., Nettekoven, M. (2003) Ligand-based combinatorial design of selective purinergic receptor (A_{2A}) antagonists using self-organizing maps. *J Comb Chem* **5**, 233–337.
26. Turner, D. B., Tyrrell, S. M., Willett, P. (1997) Rapid quantification of molecular diversity for selective database acquisition. *J Chem Inf Comput Sci* **37**, 18–22.
27. Jamois, E. A. (2003) Reagent-based and product-based computational approaches in library design. *Curr Opin Chem Biol* 7, 326–330.
28. Potter, T., Matter, H. (1998) Random or rational design? Evaluation of diverse compound subsets from chemical structure databases. *J Med Chem* **41**, 478–488.
29. McGregor, M. J., Muskal, S. M. (2000) Pharmacophore fingerprinting. 2. Application to primary library design. *J Chem Inf Comput Sci* **40**, 117–125.
30. Zheng, W., Cho, S. J., Tropsha, A. (1998) Rational combinatorial library design. 1. Focus-2D: a new approach to targeted combinatorial chemical libraries. *J Chem Inf Comput Sci* **38**, 572–584.
31. Leach, A. R., Green, D. V. S., Hann, M. M., Judd, D. B., Good, A. C. (2000) Where are the gaps? A rational approach to monomer acquisition and selection. *J Chem Inf Comput Sci* **40**, 1262–1269.
32. Gillet, V. J., Willett, P., Bradshaw, J. (1997) The effectiveness of reactant pools for generating structurally diverse combinatorial libraries. *J Chem Inf Comput Sci* **37**, 731–740.
33. Chen, H., Börjesson, U., Engkvist, O., Kogej, T., Svensson, M. A., Blomberg, N., Weigelt, D., Burrows, J. N., Lagne, T. (2009) ProSAR: a new methodology for combinatorial library design. *J Chem Inf Model* **49**, 603–614.
34. Kogej, T., Engkvist, O., Blomberg, N., Muresan, S. (2006) Multifingerprint based similarity searches for targeted class compound selection. *J Chem Inf Model* **46**, 1201–1213.
35. Bradley, E. K., Miller, J. L., Saiah, E., Grootenhuis, P. D. J. (2003) Informative library design as an efficient strategy to identify and optimize leads: application to cyclin-dependant kinase 2 antagonists. *J Med Chem* **46**, 4360–4364.
36. Python Programming Language Official Website, http://www.python.org/
37. Blomberg, N., Cosgrove, D. A., Kenny, P. W., Kolmodin, K. (2009) Design of compound libraries for fragment screening. *J Comput Aided Mol Des* **23**, 513–525.
38. Daylight Theory Manual; Daylight Chemical Information Systems, Inc. http://www.daylight.com/dayhtml/doc/theory/
39. MDL Available Chemicals Directory database 2007, Symyx Technologies, Inc., Santa Clara, CA 95051, USA.
40. GVKBio Medchem database 2007, GVK Biosciences Private Ltd., Hyderabad 500016, India.
41. Shannon, C. E., Weaver, W. (1963) *The Mathematical Theory of Communication*, University of Illinois Press, Urbana, IL, USA.
42. Godden, J. W., Stahura, F. L., Bajorath, J. (2000) Variabilities of molecular descriptors in compound databases revealed by Shannon entropy calculations. *J Chem Inf Comput Sci* **40**, 796–800.
43. Lamb, M. L., Bradley, E. K., Beaton, G., Bondy, S. S., Castellino A. J., Gibbons, P. A., Suto, M. J., Grootenhuis, P. D. J. (2004) Design of a gene family screening library targeting G-protein coupled receptors. *J Mol Graph Model* **23**, 15–21.
44. Miller, J. L., Bradley, E. K., Teig, S. L. (2003) Luddite: an information-theoretic library design tool. *J Chem Inf Comput Sci* **43**, 47–54.
45. Keating, M. T., Sanguinetti, M. C. (1996) Molecular genetic insights into cardiovascular disease. *Science* **272**, 681–685.
46. Cavalli, A., Poluzzi, E., De Ponti, F., Recanatini, M. (2002) Toward a pharmacophore for drugs including the long QT syndrome: insights from a CoMFA study of HERG K(+) channel blockers. *J Med Chem* **45**, 3844–3853.
47. Pearlstein, R. A., Vaz, R. J., Kang, J., Chen, X. -L., Preobrazhenskaya, M.,

Shchekotikhin, A. E., Korolev, A. M., Lysenkova, L. N., Miroshnikova, O. V., Hendrix, J., Rampe, D. (2003) Characterization of HERG potassium channel inhibition using CoMSiA 3D QSAR and homology modeling approaches. *Bioorg Med Chem Lett* **13**, 1829–1835.

48. Jouyban, A., Soltanpour, S., Soltani, S., Chan, H. K., Acree, W. E. (2007) Solubility prediction of drugs in water-cosolvent mixtures using Abraham solvation parameters. *J Pharm Sci* **10**, 263–277.
49. Eagan, W. J., Merz, K. M., Baldwin, J. J. (2000) Prediction of drug absorption using multivariate statistics. *J Med Chem* **43**, 3867–3877.
50. Darvas, F., Dorman, G., Papp, A. (2000) Diversity measures for enhancing ADME admissibility of combinatorial libraries. *J Chem Inf Comput Sci* **40**, 314–322.
51. Bruneau, P. (2001) Search for predictive generic model of aqueous solubility using Bayesian neural nets. *J Chem Inf Comput Sci* **41**, 1605–1616.
52. Gavaghan, C. L., Arnby, C. H., Blomberg, N., Strandlund, G., Boyer, S. (2007) Development, interpretation and temporal evaluation of a global QSAR of hERG electrophysiology screening data. *J Comput Aided Mol Des* **21**, 189–206.
53. Oprea, T. I., Davis, A. M., Teague, S. J., Leeson, P. D. (2001) Is there a difference between leads and drugs? A historical perspective. *J Chem Inf Comput Sci* **41**, 1308–1335.
54. Oprea, T. I. (2002) Current trends in lead discovery: are we looking for the appropriate properties? *J Comp Aided Mol Des* **16**, 325.
55. Reynolds, C. H., Tropsha, A., Pfahler, D. B., Druker, R., Chakravorty, S., Ethiraj, G., Zheng, W. (2001) Diversity and coverage of structural sublibraries selected using the SAGE and SCA algorithms. *J Chem Inf Comput Sci* **41**, 1470–1477.
56. Jamois, E. J., Hassan, M., Waldman, M. (2000) Evaluation of reagent-based and product-based strategies in the design of combinatorial library subsets. *J Chem Inf Comput Sci* **40**, 63–70.
57. Szardenings, A. K., Antonenko, V., Campbell, D. A., DeFrancisco, N., Ida, S., Si, L., Sharkov, N., Tien, D., Wang, Y., Navre, M. (1999) Identification of highly selective inhibitors of collagenase-1 from combinatorial libraries of diketopiperazines. *J Med Chem* **42**, 1348–1357.
58. Campbell, D. A., Look, G. C., Szardenings, A. K., Patel, P. V. (2001) US6271232B1; Campbell, D. A., Look, G. C., Szardenings, A. K., Patel, P. V. (1999) US5932579A; Campbell, D. A., Look, G. C., Szardenings, A. K., Patel, P. V. (1997) WO97/48685A1.
59. Szardenings, A. K., Harris, D., Lam, S., Shi, L., Tien, D., Wang, Y., Patel, D. V., Navre, M., Campbell, D. A. (1998) Rational design and combinatorial evaluation of enzyme inhibitor scaffolds: identification of novel inhibitors of matrix metelloproteinases. *J Med Chem* **41**, 2194–2200.
60. Martin, Y. C., Kofron, J. L., Traphagen, L. M. (2002) Do structurally similar molecules have similar biological activity? *J Med Chem* **45**, 4350–4358.
61. Pickett, S. D, McLay I. M., Clark, D. E. (2000) Enhancing the hit-to-lead properties of lead optimization libraries. *J Chem Inf Comput Sci* **40**, 263–272.
62. Gillet, V. J., Khatlib, W., Willett, P., Fleming P. J., Green, D. V. S. (2002) Combinatorial library design using a multiobjective genetic algorithm. *J Chem Inf Comput Sci* **42**, 375–385.
63. Brown, R. D., Hassan, M., Waldman, M. (2000) Combinatorial library design for diversity, cost efficiency, and drug-like character. *J Mol Graph Model* **18**, 427–437.

Section II

Structure-Based Library Design

Chapter 8

Docking Methods for Structure-Based Library Design

Claudio N. Cavasotto and Sharangdhar S. Phatak

Abstract

The drug discovery process mainly relies on the experimental high-throughput screening of huge compound libraries in their pursuit of new active compounds. However, spiraling research and development costs and unimpressive success rates have driven the development of more rational, efficient, and cost-effective methods. With the increasing availability of protein structural information, advancement in computational algorithms, and faster computing resources, in silico docking-based methods are increasingly used to design smaller and focused compound libraries in order to reduce screening efforts and costs and at the same time identify active compounds with a better chance of progressing through the optimization stages. This chapter is a primer on the various docking-based methods developed for the purpose of structure-based library design. Our aim is to elucidate some basic terms related to the docking technique and explain the methodology behind several docking-based library design methods. This chapter also aims to guide the novice computational practitioner by laying out the general steps involved for such an exercise. Selected successful case studies conclude this chapter.

Key words: Structure-based library design, drug discovery, docking, high-throughput screening, combinatorial chemistry.

1. Introduction

The finding, optimization, and bioevaluation of small molecules that can interact with therapeutically relevant targets to modulate biological processes is the core of the drug discovery process. So far, this has been mainly dominated by high-throughput screening (HTS), a hardware technology that allows the rapid screening of compound libraries to identify potentially active ones (1, 2). HTS, however, requires a ready source of large and preferably diverse set of compounds to serve as starting points for the

J.Z. Zhou (ed.), *Chemical Library Design*, Methods in Molecular Biology 685,
DOI 10.1007/978-1-60761-931-4_8, © Springer Science+Business Media, LLC 2011

screening process (3). In the pursuit of increasing the chemical space for such molecules, combinatorial chemistry, a technology that systematically mixes and matches various chemical building blocks to generate chemical libraries was developed (4). Such libraries were expected to cover the entire chemical space, which is estimated to consist of $10^{60–100}$ compounds (5, 6). This combination of HTS and combinatorial chemistry was expected to provide a large and diverse set of lead compounds, enhance shrinking drug candidate pipelines, and reduce drug discovery time frames (7, 8). Thus, over the past two decades, combinatorial chemistry and HTS have been widely used in the modern drug discovery process with reasonable success (9, 10). Notable improvements in HTS technologies (e.g., robotics, automated liquid handling devices, assay miniaturization techniques, signal detectors, and data processing software) facilitated even further the rapid screening of these libraries to identify promising compounds against validated drug targets (7).

However, after a detailed inspection of the results of these screening campaigns, it is now evident that neither hit rates (11) nor hit quality (e.g., unsuitable functional groups, poor solubility of identified hits (12)) obtained from HTS experiments have shown any significant improvements over time.

On the other hand, screening such huge libraries for every target is impractical, uneconomical, and inefficient. Part of the problem is attributed to the quality of compounds used for HTS (13) (e.g., lack of drug-like properties, compounds with properties unsuitable for biological testing).

As HTS still remains the method of choice to discover novel hit compounds, researchers focused their attention to the design and development of appropriate tools to reduce the size of the chemical libraries to be tested while increasing the quality of the compounds, what could maximize the chances of identifying hit compounds amenable to the subsequent lead optimization stages (14). Such focused libraries were also expected to decrease synthesis, repository management, and screening costs (15). It has also been observed that the number of drug-like compounds with relevant pharmacological profiles is smaller than the total chemical space, and hits for a given target are clustered in a finite region of the compound space (14). Compounds are considered to be drug-like if they contain functional groups and possess physical properties consistent with the majority of known drugs (6), along with acceptable absorption, distribution, metabolism, excretion, and toxicological (ADMET) profiles to pass through Phase 1 clinical trials (16).

Thus, it is but natural to incorporate the structural information from the target to bias compound selection prior to the experimental testing (9, 17, 18). With the advent of high-throughput protein crystallography (19), structural genomics consortium projects (20), and developments in homology modeling methods (21), an increasing number of 3D structures of targets are now available for several structure-based drug discovery applications (e.g., virtual screening (22–26), binding mode predictions (27–31), and lead optimization (32)). Structural information coded in the characteristics of binding sites, such as receptor:ligand interaction patterns, can be used to prioritize compounds for experimental screening using docking methods (33–35), and such exercises have been successful in the past (36, 37). With the continual development in docking algorithms and computational resources, structure-based docking methods will play an increasing important role in compound library design.

It is timely, then, to review the use of docking methods for structure-based library design and to understand how best to implement them in drug discovery. First, we will define and explain basic concepts and terminology related to structure-based drug design and docking. Next, docking methods for library design will be presented with brief notes explaining the practical considerations involved in such an exercise. The chapter will conclude with selected case studies highlighting the recent successes of docking methods for structure-based library design.

2. Requirements

The three major requirements for docking-based library design are basically the same as for high-throughput docking (HTD): a 3D representation of the target structure (experimental or modeled), a database of compounds in electronic format, and a suitable docking algorithm.

2.1. 3D Structure of Target

Advancements in crystallography/NMR techniques have resulted in an exponential increase in the number of protein structures in publicly available structural databases (e.g., as of September 2009, the Protein Database Bank PDB contains experimentally solved 3D structural data for ~60,000 structures (www.pdb.org)). In addition, several structural genomics consortiums aim to provide crystal structures across all protein families (38). In case when experimental structures are not available, techniques such as homology modeling are often used to build structural models of other homologous proteins (21).

The structure is thoroughly analyzed to identify putative binding sites, e.g., by the known location of co-crystallized active compounds, or by applying in silico methods to identify such sites (e.g., POCKET (39), LIGSITE (40), and SURFNET (41)). The information obtained from these sites (receptor:ligand interactions (42), physiochemical characteristics of binding site residues, nature, and size of the binding site) may then be used to restrict the size of compound libraries by adequate filtering (43).

However, caution must be exercised in using crystal structures as is, since they may contain several inaccuracies (44). Low resolution of electron density maps and crystallization conditions different than those maintained in biological assays may introduce errors in the final structure (45). Assumptions made by the crystallographer may result in errors in the orientation of side chains (e.g., asparagine and glutamine flips, histidine tautomerization) or proper location and conformation of the ligand (45). In addition, the crystal structure represents just one snapshot of a highly dynamic conformational equilibrium ensemble. The impact of protein flexibility in docking is not yet fully understood (46), which further undermines the applicability of the crystal structure as is for structure-based drug discovery.

To "prepare" the protein for a docking procedure, the following considerations are usually taken into account:

(a) Removal of the ligand or co-factors if any, from the co-crystallized protein complex.

(b) PDB structures may contain coordinates of several water molecules. Water molecules play an important role in ligand binding by mediating hydrogen bonds between the protein and the ligand or by being displaced by the ligand upon binding (47, 48). If available, several crystal structures of the target are investigated to study water positions. Only those which are highly conserved or tightly bound to the receptor are retained (49, 50).

(c) Inspection and correction of any error in the crystal structures, such as incorrect bond orders and missing residues (particularly in the binding site).

(d) Crystal structures lack hydrogens. It is necessary to check the protonation state of receptor residues, add hydrogens, and perform energy minimization.

(e) Check for asparagine and glutamine flips and for the correct histidine tautomerization states.

Sequence identity and the quality of sequence alignment play an important role in their accuracy of homology models. Exceptions to this rule exist, as in the case of Class A G protein-coupled receptors where structural rather than sequence similarity drives the modeling (21). As a word of caution, high sequence identities may mask the dissimilarities in certain flexible regions, what may render the model less useful for drug discovery applications. The choice of template and inefficient refinement methods are the other sources of errors in homology modeling (21).

2.2. Compound Collections

Many commercial vendors and academic labs provide collection of compounds or fragments (Sigma-Aldrige, Available Chemicals Directory (ACD) (http://www.symyx.com), Maybridge (www.maybridge.com), TCI America (http://www.tciamerica.com), ChemDB (51) (http://cdb.ics.uci.edu), ZINC (52), ChemBridge (www.chembridge.com); cf (53) for a review on public accessible chemical databases), in computationally readably formats like sdf or mol2 (42). Along with the experimental combinatorial library design process mentioned earlier, several software tools, such as CombiLibMaker in Sybyl (www.tripos.com) and QuaSAR-Combigen from CCG (www.chemcomp.com), were developed to enumerate and predict the 2D/3D structures of compounds using chemical fragments without the expensive and tedious experimental part. On the other hand, pharmaceutical companies have historically maintained huge compound libraries and continue to add compounds with novel chemistry to these collections. The size of such compound collections is estimated to be $\sim 10^6$ compounds (54). These compound libraries or their constitutive parts (reagents, fragments) form the source of inputs for docking methods for structure-based library design.

However, these compounds and the fragments are not without their intrinsic problems and should not be used as is. Some examples of potentially problematic compounds include those with chemically reactive groups, dyes, and fluorescent compounds which interfere with assays, frequent hitters/promiscuous binders, and inorganic complexes (55). It is important, then, to a priori filter out such compounds or reagents which are practically useless from a drug discovery point of view.

Some of the common steps toward preparing and filtering chemical compound libraries include

(a) Removing compounds with salts, counter ions, chemically reactive groups (e.g., metal chelators), undesirable atoms or functional groups, inorganic compounds, and duplicates.

(b) Generation of correct tautomers, protonation, and stereoisomeric states for each compound. Eventually, several states for a given compound could be generated and kept in the final library.

(c) Filtering compounds based on drug-like or lead-like physiochemical properties or other in-house scoring schemes (e.g., Lipinski's rule of 5 (56), lead-like filters suggested by Oprea et al. (57)).

In cases where fragments, rather than compounds, are used to design libraries, the fragments may be filtered based on the nature of the binding site and their ability to adhere to existing chemistry protocols with respect to their attachments to templates or other fragments (58). Other filtering criteria include excluding fragments with hydrolysable groups (e.g., sulfonyl halides, anhydride aliphatic ester) and potential cytotoxic groups like thiourea and cyclohexanone (55).

2.3. Docking Algorithms and Methods

The third important requirement is the selection of an appropriate docking algorithm. In brief, a docking-based virtual screening (or HTD) consists of the following steps:

(a) Positioning compounds into the binding site of the target via a process called docking.

(b) Assignment of a score to each compound which represents the likelihood of binding to the target (scoring). Please refer to (33, 34, 59, 60) for tutorials on docking and structure-based virtual screening.

(c) Prioritization of a subset of compounds based on scores and other post-screening criteria, like the ability to mimic key receptor:ligand interactions.

Docking programs routinely incorporate compound flexibility. However, incorporating receptor flexibility in docking procedures is still a major hurdle (61). Of late, several attempts have been made for this purpose (cf (46, 62) for review).

One may choose from several docking methods, but it should be noted that a thorough understanding of the principles underlying the program is important to achieve meaningful results (7). A systematic review of docking methods or programs is not the focus of this chapter, but their use in the context of structure-based library design. The interested reader may refer to (59, 63, 64) for reviews on docking programs. It should be stressed that none of the current docking programs is universally applicable (65–67). Thus, instead of using the default settings of the programs, one could develop, test, and validate a protocol that optimizes the use of the program parameters for a given target.

After HTD, one may still be left with a large number of compounds. In such cases, and on top of filtering according to key ligand:receptor interaction patterns – if available, various data-mining techniques may be applied to narrow down and identify diverse compounds as possible (68). Clustering algorithms (e.g., exclusion sphere, k-nearest neighbor, Jarvis–Patrick) provide an easy way to overview different chemical classes in the result set and choose representative compounds within each class for experimental testing (3). Neural networks, support vector machine-based approaches are used to predict target-class likeliness (e.g., for GPCRs and kinases (69)).

Docking methods for library design can be broadly classified into two strategies:

- *Sequential docking*, where pre-enumerated compounds are docked into the receptor binding site, scored, ranked, and selected for further experimental testing; the conformational space of the compounds is explored by flexible compound docking or rigid docking of pre-generated conformers of each compound.

- *Fragment-based design*, where the constituents of the compounds (scaffolds and functional groups/substituents) are docked in the binding site and then linked together to build combinatorial libraries.

The latter strategy has two flavors (70).

(a) *Seed and grow:* A pre-selected scaffold is first docked into the binding site. Each scaffold pose is scored and only top-ranking poses are considered for subsequent stages. Substituents are then attached to each selected scaffold pose, optimized, and scored (71). The top-scoring substituents are then used to build a combinatorial library. The advantage of this method is that it avoids the combinatorial explosion problem by narrowing down the number of substituents used to build libraries and including knowledge from the binding site of the target structure. This approach is depicted in **Fig. 8.1**.

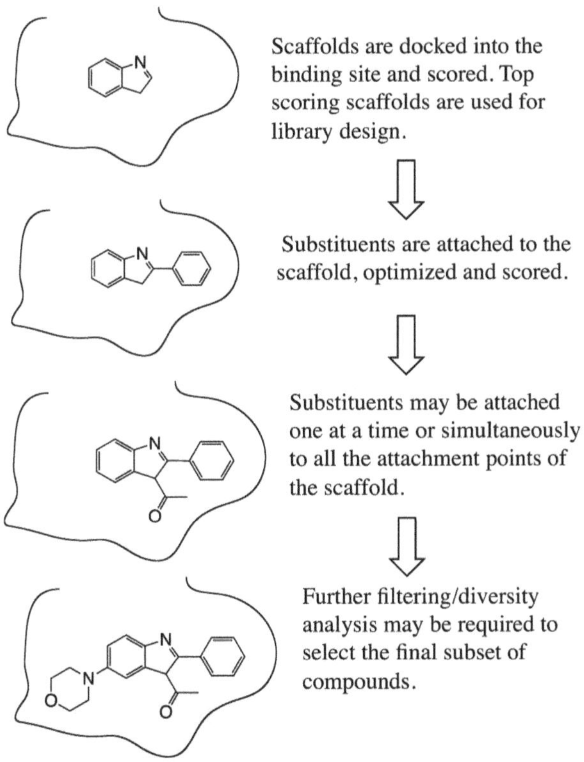

Fig. 8.1. Schematic depiction of the *seed and grow* docking approach for structure-based library design. The programs that use this approach include CombiDOCK and PRO_SELECT.

(b) *Dock and link*: The substituent groups are docked to the interacting sites in the binding pocket, scored, and then linked to each other based on chemistry constraints (70). This method, as illustrated in **Fig. 8.2**, attempts to take advantage of the known significant interactions within the binding site to bias the final compound library.

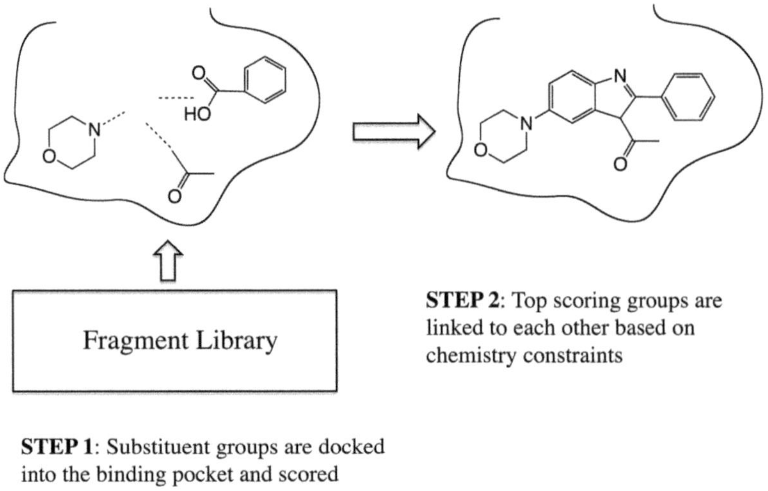

Fig. 8.2. Schematic depiction of the *dock and link* docking approach. The program BUILDER v.2 is based on this approach.

Notes:

For fragment-based methods, one needs to consider some important issues.

(a) For the *seed and grow* method, the orientation of the scaffold is highly critical. Any errors at this stage may render the results at later steps irrelevant (71).

(b) There must be a ready-to-use synthetic protocol to build these libraries based on the scaffold and fragments used.

(c) Although the *seed and grow* method reduces the number of compounds as compared to a full library enumeration, the availability of large number of fragments may still result in a huge number of compounds. Further filtering steps, diversity analysis exercises may be required to choose a final subset of compounds.

(d) It is important to have diverse fragment libraries to maximize the chances of library diversity.

(e) In the case of the *dock and link* method, though it is likely to have fragments satisfying key interactions within the binding site, the final compound may not be amenable to synthesis (71).

(f) It should be noted that docking programs may introduce errors due to the inherent inaccuracies of force fields, sampling, and scoring functions (7, 33, 72).

3. Docking Methods for Structure-Based Library Design

CombiDock is one of the first programs developed to design structure-based combinatorial libraries (73). It is based on a simple variation of the original DOCK algorithm (74). In brief, DOCK generates a negative image of the receptor binding site which is represented by spheres. The algorithm searches for internal distance matches between subsets of ligand atoms and spheres generated from the earlier stage. Based on a match, ligand atoms are placed and scored using force field or empirical functions that estimate the interaction energies. CombiDock tweaks this original algorithm such that only the scaffold atoms are used instead of the ligand. The scaffold is oriented in different conformations in the binding site and its atoms are matched against receptor binding site spheres. In the next step, all fragments/functional groups are attached at every individual attachment point and interaction scores are calculated for the scaffold and each attached fragment. The fragments with higher scores are then combined to form individual compounds. The best combinations are scanned for any intermolecular clashes with the receptor and saved. The method reduces the combinatorial process to a simple numerical addition of fragment scores to speed up library design (73).

Another tool, which combines combinatorial chemistry and fragment-based docking methods to rationally restrict the size of combinatorial libraries using structural restraints from binding site is PRO_SELECT (SELECT = Systematic Elaboration of Libraries Enhanced by Computational Techniques) (75). The underlying assumptions of this method are that the template fragment and the receptor are considered rigid and each individual substituent can be assessed independently of each other. PRO_SELECT also guides the library design process to build compounds that are accessible to specified synthetic routes, eliminating the uncertainties associated with synthetic feasibility of virtual compound libraries.

The PRO_SELECT methodology consists of three major parts which are explained in brief as follows:

I. Designing specifications for the target and the molecular templates (scaffold)
   a. The target is prepared on a protocol similar to the general steps described earlier and analyzed for possible interaction features represented as vectors, which

denote favorable position and direction for hydrogen bond interactions with the active site, and points, which denote positions of favorable hydrophobic contacts with the active site.

b. Using the structural knowledge of the receptor, template/s are chosen. It is desirable that these templates have multiple attachment points to attach several substituent groups and restricted conformational freedom to limit the number of alternative template positions within the binding pocket.

c. A design model is then developed which contains the vectors and points along with *link sites*, which are the positions on the template where a potential substituent group may be attached.

d. The templates are placed in the binding site using docking protocols based on molecular mechanics energy calculations (76, 77) or geometric positioning upon interaction sites (78).

II. Substituent/functional group selection

a. Databases of commercially available fragments (e.g., ACD) are used to search for possible substituent groups.

b. The fragments are computationally screened using PRO_LIGAND (*79*) and only those that can form good molecular interactions based on the original template position in the pocket (hydrogen bond interactions, lipophilic interactions) are selected.

c. Possible bioisoteric (functional groups possessing similar chemical properties) replacements are searched in the pursuit of novel compounds.

d. The substituents for each position are minimized using a molecular mechanics energy function where the receptor and template are held rigid, scored using the function developed by Bohm (78), and ranked.

III. Combinatorial enumeration:

a. The shortlisted compounds from the earlier stage are saved in a *list*.

b. It is recommended to reduce the size of the *list* by excluding structures with high strain energies, bad chemistries or geometries, and poor Bohm scores.

c. The structures may then be clustered based on 2D chemical functionality.

d. Finally, via combinatorial enumeration, a final compound library is generated.

DREAM++ (Docking and Reaction programs using Efficient seArch Methods) is a suite of programs (ORIENT++, REACT++, SEARCH++) developed to design chemical libraries by incorporating information from known chemical reactions and receptor active sites (80). The advantage of using well-studied organic reactions is that only synthetically accessible product compounds are generated in the final stage. The procedure begins by docking anchor parts or scaffolds into the binding site. These are then minimized, scored, and analyzed based on binding modes and other user-defined criteria. Functional groups from vendor libraries are virtually reacted with reagents using knowledge from a wide variety of organic reactions (e.g., amide bond formation, urea formation, reductive animation, alkylation, and ester formation) and are systematically combined to generate compounds. The conformational space of these compounds is explored and these steps are repeated until a complete library is produced. The generated library may then be visually inspected to study putative binding modes and offer further insights prior to selection for experimental results.

The program BUILDER v.2 (81) belongs to the *dock and link* category, where the importance is given to satisfying key interactions within the receptor binding site using fragments and then linking these fragments to form product compounds. Prior to the docking of fragments, the binding site is thoroughly investigated to identify *hot spots* or sites of potentially strong interaction with the receptor. The program DOCK (74) is used to place fragments or functional groups in the *hot spots.* By using a lattice around the protein, any two atoms of different fragments are connected via a set of lattice points. The set of such points being termed as "generic paths." These paths are generated using a modified breadth-first search algorithm (a graph search algorithm which begins at the root node and explores all the neighboring nodes). The points on these generic paths are considered to be atoms. Using three atoms in the path and their bond angles, the putative hybridization state of that atom is calculated. A pre-determined list, GOODLIST, contains a mapping of chemically reasonable functional groups (e.g., carbonyl, amide, thioester, and phenyl) for several of such three-atom combinations. Using the GOODLIST and the three-atom combinations, specific atom types are added to the atoms. BUILDER uses the SHAKE algorithm (82) to check for correct atom-type combinations, bond lengths, angles, and bump checks (steric clashes) against the receptor. Finally the paths are then reexamined to generate linker groups or bridges. Preference is given to embed a ring structure; however, other simpler and chemically synthesizable connecting groups are also considered. The bridges are expected to not have any strong contribution to binding. These bridges along with the original fragments are then attached to generate a product compound.

Another docking-based program developed by Sprous et al., OptiDock (83), attempts to exploit the common cores in a pre-enumerated combinatorial library. Instead of docking fragments or scaffolds, a subset of compounds spanning the structural space of the compound library are chosen and docked using the program FlexX (84). The binding mode for each compound is analyzed and distinctly different modes are shortlisted and the functional groups of these compounds are stripped. The core position is held constant, functional groups are attached, and interaction energies are calculated for each compound.

Recently, Zhou et al. developed a novel method termed as basis products (BPs) (85), which exploits the redundancy of fragments in a combinatorial library. The premise of this method is that all functional groups in a combinatorial library can be completely represented by a selected product subset of the library. This subset of compounds is called as basis products (BPs), which are formed by combining the smallest reactants (functional groups) of all reaction components except one. The remaining reactant is used against all viable reactants for a particular reaction while the other reactants are held constant. Thus for a two component reaction $A + B \rightarrow AB$, the entire library will consist of all the combinations of reactants A and B. In case of BPs, two capping molecules As and Bs are pre-selected with the smallest A and B, respectively. These capping molecules are then combined by changing only one component on the other side to generate two sub-libraries {AsB} and {ABs}. The sum of these libraries is much smaller than the single set of the entire library {AB}. Thus, every virtual library compound can be represented by a smaller set of BPs. Given a target, BPs can be docked using various docking programs. Based on the scores, the BPs are selected for the follow-up process, which involves designing libraries by using the reactants corresponding to the variable components of the BP hits among other strategies. To further improve the efficiency of the method, the BP's themselves may be filtered based on physiochemical properties to reduce the number of BPs for the docking process. The algorithm was tested in a comparison-type study (85) where an entire virtual library (~34,000 compounds) and a smaller subset (~1225 compounds identified by BPs and hit follow-up library) were both docked to the active site of dihydrofolate reductase and the top-ranked compounds were checked. In both cases, the top 350 ranked compounds were the same. Thus, in this case, it was shown that a smaller but focused library can achieve comparable results as compared to docking entire virtual libraries.

Several other programs like COMBISMOG (86), CombiGlide (www.schrodinger.com), and COMBIBUILD (http://mdi.ucsf.edu/CombiBUILD.html) have been developed for the purpose of library design. Please refer to **Table 8.1** for a list of programs listed in this chapter. However, it should be noted that though useful, most of the programs are neither easy to implement nor use as is (87). As a result, these methods have found limited applicability in the scientific community. On the other hand, several commercial library vendors like Cerep (www.cerep.fr), Asinex (www.asinex.com), and Enamine (www.enamine.net) offer target-focused libraries using docking-based protocols. A list of the docking methods/software tools mentioned in this chapter can be found in **Table 8.1**.

Table 8.1
Docking-based programs for library design

Method/program	Description	Refs
CombiDock	Tweaks the DOCK algorithm to identify suitable scaffold orientations in the binding pocket. Proceeds using the *seed and grow* approach to design combinatorial libraries	(73)
PRO_SELECT	Combines combinatorial chemistry and fragment-based docking methods to design structure-based libraries	(75)
DREAM++	Designs chemical libraries incorporating information from known chemical reactions and receptor binding sites	(80)
Builder v.2	Uses the *dock and link* strategy to link relevant fragments, which satisfy key receptor:ligand interactions to form product compounds	(81)
OptiDOCK	Uses the *seed and grow* strategy to first dock representative compounds spanning the chemical space of the library and subsequently use an optimal core for library enumeration	(83)
Basis products (BPs)	Exploits the redundancy of fragments in a combinatorial library and identifies a small subset of compounds (BPs) which represent the entire virtual library. BPs are docked, scored, and used for final library enumeration	(85)
CombiGlide	Combines docking algorithms and core hopping technologies to design focused libraries	www.schrodinger.com
CombiSMoG	Uses a Monte Carlo ligand growth algorithm and knowledge-based potentials to combine combinatorial and rational strategies for generating biased compound libraries	(86)

4. Applications

This section will highlight a few applications of docking-based methods for library design.

The CombiDOCK algorithm was applied to design a structure-based library for cathepsin D protease using a hydroxyethylamine scaffold. This scaffold has three attachment points. Ten fragments for each site were chosen and incorporated in the final library design. The 1000 compounds were filtered to check inaccuracies in bond geometries to give ~750 compounds which were synthesized and assayed for experimental testing. Results indicated that this library had an enrichment factor (EF) of 2.5, whereas a completely random ranking would result in an EF of 1.0. (73) The EF is the ratio between the probability of finding a true ligand in a filtered sub-library compared to the probability of finding a ligand at random.

The PRO_SELECT method was applied to design an inhibitor library for thrombin, a key serine protease. The crystal structure of thrombin includes a covalently bound inhibitor, tri-peptide PPACK. L-proline, the centrally located portion of PPACK was chosen as the template and its alternate locations were generated by docking/modeling a noncovalently bound analogue of PPACK. Analysis of the binding site revealed the requirements for a hydrogen bond donor and a hydrophobic group at either ends of the template. A 3D database search of potential fragment binders based on the analysis of the binding site resulted in over 400,000 hits. PRO_SELECT method was able to drastically reduce the number of fragments to 17, which were then used to build a chemical library. Over 30 molecules were then synthesized, of which at least 50% showed micromolar activities (75).

In another study Head et al. used a docking-based method to design a library of potentially novel inhibitors for caspases 3 and 8, a key regulator of apoptosis (88). The authors chose thiomethylketone as a scaffold for this study, as it is a common denominator of a class of compounds inhibiting caspases 3 and 8. Two attachment points on the thiomethylketone scaffold were identified. The ketone group is postulated to covalently bind with the catalytic cysteine. Thus from **Fig. 8.3** it is seen that the R' group points away from the S2 binding pocket. Hence a small number of reagents (8) were fixed for R' based on availability and ease of synthesis. To identify potential functional groups for the other attachment point (R), roughly 7000 monoacid reagents from the ACD database were selected for combinatorial docking. First, a simplified thiomethylketone with R and R' set to methyl was docked in the binding pocket to identify initial template

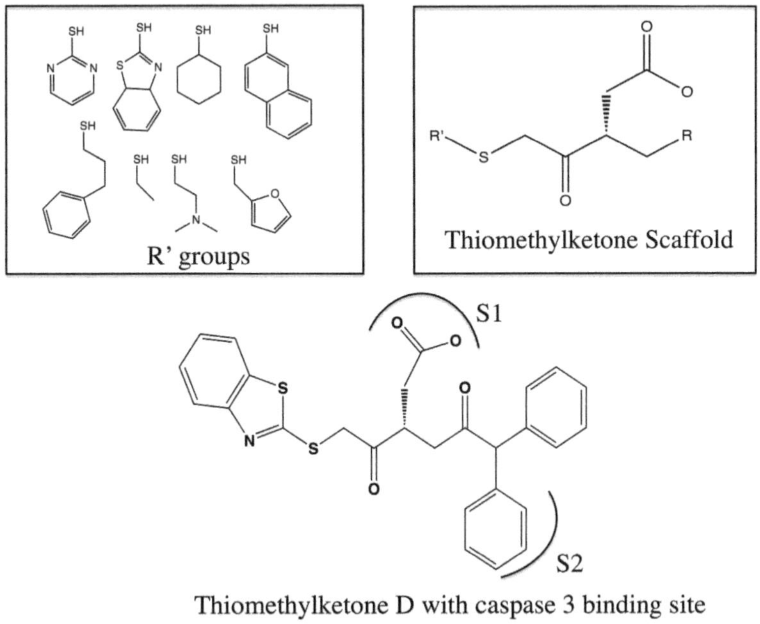

Fig. 8.3. (*Bottom*): Thiomethylketone D of (88) is used as an example of a caspase 3 inhibitor designed via a docking-based library generation protocol. S1 and S2 denote the interaction sites within the binding pocket of caspase 3. (*Top right*): The thiomethylketone scaffold that is used as the starting point for library design. (*Top left*): The eight R-groups used to attach to the R' attachment point of the scaffold.

locations. Next, the eight reagents for the R point and 7000 monoacids for the R' points were combinatorially attached to the templates, docked, and scored. Two criteria were used to obtain the final reagents for the R group: (1) docking scores and (2) distance filters based on the experimental data of isatin-based compounds and crystal structures of other caspases. Based on these results approximately 150 reagents were selected per caspase and roughly 10% of these reagents underwent full conformational sampling. As the array size for synthesis was 96, only 12 reagents for the R group (seven for caspase 3, three for caspase 8, and three common for both) were selected based on visual inspection of the predicted binding modes. Sixty-one compounds were synthesized and tested. Five of the 61 compounds tested against caspase 3 and two compounds against caspase 8 showed micromolar activity. Interestingly, a homology model of caspase 8 was used for this study, which clearly indicates the usefulness of homology modeling in structure-based library design.

Decornez et al. used a generalized kinase model and a combination of 2D (fingerprint based similarity) and 3D methods (docking) to develop a kinase family focused library (15). The authors used ~ 2800 kinase inhibitors compounds as a reference for a 2D search of their in-house database of ~260 K compounds which resulted in 3135 compounds. As 2D methods are grossly inadequate to incorporate receptor information, a docking protocol was developed using the crystal structure of PKA (PDB code 1BX6) and the software Glide (www.schrodinger.com). Since the goal of the project was to design a generic kinase-specific library, the authors mutated several residues of the crystal structure to avoid any bias in the eventual compound library. The ~3100 compounds were then docked, scored, and the top 170 compounds with significant 2D similarity to known inhibitors and 3D binding characteristics were submitted for biochemical screening. The identified hits were similar or analogues of p38, tyrosine kinase, and PKC kinases.

Zhao et al. implemented a structure-based docking protocol to narrow down 500 compounds from a database of ~57 K compounds in their pursuit of FKBPs inhibitors (89). A novel scaffold was designed using the information obtained from the binding mode analysis of a known weak binder. To avoid any scoring function shortcomings, three scoring functions were used to select the 500 compounds. Of these, 43 were synthesized and tested to identify one potent inhibitor in a mouse peripheral synthetic nerve model.

5. Conclusions

Despite the initial promise, advancements in HTS methods and combinatorial chemistry have so far failed to improve the success rates of drug discovery programs. Since the experimental screening of these gigantic libraries is costly and time consuming, it is of utmost importance to rationally, efficiently, and economically explore the available chemical space of compounds in order to design smaller and focused compound libraries for experimental evaluation. Several docking-based methods make use of the increasing availability of structural information of drug targets to a priori filter out those compounds that are unlikely to bind to the target. This chapter highlights several of such docking methods used in library design, together with their application to actual cases.

References

1. Mayr, L. M., Fuerst, P. (2008) The future of high-throughput screening. *J Biomol Screen* **13**, 443–448.
2. Entzeroth, M. (2003) Emerging trends in high-throughput screening. *Curr Opin Pharmacol* **3**, 522–529.
3. Schnecke, V., Bostrom, J. (2006) Computational chemistry-driven decision making in lead generation. *Drug Discov Today* **11**, 43–50.
4. Boldt, G. E., Dickerson, T. J., Janda, K. D. (2006) Emerging chemical and biological approaches for the preparation of discovery libraries. *Drug Discov Today* **11**, 143–148.
5. Bohacek, R. S., McMartin, C., Guida, W. C. (1996) The art and practice of structure-

based drug design: a molecular modeling perspective. *Med Res Rev* **16**, 3–50.
6. Walters, W. P., Stahl M. T., Murcko, M. A. (1998) Virtual screening – an overview. *Drug Discov Today* **3**, 160–178.
7. Phatak, S. S., Stephan, C. C., Cavasotto, C. N. (2009) High-throughput and *in silico* screenings in drug discovery. *Expert Opin. Drug Discov* **4**, 947–959.
8. Keseru, G. M., Makara, G. M. (2006) Hit discovery and hit-to-lead approaches. *Drug Discov Today* **11**, 741–748.
9. Macarron, R. (2006) Critical review of the role of HTS in drug discovery. *Drug Discov Today* **11**, 277–279.
10. Fox, S., Farr-Jones, S., Sopchak, L., Boggs, A., Nicely, H. W., Khoury, R., Biros, M. (2006) High-throughput screening: update on practices and success. *J Biomol Screen* **11**, 864–869.
11. Keseru, G. M., Makara, G. M. (2009) The influence of lead discovery strategies on the properties of drug candidates. *Nat Rev Drug Discov* **8**, 203–212.
12. Lipkin, M. J., Stevens, A. P., Livingstone, D. J., Harris, C. J. (2008) How large does a compound screening collection need to be? *Comb Chem High Throughput Screening* **11**, 482–493.
13. Nestler, H. P. (2005) Combinatorial chemistry and fragment screening – Two unlike siblings? *Curr Drug Discov Tech* **2**, 1–12.
14. Diller, D. J., Merz, K. M., Jr. (2001) High throughput docking for library design and library prioritization. *Proteins* **43**, 113–124.
15. Decornez, H., Gulyas-Forro, A., Papp, A., Szabo, M., Sarmay, G., Hajdu, I., Cseh, S., Dorman, G., Kitchen, D. B. (2009) Design, selection, evaluation of a general kinase-focused library. *ChemMedChem* **4**, 1273–1278.
16. Lipinski, C. A. (2000) Drug-like properties and the causes of poor solubility and poor permeability. *J Pharmacol Toxicol Methods* **44**, 235–249.
17. Schnur, D. M. (2008) Recent trends in library design: 'rational design' revisited. *Curr Opin Drug Discov Devel* **11**, 375–380.
18. Villar, H. O., Koehler, R. T. (2000) Comments on the design of chemical libraries for screening. *Mol Divers* **5**, 13–24.
19. Manjasetty, B. A., Turnbull, A. P., Panjikar, S., Bussow, K., Chance, M. R. (2008) Automated technologies and novel techniques to accelerate protein crystallography for structural genomics. *Proteomics* **8**, 612–625.
20. Gileadi, O., Knapp, S., Lee, W. H., Marsden, B. D., Muller, S., Niesen, F. H., Kavanagh, K. L., Ball, L. J., von Delft, F., Doyle, D. A., Oppermann, U. C., Sundstrom, M. (2007) The scientific impact of the Structural Genomics Consortium: a protein family and ligand-centered approach to medically-relevant human proteins. *J Struct Funct Genomics* **8**, 107–119.
21. Cavasotto, C. N., Phatak, S. S. (2009) Homology modeling in drug discovery: current trends and applications. *Drug Discov Today* **14**, 676–683.
22. Cavasotto, C. N., Orry, A. J., Murgolo, N. J., Czarniecki, M. F., Kocsi, S. A., Hawes, B. E., O'Neill, K. A., Hine, H., Burton, M. S., Voigt, J. H., Abagyan, R. A., Bayne, M. L., Monsma, F. J., Jr. (2008) Discovery of novel chemotypes to a G-protein-coupled receptor through ligand-steered homology modeling and structure-based virtual screening. *J Med Chem* **51**, 581–588.
23. Hong, T. J., Park, H., Kim, Y. J., Jeong, J. H., Hahn, J. S. (2009) Identification of new Hsp90 inhibitors by structure-based virtual screening. *Bioorg Med Chem Lett* **19**, 4839–4842.
24. Brozic, P., Turk, S., Lanisnik Rizner, T., Gobec, S. (2009) Discovery of new inhibitors of aldo-keto reductase 1C1 by structure-based virtual screening. *Mol Cell Endocrinol* **301**, 245–250.
25. Park, H., Bhattarai, B. R., Ham, S. W., Cho, H. (2009) Structure-based virtual screening approach to identify novel classes of PTP1B inhibitors. *Eur J Med Chem* **44**, 3280–3284.
26. Heinke, R., Spannhoff, A., Meier, R., Trojer, P., Bauer, I., Jung, M., Sippl, W. (2009) Virtual screening and biological characterization of novel histone arginine methyltransferase PRMT1 inhibitors. *ChemMedChem* **4**, 69–77.
27. Wang, Q., Wang, J., Cai, Z., Xu, W. (2008) Prediction of the binding modes between BB-83698 and peptide deformylase from Bacillus stearothermophilus by docking and molecular dynamics simulation. *Biophys Chem* **134**, 178–184.
28. Padgett, L. W., Howlett, A. C., Shim, J. Y. (2008) Binding mode prediction of conformationally restricted anandamide analogs within the CB1 receptor. *J Mol Signal* **3**, 5.
29. Zampieri, D., Mamolo, M. G., Vio, L., Banfi, E., Scialino, G., Fermeglia, M., Ferrone, M., Pricl, S. (2007) Synthesis, antifungal and antimycobacterial activities of new bis-imidazole derivatives, and prediction of their binding to P450(14DM) by molecular docking and MM/PBSA method. *Bioorg Med Chem* **15**, 7444–7458.
30. Monti, M. C., Casapullo, A., Cavasotto, C. N., Napolitano, A., Riccio, R. (2007)

Scalaradial, a dialdehyde-containing marine metabolite that causes an unexpected non-covalent PLA2 Inactivation. *Chembiochem* **8**, 1585–1591.
31. Diaz P., Phatak, S. S., Xu, J., Fronczek, F. R., Astruc-Diaz, F., Thompson, C. M., Cavasotto, C. N., Naguib, M. (2009) 2,3-Dihydro-1-benzofuran derivatives as a series of potent selective cannabinoid receptor 2 agonists: design, synthesis, and binding mode prediction through ligand-steered modeling. *ChemMedChem* **4**, 1615–1629.
32. Andricopulo, A. D., Salum, L. B., Abraham, D. J. (2009) Structure-based drug design strategies in medicinal chemistry. *Curr Topics Med Chem* **9**, 777–790.
33. Cavasotto, C. N., Orry, A. J. (2007) Ligand docking and structure-based virtual screening in drug discovery. *Curr Top Med Chem* **7**, 1006–1014.
34. Kitchen, D. B., Decornez, H., Furr, J. R., Bajorath, J. (2004) Docking and scoring in virtual screening for drug discovery: methods and applications. *Nat Rev Drug Discov* **3**, 935–949.
35. Cavasotto, C. N., Ortiz, M. A., Abagyan, R. A., Piedrafita, F. J. (2006) In silico identification of novel EGFR inhibitors with antiproliferative activity against cancer cells. *Bioorg Med Chem Lett* **16**, 1969–1974.
36. Klebe, G. (2006) Virtual ligand screening: strategies, perspectives and limitations. *Drug Discov Today* **11**, 580–594.
37. Zoete, V., Grosdidier, A., Michielin, O. (2009) Docking, virtual high throughput screening and in silico fragment-based drug design. *J Cell Mol Med* **13**, 238–248.
38. Marsden, R. L., Orengo, C. A. (2008) Target selection for structural genomics: an overview. *Methods Mol Biol* **426**, 3–25.
39. Levitt, D. G., Banaszak, L. J. (1992) POCKET: a computer graphics method for identifying and displaying protein cavities and their surrounding amino acids. *J Mol Graph* **10**, 229–234.
40. Hendlich, M., Rippmann, F., Barnickel, G. (1997) LIGSITE: automatic and efficient detection of potential small molecule-binding sites in proteins. *J Mol Graph Model* **15**, 359–363, 389.
41. Laskowski, R. A. (1995) SURFNET: a program for visualizing molecular surfaces, cavities, intermolecular interactions. *J Mol Graph* **13**, 323–330, 307–328.
42. Balakin, K. V., Kozintsev, A. V., Kiselyov, A. S., Savchuk, N. P. (2006) Rational design approaches to chemical libraries for hit identification. *Curr Drug Discov Technol* **3**, 49–65.
43. Orry, A. J., Abagyan, R. A., Cavasotto, C. N. (2006) Structure-based development of target-specific compound libraries. *Drug Discov Today* **11**, 261–266.
44. Brown, E. N., Ramaswamy, S. (2007) Quality of protein crystal structures. *Acta Crystallogr D Biol Crystallogr* **63**, 941–950.
45. Davis, A. M., St-Gallay, S. A., Kleywegt, G. J. (2008) Limitations and lessons in the use of X-ray structural information in drug design. *Drug Discov Today* **13**, 831–841.
46. Cavasotto, C. N., Singh, N. (2008) Docking and high throughput docking: successes and the challenge of protein flexibility. *Curr Comput Aided Drug Design* **4**, 221–234.
47. Sousa, S. F., Fernandes, P. A., Ramos, M. J. (2006) Protein-ligand docking: current status and future challenges. *Proteins* **65**, 15–26.
48. Li, Z., Lazaridis, T. (2007) Water at biomolecular binding interfaces. *Phys Chem Chem Phys* **9**, 573–581.
49. Mancera, R. L. (2007) Molecular modeling of hydration in drug design. *Curr Opin Drug Discov Devel* **10**, 275–280.
50. Corbeil, C. R., Moitessier, N. (2009) Docking ligands into flexible and solvated macromolecules. 3. Impact of input ligand conformation, protein flexibility, and water molecules on the accuracy of docking programs. *J Chem Inf Model* **49**, 997–1009.
51. Chen, J., Swamidass, S. J., Dou, Y., Bruand, J., Baldi, P. (2005) ChemDB: a public database of small molecules and related chemoinformatics resources. *Bioinformatics* **21**, 4133–4139.
52. Irwin, J. J., Shoichet, B. K. (2005) ZINC–a free database of commercially available compounds for virtual screening. *J Chem Inf Model* **45**, 177–182.
53. Williams, A. J. (2008) Public chemical compound databases. *Curr Opin Drug Discov Develop* **11**, 393–404.
54. Drie, J. H. (2005) Pharmacophore-based virtual screening: a practical perspective, in (Alvarez, J., Shoichet, B., eds.) *Virtual Screening in Drug Discovery.* CRC Press, Boca Raton, FL, pp. 157–205.
55. Oprea, T. I., Bologa, C. G., Olah, M. M. (2005) Compound selection for virtual screening, in *Virtual screening in Drug Discovery* (Alvarez, J., Shoichet, B., eds.), CRC Press, Boca Raton, FL, pp. 89–106.
56. Lipinski, C. A., Lombardo, F., Dominy, B. W., Feeney, P. J. (1997) Experimental and computational approaches to estimate solubility and permeability in drug discovery and development settings. *Adv Drug Del Rev* **23**, 3–25.

57. Oprea, T. I. (2002) Current trends in lead discovery: are we looking for the appropriate properties? *J Comput Aided Mol Des* **16**, 325–334.
58. Hubbard, R. E. (2008) Fragment approaches in structure-based drug discovery. *J Synchrotron Radiat* **15**, 227–230.
59. Kroemer, R. T. (2007) Structure-based drug design: docking and scoring. *Curr Protein Pept Sci* **8**, 312–328.
60. Barril, X., Hubbard, R. E., Morley, S. D. (2004) Virtual screening in structure-based drug discovery. *Mini Rev Med Chem* **4**, 779–791.
61. Teague, S. J. (2003) Implications of protein flexibility for drug discovery. *Nat Rev Drug Discov* **2**, 527–541.
62. B-Rao, C., Subramanian, J., Sharma, S. D. (2009) Managing protein flexibility in docking and its applications. *Drug Discov Today* **14**, 394–400.
63. Dias, R., de Azevedo, W. F., Jr. (2008) Molecular docking algorithms. *Curr Drug Targets* **9**, 1040–1047.
64. Sperandio, O., Miteva, M. A., Delfaud, F., Villoutreix, B. O. (2006) Receptor-based computational screening of compound databases: the main docking-scoring engines. *Curr Protein Pept Sci* **7**, 369–393.
65. Stahl, M., Rarey, M. (2001) Detailed analysis of scoring functions for virtual screening. *J Med Chem* **44**, 1035–1042.
66. Perola, E., Walters, W. P., Charifson, P. S. (2005) An analysis of critical factors affecting docking and scoring, in (Alvarez, J., Shoichet, B., eds.) *Virtual screening in drug discovery*. CRC Press, Boca Raton, FL, pp. 47–85.
67. Warren, G. L., Andrews, C. W., Capelli, A. M., Clarke, B., LaLonde, J., Lambert, M. H., Lindvall, M., Nevins, N., Semus, S. F., Senger, S., Tedesco, G., Wall, I. D., Woolven, J. M., Peishoff, C. E., Head, M. S. (2006) A critical assessment of docking programs and scoring functions. *J Med Chem* **49**, 5912–5931.
68. Waszkowycz, B. (2008) Towards improving compound selection in structure-based virtual screening. *Drug Discov Today* **13**, 219–226.
69. Manallack, D. T., Pitt, W. R., Gancia, E., Montana, J. G., Livingstone, D. J., Ford, M. G., Whitley, D. C. (2002) Selecting screening candidates for kinase and G protein-coupled receptor targets using neural networks. *J Chem Inf Comput Sci* **42**, 1256–1262.
70. Schneider, G. (2002) Trends in virtual combinatorial library design. *Curr Med Chem* **9**, 2095–2101.
71. Beavers, M. P., Chen, X. (2002) Structure-based combinatorial library design: methodologies and applications. *J Mol Graph Model* **20**, 463–468.
72. Coupez, B., Lewis, R. A. (2006) Docking and scoring–theoretically easy, practically impossible? *Curr Med Chem* **13**, 2995–3003.
73. Sun, Y., Ewing, T. J., Skillman, A. G., Kuntz, I. D. (1998) CombiDOCK: structure-based combinatorial docking and library design. *J Comput Aided Mol Des* **12**, 597–604.
74. Kuntz, I. D., Blaney, J. M., Oatley, S. J., Langridge, R., Ferrin, T. E. (1982) A geometric approach to macromolecule-ligand interactions. *J Mol Biol* **161**, 269–288.
75. Murray, C. W., Clark, D. E., Auton, T. R., Firth, M. A., Li, J., Sykes, R. A., Waszkowycz, B., Westhead, D. R., Young, S. C. (1997) PRO_SELECT: combining structure-based drug design and combinatorial chemistry for rapid lead discovery. 1. Technology. *J Comput Aided Mol Des* **11**, 193–207.
76. Blaney, J. M., Dixon, J. S. (1993) A good ligand is hard to find: automated docking methods. *Perspect Drug Discov Des* **1**, 301–319.
77. Kuntz, I. D., Meng, E. C., Shoichet, B. (1994) Structure-based molecular design. *Acc Chem Res* **27**, 117–123.
78. Bohm, H. J. (1994) The development of a simple empirical scoring function to estimate the binding constant for a protein-ligand complex of known three-dimensional structure. *J Comput Aided Mol Des* **8**, 243–256.
79. Clark, D. E., Frenkel, D., Levy, S. A., Li, J., Murray, C. W., Robson, B., Waszkowycz, B., Westhead, D. R. (1995) PRO-LIGAND: an approach to de novo molecular design. 1. Application to the design of organic molecules. *J Comput Aided Mol Des* **9**, 13–32.
80. Makino, S., Ewing, T. J., Kuntz, I. D. (1999) DREAM++: flexible docking program for virtual combinatorial libraries. *J Comput Aided Mol Des* **13**, 513–532.
81. Roe, D. C., Kuntz, I. D. (1995) BUILDER v.2: improving the chemistry of a de novo design strategy. *J Comput Aided Mol Des* **9**, 269–282.
82. Van Gunsteren, W. F., Berendsen, H. J. C. (1977) Algorithms for macromolecular dynamics and constraint dynamics. *Mol Phys* **34**, 1311–1327.
83. Sprous, D. G., Lowis, D. R., Leonard, J. M., Heritage, T., Burkett, S. N., Baker, D. S., Clark, R. D. (2004) OptiDock: virtual

HTS of combinatorial libraries by efficient sampling of binding modes in product space. *J Comb Chem* **6**, 530–539.
84. Rarey, M., Lengauer, T. (2000) A recursive algorithm for efficient combinatorial library docking. *Perspect Drug Discov Des* **20**, 63–81.
85. Zhou, J. Z., Shi, S., Na, J., Peng, Z., Thacher, T. (2009) Combinatorial library-based design with Basis Products. *J Comput Aided Mol Des* DOI 10.1007/s10822-009-9297-9.
86. Grzybowski, B. A., Ishchenko, A. V., Kim, C. Y., Topalov, G., Chapman, R., Christianson, D. W., Whitesides, G. M., Shakhnovich, E. I. (2002) Combinatorial computational method gives new picomolar ligands for a known enzyme. *Proc Natl Acad Sci USA* **99**, 1270–1273.
87. Zhou, J. Z. (2008) Structure-directed combinatorial library design. *Curr Opin Chem Biol* **12**, 379–385.
88. Head, M. S., Ryan, M. D., Lee, D., Feng, Y., Janson, C. A., Concha, N. O., Keller, P. M., deWolf, W. E., Jr. (2001) Structure-based combinatorial library design: discovery of non-peptidic inhibitors of caspases 3 and 8. *J Comput Aided Mol Des* **15**, 1105–1117.
89. Zhao, L., Huang, W., Liu, H., Wang, L., Zhong, W., Xiao, J., Hu, Y., Li, S. (2006) FK506-binding protein ligands: structure-based design, synthesis, and neurotrophic/neuroprotective properties of substituted 5,5-dimethyl-2-(4-thiazolidine)carboxylates. *J Med Chem* **49**, 4059–4071.

Chapter 9

Structure-Based Library Design in Efficient Discovery of Novel Inhibitors

Shunqi Yan and Robert Selliah

Abstract

Structure-based library design employs both structure-based drug design (SBDD) and combinatorial library design. Combinatorial library design concepts have evolved over the past decade, and this chapter covers several novel aspects of structure-based library design together with successful case studies in the anti-viral drug design HCV target area. Discussions include reagent selections, diversity library designs, virtual screening, scoring/ranking, and post-docking pose filtering, in addition to the considerations of chemistry synthesis. Validation criteria for a successful design include an X-ray co-crystal complex structure, in vitro biological data, and the number of compounds to be made, and these are addressed in this chapter as well.

Key words: Structure-based drug design, structure-based library design, library design, focused library design, diversity library, combinatorial library, docking, reagent selections, HCV NS5B, thiazolone.

1. Introduction

Structure-based library design engages in dual approaches of structure-based drug design (SBDD) and combinatorial library design (**Fig. 9.1**). Design concepts of combinatorial library have been evolving since its conception more than a decade ago. Early efforts mainly focused on the capability to synthesize large number of compounds through combinatorial chemistry with the confidence that high-throughput screening (HTS) (1) of every possible compound in a large library would lead to potential druggable hits and leads, and eventually development of candidates after subsequent lead optimizations. Needless to say, this approach

J.Z. Zhou (ed.), *Chemical Library Design*, Methods in Molecular Biology 685,
DOI 10.1007/978-1-60761-931-4_9, © Springer Science+Business Media, LLC 2011

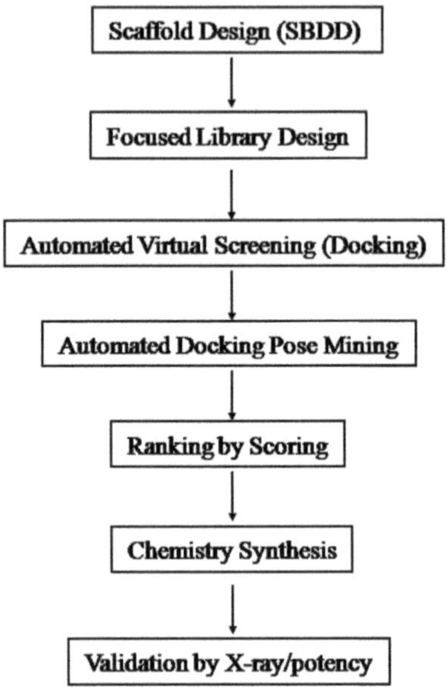

Fig. 9.1. Structure-based combinatorial library design.

oversimplified the complex processes of drug discovery. Drugs reported to originate solely from combinatorial chemistry thus far are rare.

The advent of more accurate and rapid tools in chemoinformatics and virtual screening makes it possible to design and synthesize a small subset of representative compounds (focused library) of a larger library. Out of various improved methods these two diversity- or structure-based approaches are frequently exercised in the design of a focused library. Once the 3D coordinates of a protein target are determined by either X-ray crystal structures or NMR, a structure-based library design is a more productive and viable approach.

This chapter covers several aspects of structure-based library designs coupled with the successful case studies in the anti-viral HCV area. Discussions include reagent selections, diversity library designs, virtual screening, scoring/ranking, and post-docking pose filtering, in addition to the considerations of chemistry synthesis (**Fig. 9.1**). Validation criteria for a successful design include an X-ray co-crystal complex structure and in vitro biological data, and these are addressed in this chapter as well.

2. Materials

A number of computational methodologies have been used in this experiment. Reagent selections for library designs were exported from ACD database (2). GOLD (3) and LigandFit (4) were used for docking, where MOE software (5) was used for reagent filtering and library enumeration. Post-docking pose filters and reactive group filtering were carried out using a MOE SVL script (5). Diversity analysis and physical property calculation were performed by Ceris2 (4). HKL package was applied for X-ray data process (6) and the X-ray structure determination and refinement were carried out by CNX (4). X-ray co-crystal complex structures discussed in this chapter were deposited in PDB database with codes 2O5D, 2HWH, 2HWI, and 2I1R.

The chemical synthesis of the compounds is applicable for library production. The reactions start with a condensation reaction of a readily available reagent, rhodinine, with an aldehyde to afford a thiazolonone intermediate, which can undergo a coupling reaction with an amine, amino acid, or a variety of other amino-containing derivatives to give final products with good yields (7–9).

3. Methods

3.1. Introduction

SBDD begins with designs of novel scaffolds based on the structural binding information of hit or lead compounds in complex with a target which is usually an enzyme or a protein. Previously hard-to-crystallized protein and their corresponding protein–ligand complex co-crystal structures have been routinely determined nowadays due to the significant technology improvement in crystallography, molecular biology, and protein science in the last decade. Given an X-ray complex structure of a protein–ligand co-crystal, various computational tools such as virtual screening (10, 11), de novo design (11), and scaffold hopping are utilized to design brand new and better molecules with potentially novel IP space coverage. Such designs are often accomplished through exploring better or comparable ligand–protein interaction as predicted computationally. Alternatively, a 3D structure of a target can be also approximated with reasonable confidence from a homology model if identity level of amino acid sequences between the target and a known structure from either in-house X-ray determinations or PDB database is high. Recently more and more protein structures have been solved with high resolutions (<2.5 Å) by NMR experiments. Structures from NMR are sometimes advantageous for SBDD in terms of the flexibility of protein since the experiments are generally performed in solution at room temperature and therefore, the dynamic states of protein resemble more of the genuine states of protein in physiological conditions than a X-ray structure derived from a solid state of protein at flash-frozen condition with liquid nitrogen.

Docking methods are commonly used to determine whether the newly designed scaffold fits to the target in a desirable binding mode – similar to the one(s) found by an X-ray complex structure. A scaffold is rarely selected for further pursuit, for example, chemistry synthesis, if initial docking evaluations show unexpected results. Various docking programs, such as Glide (12), GOLD (3), FlexX (13–15), Surflex-Dock (16), Ligandfit (4), AutoDock (17), and DOCK (18), are commercially available. It is the specific user's job to validate which docking method is applicable to their target (**Fig. 9.2**). A docking program is by and large acceptable if it reproduces the X-ray complex structure with a moderately low RMSD (<1.5 Å) from the superimposition of ligand structures between X-ray and a docking study (**Fig. 9.2**). It is frequently practiced as well to apply two different docking programs for cross-validation to increase the confidence level of docking results.

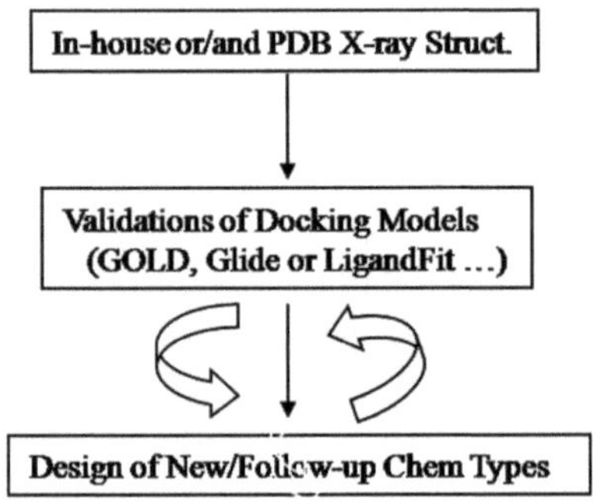

Fig. 9.2. Validation of docking programs for scaffold designs.

3.2. Focused Library Design

Rational design of a target-based focus library becomes critical when a new scaffold carries different substituents, i.e., R_1 and R_2, which originate from functional handles and derivatives such as amine, carboxyl acid, aldehyde, hydrazine (**Fig. 9.3**). As one can imagine, the resulting virtual library for product **C** can easily amount to millions of individual compounds based on availability of R groups. The critical questions for a medicinal chemist are which compound sets are to be made first? Computational design of focused library aims to trim a large virtual library to a manageable and viable "realistic" library that can be synthesized and tested for desired activity of novel scaffolds (**Fig. 9.3**).

Computational library design process begins with reagent selections, followed by diversity analysis and virtual library enumeration, and ends with selection of a final set of molecular structures to be synthesized (**Fig. 9.4**). Two databases, Available Chemical Database (ACD) (2) and Chemicals Available for

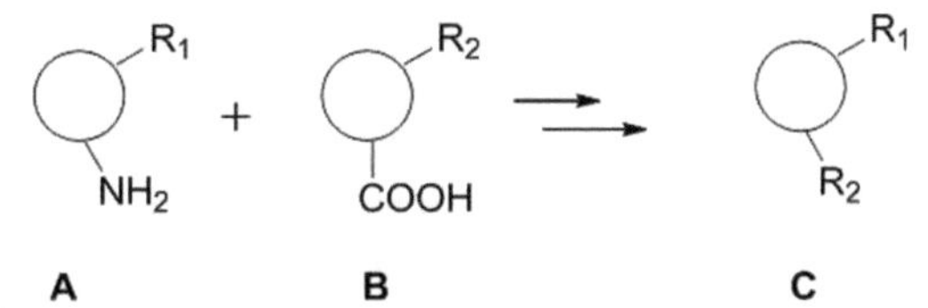

Fig. 9.3. A schematic reaction.

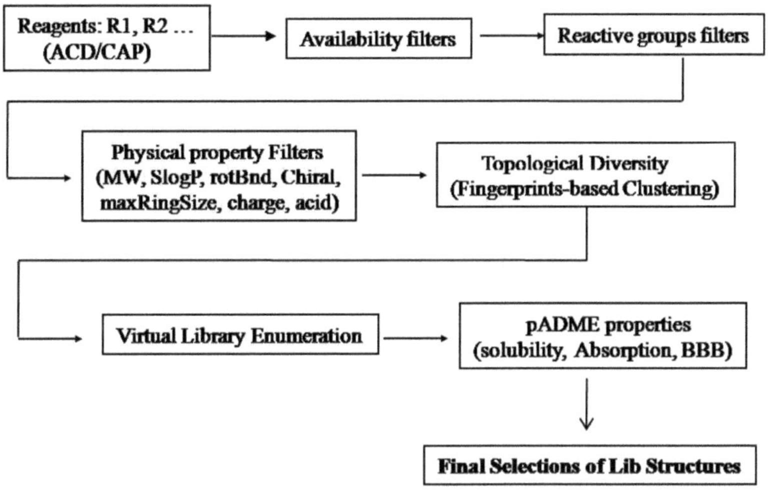

Fig. 9.4. Focused library design.

Purchase (CAP) (4), are commonly used by medicinal chemists to select reagents for their chemical synthesis and these are also used by computational chemists for the selection of building blocks; both databases allow for export of all of the desirable reagents into a single SDF structural file. Other non-structural information of the reagents such as the origins of countries, price, and time for delivery are normally available in the databases and can be exported into the same SDF file as well. With structures and non-structural information together in one file, the task of weeding out some undesirable reagents and building blocks becomes easy. Scripting language implemented in MOE software package is used interactively for these tasks. Price of reagents, availability, delivery time are collectively classified as availability filters and these filters are applied to select out non-optimal features. The list of reagents is further narrowed down by applying (a) reactive groups to remove building blocks that contain undesirable reactive functional groups which may interfere with desired chemical reactions and (b) physical property filters to remove the ones that have undesirable physical properties such as high molecular weight (MW > 300), too many rotational bonds (rot > 5), and too many chiral centers ($n > 2$). Topological filter can further reduce reagent list if it is necessary. Virtual library enumeration is enumerated using the refined building block list using commercial software such as MOE (5) or CombiLibMaker in Sybyl (16). The virtual library thus obtained may be further trimmed using ADME filters. The final focus library is thus prepared for virtual screening against a protein target (**Fig. 9.4**) (4, 5, 16).

3.3. Virtual Screening, Scoring, and Ranking of Focused Library

Virtual screening methods have been routinely and extensively applied in the generation of lead compounds from commercially available chemical libraries (19–22). Conventional virtual screening programs for combinatorial libraries include CombiGlide (12) and FlexX^{c} (23). Both methods work similarly in a way by first anchoring the core structure (or scaffold) in a predetermined ideal location and then making side chain R groups flexible to identify a focused set of molecules with most favorable R groups (12, 23). Docking molecules in this way renders different R groups invisible to each other during docking and can thus generate a focused library by eliminating energetically unfavorable R-group and conformations upon binding to a target.

This shortcoming of docking in a combinatorial fashion is readily overcome by docking all molecules individually using conventional docking programs, i.e., GOLD (23), Glide (12), FlexX (23), or Surflex-Dock (16), followed by post-docking pose mining. The post-docking filters are realized through some straightforward scripting language such as MOE SVL (5). A typical script program allows users to identify all structures in a library that bind to a target in a way as desired. Specifically, this program will read the criteria definition parameters from a file for pharmacophore matching (**Table 9.1**) and automatically select all molecules in the docking pose database with desirable and anticipated poses. For example, in **Table 9.1**, column 1 is label; column 2 represents SMARTS patterns of fragments, which is usually the core structure of ligand, potentially for either hydrogen bonding or hydrophobic contacts; column 3 is the coordinates of the core interaction points in a receptor, and the last column denotes distance criteria between column 2 and column 3. Molecules in the

Table 9.1
Parameter definition for post-docking filter

Label,	smarts_pattern,	x_y_z_coordinates,	distance_ threshold
['dl_NH',	'[NH]([CH2])cn',	[165.18, −27.16, 27.46],	2.5]
['d2_OC',	'O(=C([NH])[CH2][NH])',	[165.68, −30.03, 29.63],	2.0]
['d3_nap',	'[cX3][nX3]nc([NH])c[#6]',	[162.06, −25.15, 28.78],	2.0]
['d4_Me',	'[nX3]nc([NH])cc',	[162.6, −25.67, 29.98],	2.0]
['d5_nR',	'[cX3]([cX3])[cX3]([NH])n[nX3]',	[162.92, −25.61, 27.74],	2.0]

initial virtual library are selected for further analysis only if all of the distance criteria in column 4 of **Table 9.1** are satisfied.

Post-docking pharmacophore-based filtering, followed by various scoring functions (**Fig. 9.1**), is greatly advantageous in comparison with results when only docking scoring functions are used. Docking scoring functions (3–5, 12, 16, 17, 23) are in reality met with considerable limitations in reasonably prioritizing compounds in accordance with their corresponding binding affinities or enzymatic potency (24–27). One of the reasons for this lack of correlation is the artificially high ranking for an incorrect docking pose (28, 29). However, it is well documented that docking methods are able to reproduce the bound conformation of a ligand in a protein–ligand complex determined by X-ray crystallography (29–32). Therefore, once the molecules with the correct poses are identified by post-docking filters, the problem of scoring wrong poses is avoided and multiple scoring functions can thus be better suited to rank molecules in the focused library with better chance of success (**Fig. 9.1**).

A set of molecules that rank high after this process would be synthesized and subject to biological tests, i.e., in vitro enzymatic assay or binding affinity experiments, in order to confirm design rationales. Simultaneously, X-ray co-crystal structures of these ligands in complex with the target are to be determined to further corroborate modeling results. Positive results from such approaches are decisive for selection of next set of compounds for synthesis and the future directions of lead optimization.

3.4. Structure-Based Library Design in Discovery of HCV NS5B Polymerase Inhibitors

3.4.1. Background

Hepatitis C virus (HCV) was discovered in 1989 and has been regarded as the key causative agent for non-A, non-B virus hepatitis (33–35). It is estimated that there are over 170 million people worldwide and about 4 million individuals in United States with chronic HCV infection (36). Majority of the infected persons (80%) develop chronic hepatitis, where about 10–25% of them could advance to serious HCV-related liver diseases such as fibrosis, cirrhosis, and hepatocellular carcinoma (37). Only a fraction of patients respond to current FDA-approved standard therapy with a sustained viral load reduction (38), and many of them could not tolerate the treatment because of the various severe side effects (39). Therefore, HCV still represents an unmet medical need which requires discovery and development of more effective and well-tolerated therapies.

HCV NS5B polymerase is a non-structural proteins encoded in HCV genome. This polymerase plays a crucial role in replicating HCV virus and causing infectivity (38) and thus is a key target for drug discovery against HCV (40). Various series of non-nucleoside molecules with different scaffolds have been published recently as HCV NS5B inhibitors (7–9, 41, 42). A couple of scaffolds including Merck's indole scaffold, Pfizer's dihydropyran-2-one derivatives, and Shire's phenylalanine appear to bind to allosteric sites of NS5b (43–46). These binding sites are located on the surface of the thumb sub-domains remote from the NS5B active site. Inhibitors located in such binding sites are believed to show more favorable on-target specific efficacy but less unwanted side effects due to off-target binding.

3.4.2. SBDD of a Novel Thiazolone Scaffold as HCV NS5b Inhibitor

In our HCV programs, we aimed to discover novel scaffolds efficiently to explore the allosteric site of HCV NS5B by means of structure-based approach including the focused library design (7). Our main strategies for new scaffolds are to maintain key pharmacophores of the initial hit, establish sizable chemistry space, and most importantly identify directions for future diversification and optimization. We started with a hit *1* from high-throughput screening which has an IC_{50} value of 2.0 μM (**Fig. 9.5**). An X-ray complex structure indicated that *1* binds to a location in the allosteric site (**Fig. 9.6**). Key inhibitor–protein interactions include the following: (1) both –C=O and N on the thiazolone ring hydrogen bond with backbone –NHs of Tyr477 and Ser476, (2) sulfonamide oxygen atom engages in hydrogen-bonding interaction with basic side chain $–NH_3^+$ of Arg 501, (3) the aromatic furan and phenyl rings interact with the protein by hydrophobic contacts (**Fig. 9.6**). Furthermore, such binding information enable us to envision that a novel structure *2* once incorporated with a suitable *(S)* amino acid possesses not only pharmacophore equivalent as *1* but additional chemistry opportunity for exploring more space in the pocket (**Figs. 9.5** and **9.6**).

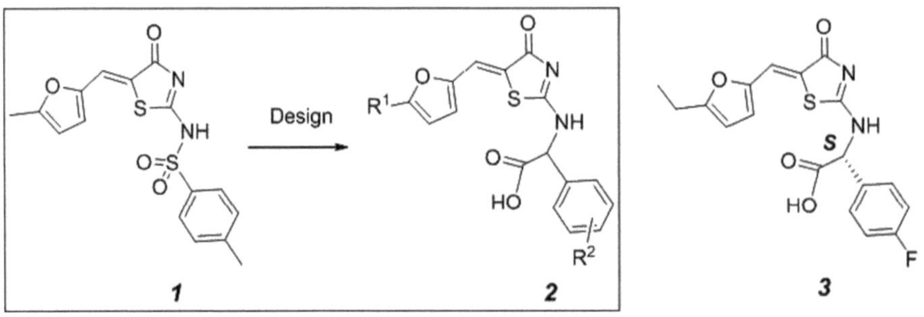

Fig. 9.5. SBDD of a novel scaffold ***2*** as NS5B inhibitor.

The scaffold *2* was confirmed by GOLD docking to have a binding mode which is similar to *1*. Besides, the carboxyl group on *2* picks up additional interaction with side chain of Lys 533 and can be further functionalized to explore more space in the pocket (**Figs. 9.6** and **9.7**).

Starting with the desirable scaffold *2* at hand, we decided to employ the approaches outlined in **Fig. 9.1** for focus library

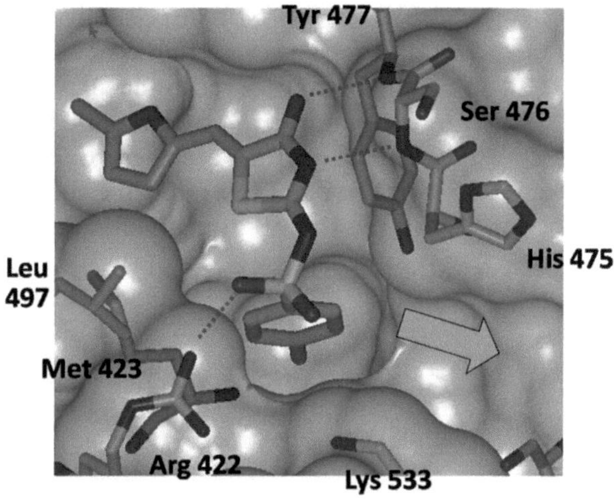

Fig. 9.6. X-ray complex structure of ***1*** with NS5B.

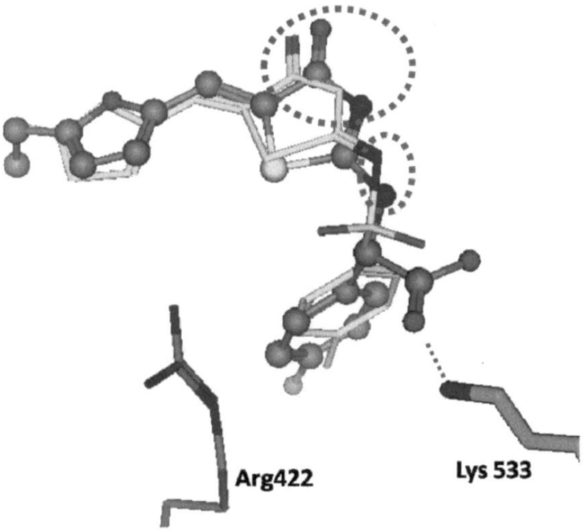

Fig. 9.7. Alignment of ***1*** (in sticks) from X-ray with ***2*** (sticks and balls) from docking.

design and selecting compounds to synthesize, where reactions to make final compounds such as *2* require amino acids as chemical reagents *(7)*. A substructure search of amino acids in ACD database (2) produced 2862 hits and the number was reduced to 1175 after application of a topological diversity selection using MOE package (5). A virtual library was then enumerated and underwent GOLD virtual screening with the standard parameters (3). During the virtual screening, the bound conformation of *1* from X-ray was used as shape template similarity constraint and the constraint weight was set to 10. Each molecule in the focused library was allowed to have 10 docking poses and totally 11,750 poses were collected and filtered according to the predetermined pharmacophore-based criteria using an in-house MOE

SVL script. Not surprisingly, only 60 molecules passed this filter and were then re-ranked by GOLD scoring function (3). One of the 10 top-scored molecules was proposed for synthesis and the compound *3* was determined to have an IC_{50} value of 3.0 μM. Subsequent X-ray structure of *3* in complex with NS5B was solved at 2.0 Å and this molecule shows a binding mode just as predicted in the thumb domain (7) (**Fig. 9.8**).

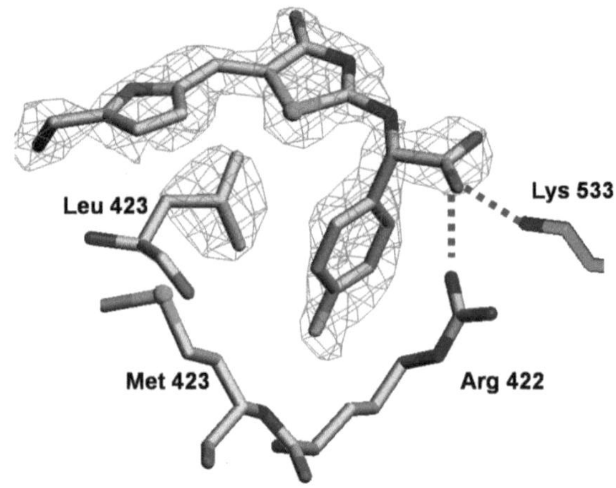

Fig. 9.8. X-ray complex structure of ***3*** with NS5B at a 2.0 Å resolution.

3.4.3. Further SBDD of Follow-Up Focused Library

Structural analysis of the binding mode of *3* in the pocket further led to the identification of more new scaffolds *4* and *5* as HCVNS5B inhibitors (**Fig. 9.9**) (9). A small focused library was enumerated and selected for synthesis after virtual screening. In general, carboxyl compounds with more flexibility resulting from addition of one methylene ($-CH_2-$) group have comparable potency with original compound *3*. The most potent compound *6* has an IC_{50} of 8.5 μM (**Table 9.2**). Compound *11* shows similar enzymatic potency to *6*, while mono-substituted molecules *7–10*, regardless of the chiral centers, showed much weaker potency (**Table 9.2**). Molecules with tetrazole moiety, a commonly used carboxyl group –COOH bioisostere, were pre-

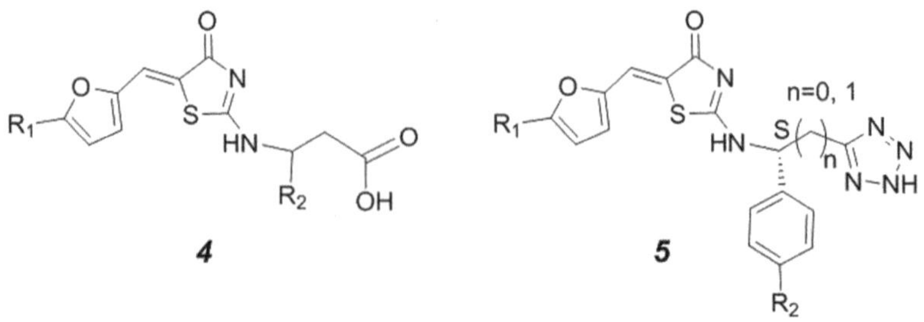

Fig. 9.9. New designs of novel scaffolds.

Table 9.2
Enzymatic potency (IC_{50} in μM) for new molecules. IC_{50} (μM) values of novel scaffolds

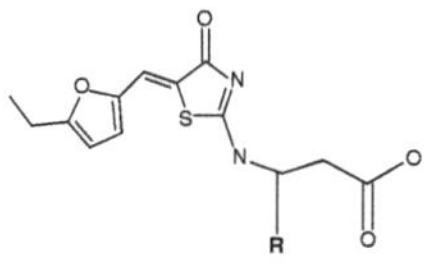

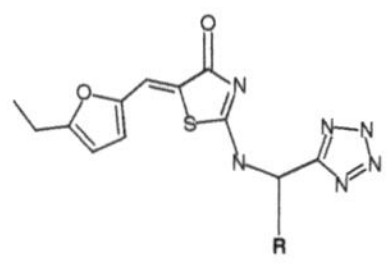

Entry	R:	IC_{50}(μM)	Entry	R:	IC_{50}(μM)	Entry	R:	IC_{50}(μM)
6	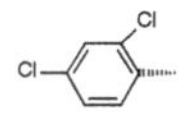	8.5	9	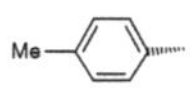	27.0	12	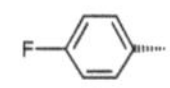	9.7
7	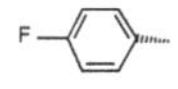	44.0	10	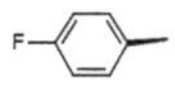	13.0	13	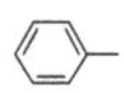	19.0
8	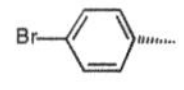	16.5	11	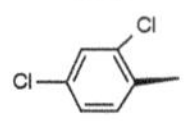	9.0	14	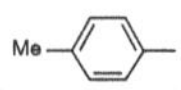	**14.0**

dicted to fit well into target as well and a few of them were synthesized. As seen from **Table 9.2**, tetrazole compounds *12–14* are moderately potent with IC_{50} values of 9.7, 19.0, and 14.0 μM, respectively. Extending the tetrazole group by one more $-CH_2-$ group is tolerated by protein. To prove the design rationale for future structure-based designs, co-crystals structure of *12* in complex with HCV NS5B was successfully established at a resolution of 2.2 Å. The electron density was clear for inhibitor *12*, which binds to the "thumb" sub-domain as expected (**Fig. 9.10**). Overall interactions of *12* with protein are comparable with those of *3* (**Fig. 9.10**).

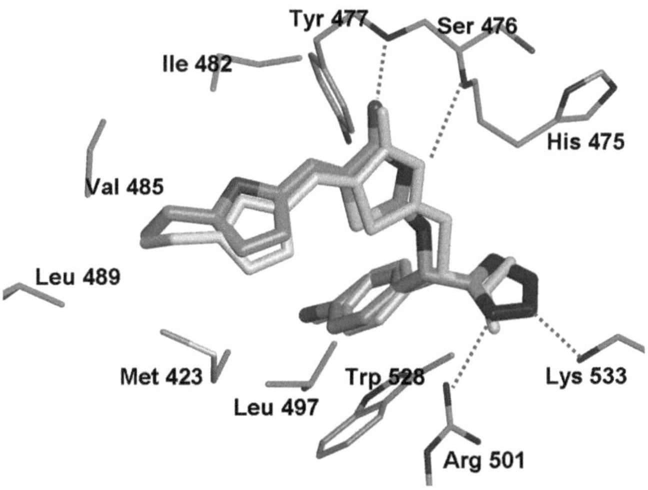

Fig. 9.10. X-ray complex structure of *12* with HCV NS5B (***3*** in *yellow* sticks).

3.4.4. Further Designs of Thiazolone-Acylsulfonamide as NS5B Inhibitors

All of the scaffolds discussed above make hydrogen-bonding interaction with Ser476, Tyr477, and Arg501 and, in the same region, engage in similar hydrophobic contacts with Met423, Ile482, Val485, Leu489, Leu497, and Trp528 (**Fig. 9.10**). In the vicinity of the inhibitor–protein interaction pocket there appears to be more space open for additional interactions. In particular, this new site has two basic residues, His475 and Lys533, as gate-keepers near its entrance (**Fig. 9.10**). A molecule with an appropriate moiety to interact with these two residues was predicted to be able to reach this extra pocket. Our continued SBDD effort was to design such a new scaffold. We envisioned that an acyl-sulfonamide *15* that has a comparable p$K_a$with –COOH could serve as a candidate to hydrogen bond with the basic side chain of Lys533 and additional aromatic moiety linked to sulfonyl group picking up π–π stacking with His475 (**Fig. 9.11**) (8).

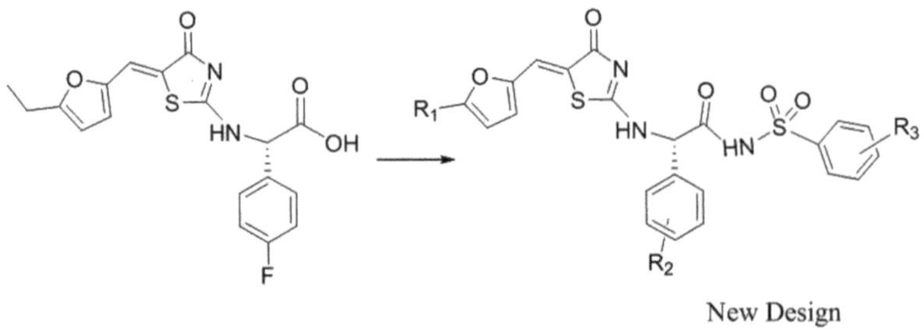

3 *15*

Fig. 9.11. Design of acylsulfonamide scaffold.

To validate the design principle, a very small focused set of library compounds, seven compounds in total, were synthesized and subsequently evaluated for the inhibiting the activity of HCV NS5B. All compounds were reasonably active with IC_{50} values in the range of 6–20 μM. One of the compounds was successfully soaked into NS5B protein crystal and its complex structure with protein was obtained at 2.2 Å. The s'electron density was clear for the inhibitor and the compound fits nicely to the same allosteric site as the *3* and *15* with additional interactions with basic side chains of Arg422 and Lys533 as predicted by GOLD (**Fig. 9.12**) *(8)*. It is also interesting to find that 4-NO_2-Ph makes a face–face π stacking with His475 (**Fig. 9.12**). New scaffolds like this open fresh opportunity for SBDD targeting this allosteric site of HCV NS5B.

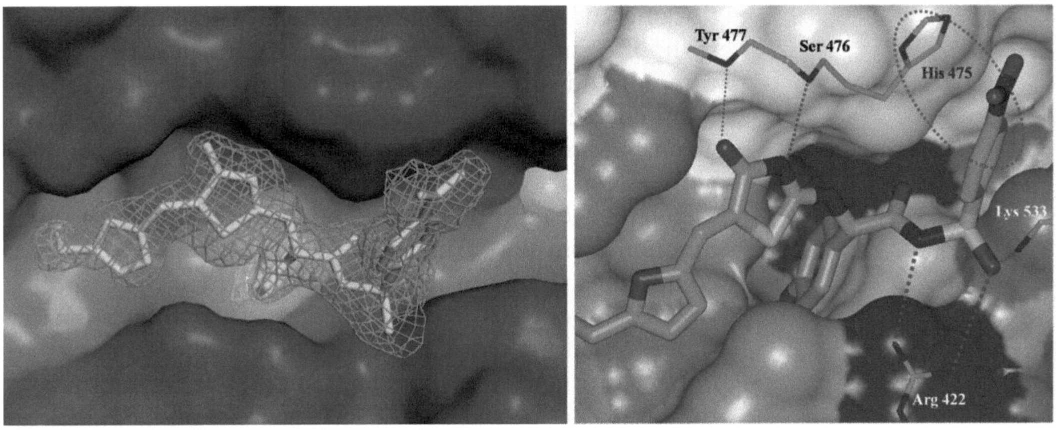

Fig. 9.12. Electron-density map and interactions of acylsulfonamide with NS5B allosteric site.

4. Notes

1. The key to a success is to diligently do various cycles of filtering such as availability, reaction groups, and diversity selection of reagents before library enumerations. It is also necessary to carry out automatic pharmacophore-based post-docking pose filtering prior to using any docking scoring functions.
2. Induced fit docking (IFD) should be carried out periodically to check whether inclusion of flexibility of receptor improves docking results or not. Regular docking, while very fast, treats all amino acids rigid which does not reflect the true nature of protein flexibility and consequently true positives may be missed.
3. Molecular dynamics (MD) should be performed for binding pockets defined mostly by side chains of flexible protein residues to generate an ensemble of binding sites. Such an ensemble can be used for subsequent docking or virtual screening in a parallel fashion.
4. A SBDD design should be confirmed by a later X-ray complex structure which in turn serves to initiate a cycle of iterative structural-based drug design (SBDD). SBDD starts from X-ray or NMR complex structure of ligand with protein and a design, if synthesized, validated, and confirmed by X-ray, creates a starting point for a new level of SBDD efforts.
5. Do not use any scoring functions blindly without validation in any specific drug targets. Most SBDD efforts involve both

docking and scoring. Docking generates a number of poses with different conformation of ligands in a binding site of a receptor and subsequent scoring function is applied to rank them energetically based on the interaction of a pose and a given binding site. One can validate a scoring function by performing a so-called enrichment ratio (ER) study, which calculates the ratio of active compounds selected by scoring function from a docking divided by the number of active compounds if chosen randomly. While there is no specific value for a good ER, a value of less than 1.0 unquestionably suggests that the scoring does not do any better than a random selection. Thus, a greater value of ER corresponds to the better performance of a scoring function in a docking experiment.

6. Reagents with multiple reactive chemical groups should be avoided in library enumeration because their presence most likely requires specific protections of certain functional groups which complicates chemical reactions and makes library production unpractical.

References

1. Hertzberg, R. P., Pope, A. J. (2000) High-throughput screening: new technology for the 21st century. *Curr Opin Chem Biol* **4**, 445–451.
2. MDL Information Systems. http://www.mdli.com.
3. Cambridge Crystallographic Data Centre, UK. http://gold.ccdc.cam.ac.uk.
4. Accelyrs, San Diego, CA, USA. http://www.accelrys.com.
5. Chemical Computing Group, Montreal, Quebec, CA. http://www.chemcomp.com.
6. HKL Research, Inc. http://www.hkl-xray.com.
7. Yan, S., Appleby, T., Larson, G., Wu, J. Z., Hamatake, R., Hong, Z., Yao, N. (2006) Structure-based design of a novel thiazolone scaffold as HCV NS5B polymerase allosteric inhibitors. *Bioorg Med Chem Lett* **16**, 5888–5891.
8. Yan, S., Appleby, T., Larson, G., Wu, J. Z., Hamatake, R. K., Hong, Z., Yao, N. (2007) Thiazolone-acylsulfonamides as novel HCV NS5B polymerase allosteric inhibitors: convergence of structure-based drug design and X-ray crystallographic study. *Bioorg Med Chem Lett* **17**, 1991–1995.
9. Yan, S., Larson, G., Wu, J. Z., Appleby, T., Ding, Y., Hamatake, R., Hong, Z., Yao, N. (2007) Novel thiazolones as HCV NS5B polymerase allosteric inhibitors: Further designs, SAR, and X-ray complex structure. *Bioorg Med Chem Lett* **17**, 63–67.
10. Lyne, P. D. (2002) Structure-based virtual screening: an overview. *Drug Discov Today* **7**, 1047–1055.
11. Jain, S. K., Agrawal, A. (2004) De novo drug design: an overview. *India J Phar Sci* **66**, 721–728.
12. Schrodinger, LLC., Portland, OR, USA. http://www.schrodinger.com.
13. Rarey, M., Kramer, B., Lengauer, T. (1999) Docking of hydrophobic ligands with interaction-based matching algorithms. *Bioinformatics* **15**, 243–250.
14. Kramer, B., Rarey, M., Lengauer, T. (1997) CASP2 experiences with docking flexible ligands using FlexX. *Proteins* **Suppl 1**, 221–225.
15. Kramer, B., Rarey, M., Lengauer, T. (1999) Evaluation of the FLEXX incremental construction algorithm for protein-ligand docking. *Proteins* **37**, 228–241.
16. Tripos, St. Louis, MO, USA. http://www.tripos.com.
17. Molecular Graphics Laboratory, The Scripps Research Institute, San Diego, CA, US. http://autodock.scripps.edu.
18. DesJarlais, R. L., Sheridan, R. P., Dixon, J. S., Kuntz, I. D., Venkataraghavan, R. (1986)

Docking flexible ligands to macromolecular receptors by molecular shape. *J Med Chem* **29**, 2149–2153.
19. Aronov, A. M., Munagala, N. R., Kuntz, I. D., Wang, C. C. (2001) Virtual screening of combinatorial libraries across a gene family: in search of inhibitors of Giardia lamblia guanine phosphoribosyltransferase. *Antimicrob Agents Chemother* **45**, 2571–2576.
20. Ghosh, S., Nie, A., An, J., Huang, Z. (2006) Structure-based virtual screening of chemical libraries for drug discovery. *Curr Opin Chem Biol* **10**, 194–202.
21. Green, D. V. (2003) Virtual screening of virtual libraries. *Prog Med Chem* **41**, 61–97.
22. Shoichet, B. K. (2004) Virtual screening of chemical libraries. *Nature* **432**, 862–865.
23. BioSolveIT GmhB, Germany. http://www.biosolveit.de.
24. Coupez, B., Lewis, R. A. (2006) Docking and scoring–theoretically easy, practically impossible? *Curr Med Chem* **13**, 2995–3003.
25. Kontoyianni, M., Madhav, P., Suchanek, E., Seibel, W. (2008) Theoretical and practical considerations in virtual screening: a beaten field? *Curr Med Chem* **15**, 107–116.
26. Kontoyianni, M., Sokol, G. S., McClellan, L. M. (2005) Evaluation of library ranking efficacy in virtual screening. *J Comput Chem* **26**, 11–22.
27. Kontoyianni, M., McClellan, L. M., Sokol, G. S. (2004) Evaluation of docking performance: comparative data on docking algorithms. *J Med Chem* **47**, 558–565.
28. Verdonk, M. L., Berdini, V., Hartshorn, M. J., Mooij, W. T., Murray, C. W., Taylor, R. D., Watson, P. (2004) Virtual screening using protein-ligand docking: avoiding artificial enrichment. *J Chem Inf Comput Sci* **44**, 793–806.
29. Stahl, M., Bohm, H. J. (1998) Development of filter functions for protein-ligand docking. *J Mol Graph Model* **16**, 121–132.
30. Stahura, F. L., Xue, L., Godden, J. W., Bajorath, J. (1999) Molecular scaffold-based design and comparison of combinatorial libraries focused on the ATP-binding site of protein kinases. *J Mol Graph Model* **17**, 1–9, 51–52.
31. Godden, J. W., Stahura, F., Bajorath, J. (1998) Evaluation of docking strategies for virtual screening of compound databases: cAMP-dependent serine/threonine kinase as an example. *J Mol Graph Model* **16**, 139–143, 65.
32. Vigers, G. P., Rizzi, J. P. (2004) Multiple active site corrections for docking and virtual screening. *J Med Chem* **47**, 80–89.
33. Choo, Q. L., Weiner, A. J., Overby, L. R., Kuo, G., Houghton, M., Bradley, D. W. (1990) Hepatitis C virus: the major causative agent of viral non-A, non-B hepatitis. *Br Med Bull* **46**, 423–441.
34. Choo, Q. L., Kuo, G., Weiner, A. J., Overby, L. R., Bradley, D. W., Houghton, M. (1989) Isolation of a cDNA clone derived from a blood-borne non-A, non-B viral hepatitis genome. *Science* **244**, 359–362.
35. Weiner, A. J., Kuo, G., Bradley, D. W., Bonino, F., Saracco, G., Lee, C., Rosenblatt, J., Choo, Q. L., Houghton, M. (1990) Detection of hepatitis C viral sequences in non-A, non-B hepatitis. *Lancet* **335**, 1–3.
36. (2000) Hepatitis C-Global prevalence (update). *Weekly Epidemiol Rec* **75**, 18–19.
37. Memon, M. I., Memon, M. A. (2002) Hepatitis C: an epidemiological review. *J Viral Hepat* **9**, 84–100.
38. Kolykhalov, A. A., Mihalik, K., Feinstone, S. M., Rice, C. M. (2000) Hepatitis C virus-encoded enzymatic' activities and conserved RNA elements in the 3′ nontranslated region are essential for virus replication in vivo. *J Virol* **74**, 2046–2051.
39. Scott, L. J., Perry, C. M. (2002) Interferon-alpha-2b plus ribavirin: a review of its use in the management of chronic hepatitis C. *Drugs* **62**, 507–556.
40. De Clercq, E. (2002) Strategies in the design of antiviral drugs. *Nat Rev Drug Discov* **1**, 13–25.
41. Rong, F., Chow, S., Yan, S., Larson, G., Hong, Z., Wu, J. (2007) Structure-activity relationship (SAR) studies of quinoxalines as novel HCV NS5B RNA-dependent RNA polymerase inhibitors. *Bioorg Med Chem Lett* **17**, 1663–1666.
42. Yan, S., Appleby, T., Gunic, E., Shim, J. H., Tasu, T., Kim, H., Rong, F., Chen, H., Hamatake, R., Wu, J. Z., Hong, Z., Yao, N. (2007) Isothiazoles as active-site inhibitors of HCV NS5B polymerase. *Bioorg Med Chem Lett* **17**, 28–33.
43. Wang, M., Ng, K. K., Cherney, M. M., Chan, L., Yannopoulos, C. G., Bedard, J., Morin, N., Nguyen-Ba, N., Alaoui-Ismaili, M. H., Bethell, R. C., James, M. N. (2003) Non-nucleoside analogue inhibitors bind to an allosteric site on HCV NS5B polymerase. Crystal structures and mechanism of inhibition. *J Biol Chem* **278**, 9489–9495.
44. Di Marco, S., Volpari, C., Tomei, L., Altamura, S., Harper, S., Narjes, F., Koch, U., Rowley, M., De Francesco, R., Migliaccio, G., Carfi, A. (2005) Interdomain communication in hepatitis C virus polymerase abolished by small molecule inhibitors bound

to a novel allosteric site. *J Biol Chem* **280**, 29765–29770.
45. Biswal, B. K., Cherney, M. M., Wang, M., Chan, L., Yannopoulos, C. G., Bilimoria, D., Nicolas, O., Bedard, J., James, M. N. (2005) Crystal structures of the RNA-dependent RNA polymerase genotype 2a of hepatitis C virus reveal two conformations and suggest mechanisms of inhibition by non-nucleoside inhibitors. *J Biol Chem* **280**, 18202–18210.
46. Biswal, B. K., Wang, M., Cherney, M. M., Chan, L., Yannopoulos, C. G., Bilimoria, D., Bedard, J., James, M. N. (2006) Non-nucleoside inhibitors binding to hepatitis C virus NS5B polymerase reveal a novel mechanism of inhibition. *J Mol Biol* **361**, 33–45.

Chapter 10

Structure-Based and Property-Compliant Library Design of 11β-HSD1 Adamantyl Amide Inhibitors

Genevieve D. Paderes, Klaus Dress, Buwen Huang, Jeff Elleraas, Paul A. Rejto, and Tom Pauly

Abstract

Multiproperty lead optimization that satisfies multiple biological endpoints remains a challenge in the pursuit of viable drug candidates. Optimization of a given lead compound to one having a desired set of molecular attributes often involves a lengthy iterative process that utilizes existing information, tests hypotheses, and incorporates new data. Within the context of a data-rich corporate setting, computational tools and predictive models have provided the chemists a means for facilitating and streamlining this iterative design process. This chapter discloses an actual library design scenario for following up a lead compound that inhibits 11β-hydroxysteroid dehydrogenase type 1 (11β-HSD1) enzyme. The application of computational tools and predictive models in the targeted library design of adamantyl amide 11β-HSD1 inhibitors is described. Specifically, the multiproperty profiling using our proprietary PGVL (Pfizer Global Virtual Library) Hub is discussed in conjunction with the structure-based component of the library design using our in-house docking tool AGDOCK. The docking simulations were based on a piecewise linear potential energy function in combination with an efficient evolutionary programming search engine. The library production protocols and results are also presented.

Key words: Multiproperty lead optimization, library design, adamantyl amide, targeted library, 11β-hydroxysteroid dehydrogenase type 1, 11β-HSD1, PGVL, Pfizer Global Virtual Library, structure-based, AGDOCK, piecewise linear, evolutionary programming.

1. Introduction

Glucocorticoids (GC) are steroid hormones that regulate various physiological processes via stimulation of the nuclear glucocorticoid receptors (1). Chronically elevated levels of active GC hormones (e.g., cortisol) have been associated with many

J.Z. Zhou (ed.), *Chemical Library Design*, Methods in Molecular Biology 685,
DOI 10.1007/978-1-60761-931-4_10,

diseases, including diabetes, obesity, dyslipidemia, and hypertension. In mammalian tissues, GC hormonal regulation is controlled by two isozymes of 11β-hydroxysteroid dehydrogenase that catalyze the interconversion of inert cortisone and active cortisol, namely, 11β-HSD1, which is present predominantly in the liver, adipose tissue, and brain, and 11β-HSD2, which is mainly expressed in the kidney and placenta (2, 3). 11β-HSD1 is a bidirectional, NADPH-dependent enzyme that catalyzes the conversion of inactive 11-keto GCs (cortisone in humans and 11-dehydrocorticosterone in rodents) into hormonally active 11β-hydroxy GCs (cortisol in human and corticosterone in rodents), whereas 11β-HSD2 is a unidirectional dehydrogenase that catalyzes the reverse reaction (cortisol to cortisone) using NAD+ solely as a cofactor (3, 4). In recent years, clinical studies in animal models (5–7) and in humans (8–12) provided evidence for the role of 11β-HSD1 enzyme activity in obesity, diabetes, and insulin insensitivity. In line with these findings, inhibition of 11β-HSD1 by the steroid carbenoxolone (CBX) showed improved insulin sensitivity in human (13–14). Thus, 11β-HSD1 is considered a promising target for the treatment of glucocorticoid-related diseases and has given rise to several classes of nonsteroidal 11β-HSD1 inhibitors (15–18), including the adamantyl triazoles and amides (19–21).

The identification of an adamantyl amide inhibitor of human 11β-HSD1 (**Fig. 10.1**) in our laboratories has prompted us to design a targeted library of close analogs using the Pfizer Global Virtual Library (PGVL) Hub, a desktop tool for designing libraries and accessing Pfizer internal tools, models, and resources. With PGVL Hub, we were able to input our customized alcohol-containing adamantyl amide template and select the appropriate reaction protocol with its corresponding set of amine monomers (**Fig. 10.2**). The reaction protocol involves the transformation of alcohols to amines via mesylation followed by amine substitution (22, 23). The initial set of amine monomers from in-house and commercial sources gave us ~13,000 amines. Reduction in the virtual chemistry space to ~1,000 was achieved by selecting only available secondary amines having molecular weights

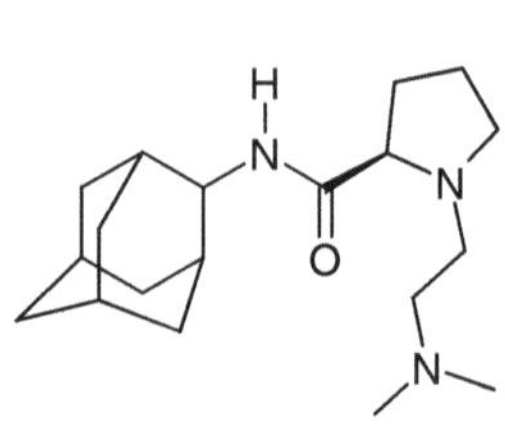

Fig. 10.1. Adamantyl amide inhibitor of human 11β-HSD1 (hu11β-HSD1 Ki(app) = 1.8 nM, EC_{50} = 171 nM, kinetic solubility = 376 μM, HLM = 7.6 μL/min/mg, HHep = 3.0 μL/min/million).

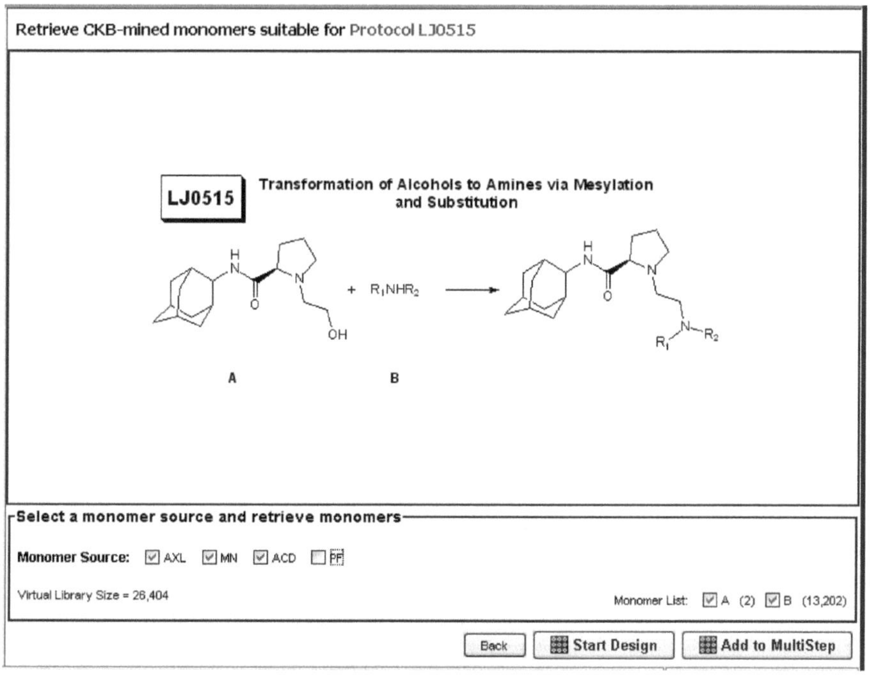

Fig. 10.2. Reaction protocol for the transformation of alcohols to amines via mesylation and amine substitution, as shown in PGVL Hub.

less than 200. Since the objective of the library design was to improve the cellular potency with retention of good solubility, stability in human liver microsomes (HLM) and human hepatocytes (HHep), in silico property calculation and profiling were performed on the virtual enumerated products, resulting in ~300 predicted property-compliant virtual products.

In order to ensure the retention of enzyme activity, the virtual products were subjected to "fixed anchor" docking using AGDOCK, wherein the adamantyl amide moiety was fixed to a specified crystal-bound coordinate during the docking simulations. At the time of our library design, the human 11β-HSD1 (hu11β-HSD1) crystal structure was not available. Thus, we utilized our available in-house guinea pig 11β-HSD1 (gp11β-HSD1) protein crystal structure for docking our virtual library and selected its bound adamantyl ligand, which showed activity in hu11β-HSD1, for defining the coordinates of our "fixed anchor" structure. The docking simulations were carried out using a piecewise linear intermolecular function (24–27) and a stochastic search algorithm based on evolutionary programming (28, 29). Evaluation of the dock hits led to the selection of the top-ranking virtual compounds based on their estimated high-throughput docking scores. The resulting structure-based and property-compliant, 88-compound library design was then submitted to production for combinatorial synthesis. Initial screening at 0.2 μM concentration followed by purification of the 37 selected best hits (>90% inhibition) yielded a compound with improved cell potency and solubility, high stability in HLM and HHep, and retention of enzyme activity. Subsequent elucidation and publication of the X-ray crystal structures of guinea pig (30), human (31), and murine (32) 11β-HSD1 enabled us to crystallize later an adamantyl amide analog in hu11β-HSD1, which confirmed the similarity in the binding modes of the adamantyl amide anchor structure in human and in guinea pig (**Fig. 10.3**), thereby lending credence to use of the gp11β-HSD1 crystal complex as reference for docking.

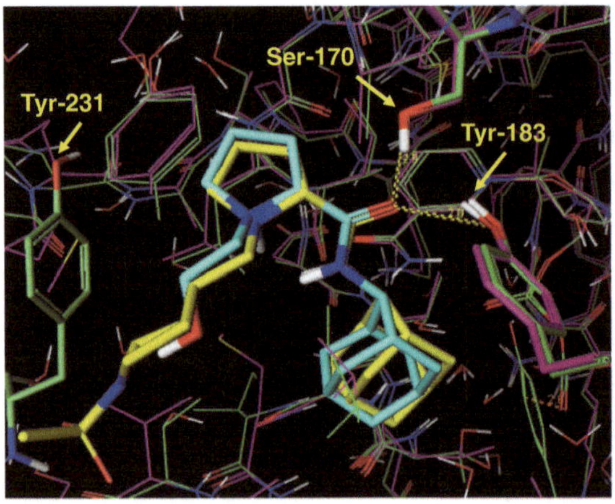

Fig. 10.3. Adamantyl amide analogs exhibit similar binding modes in guinea pig (green) and in human (pink) 11β-HSD1 cocrystal complexes, with Ser-170 and Tyr-183 forming hydrogen bond interactions with the bound ligands. A nonconserved residue (Tyr-231 in guinea pig and Asn-123 in human) differentiates the active sites for these analogs.

1.1. PGVL Overview

PGVL is defined as a set of virtual molecules that can be synthesized from the available monomers and existing templates using validated reaction protocols at Pfizer. It covers a vast virtual chemistry space in the order of 10^{13} compounds. PGVL Hub is the corresponding desktop interface used for a quick navigation of the virtual chemistry space and contains the basic features of an earlier library design tool called LiBrain (33). Searching PGVL for compounds similar to a given lead or HTS hit can be carried out using a "Lead Centric Mining" tool within PGVL Hub or a desktop application called the Bayesian Idea Generator (34). For library designs, virtual searching and screening are simply conducted on specific subsets of PGVL, as defined by reaction types that utilize a set of registered chemistry protocols along with their specific sets of mined reactant monomers. One of the most useful features in PGVL Hub is its ability to access Pfizer's internal computational tools and models. Thus, calculation of physicochemical properties (e.g., thermodynamic solubility) and use of predicted biological endpoints from these models (e.g., in silico HLM model (35)) become an integral part of the virtual screening process.

1.2. AGDOCK Theory for Docking Simulations

AGDOCK is a Pfizer application for rapid and automated computational prediction of the binding geometries (conformation and orientation) of compounds in a given protein active site, as defined by the input defining ligand. It operates in three modes, namely noncovalent docking, covalent docking (25, 36), and partially fixed or fixed anchor docking (36, 37). The default mode is noncovalent docking with full ligand conformational flexibility that explores a large number of degrees of freedom. Significant reduction in the number of degrees of freedom is achieved with the latter two modes in which part of the ligand is fixed within the active site of the protein, either through covalent bond formation with the receptor (covalent docking) or by imposition of positional constraints on an anchor fragment (fixed anchor docking) that is primarily responsible for molecular recognition. AGDOCK employs two search engines, evolutionary programming (24–27) and simulated annealing (38), both of which allow for a full search of the ligand conformation and orientation within the active site. It also supports two intermolecular potentials, AMBER (39) and piecewise linear potential (24–27), and an intramolecular potential consisting of van der Waals and torsional terms derived from the DREIDING force field (40). The intermolecular potential developed for AGDOCK incorporates both steric and hydrogen bond contributions which are calculated from the sum of pairwise interactions between the ligand and the protein heavy atoms using piecewise linear potentials. This energy function along with an evolutionary search technique enables the structure prediction of the protein-ligand complex.

2. Materials

2.1. PGVL Monomers

1. Reactant A: R-1(2-hydroxy-ethyl)-pyrrolidine-2-carboxylic acid adamantan-2-ylamide (**Fig. 10.2**)
2. Reactant B: 88 cyclic and acyclic secondary amines with molecular weights ranging from 71.2 to 162.1 Da

2.2. Reagents and Solvents

1. Anhydrous 1,2-dichloroethane (DCE)
2. Triethylamine (TEA)
3. 4-*N*,*N*-Dimethylaminopyridine (DMAP)
4. Methanesulfonyl chloride
5. Anhydrous dichloromethane (DCM)

6. Anhydrous *N*,*N*-dimethylformamide (DMF)
7. Dimethylsulfone (DMSO)
8. 95:5 Methanol/water mixture (MeOH/water)

2.3. Input Files for Docking Simulations

1. Structure file (in SDF or PDB format) containing the crystal-bound conformation of reference ligand (Reactant A) in the gp11β-HSD1 cocrystal complex
2. Structure file (in PDB format) containing the coordinates of the protein crystal structure derived from the gp11β-HSD1 cocrystal complex (**Fig. 10.4a**)
3. Structure file (in SDF or PDB format) of the anchor or core structure (**Fig. 10.4b**) which will be used in specifying the fixed coordinates of the common fragment in the virtual library of adamantyl amide analogs
4. Structure file (in SDF format) of the virtual library compounds to be subjected to fixed anchor or partially fixed docking

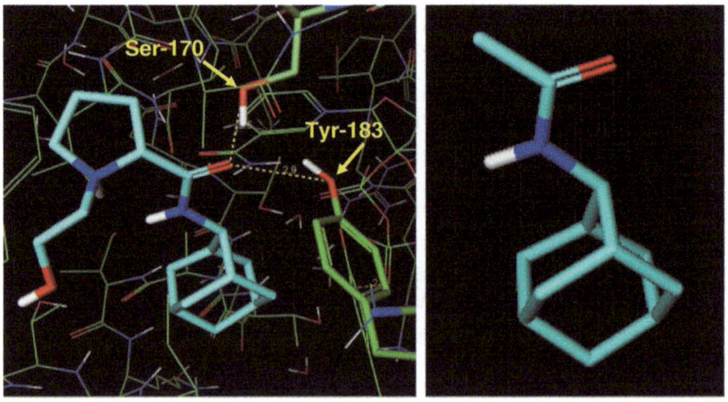

a) b)

Fig. 10.4. (**a**) Crystal structure of gp11β-HSD1 complex with Reactant A which was used as defining ligand in docking simulations. (**b**) Adamantyl amide core structure used in fixed anchor docking.

2.4. Computational Tools and Resources

1. PGVL Hub for reactant monomer retrieval, virtual product enumeration, molecular property calculation, product property profiling, and exporting virtual product structures for subsequent docking
2. Molecular property calculators and predictors (e.g., aqueous solubility model)
3. AGDOCK tool for docking the virtual library
4. PLCALC tool for calculating the protein-ligand interaction free energy (HT) scores

5. A script for ranking and extracting the best docked poses along with their HT scores and other parameters into an Excel table
6. MoViT tool for viewing the dock poses

3. Methods

3.1. PGVL Library Design

The library design was conducted with PGVL Hub which allows the retrieval of the appropriate reaction protocol along with their corresponding sets of reactant monomers (**Fig. 10.2**). There are basically four monomer sources, a commercial domain (ACD), and three in-house domains (AXL, MN, and PF). With the selection of the in-house and commercial domains, the virtual library size is 26,404 (2 "Reactant A" × 13,202 "Reactant B"). In this design, we selected only the in-house monomers which gave us a virtual library size of 11,664 (2A × 5,832B). By specifying a single alcohol-containing template for "Reactant A" which is needed for generating close analogs of the adamantyl amide lead compound, the virtual library size was reduced to 5,832 (1A × 5,832 B). Further reduction in chemistry space was achieved through filtering done both at the monomer and virtual product levels, as outlined in the subsequent library design steps.

1. Calculate the molecular weight and "structural alerts" (substructures containing undesirable or reactive functionalities) for 5,832 amines (*see* **Note 1**).
2. Perform a substructure search for secondary amines.
3. Select only secondary amines with molecular weight (MW) less than 200 and with no structural alerts. This step drastically reduced the number of amines from 5,832 to 1,019.
4. Enumerate the virtual products for the alcohol template and the selected amines using the PGVL Hub virtual product enumerator.
5. Calculate the following molecular properties within PGVL Hub using global computational tools and models: (a) Rule-of-Five (RO5), MW, cLogP, number of hydrogen-bond donors (HBD), number of N and O atoms (NO), and number of RO5 violations; (b) topological polar surface area (TPSA); (c) number of rotatable bonds (NRB); (d) LogD (*see* **Note 2**); and (e) aqueous solubility (*see* **Note 3**).
6. Impose the desired molecular property profile for the virtual products by setting computed property thresholds using the PGVL Hub Decision Maker feature, as shown in **Fig. 10.5**.

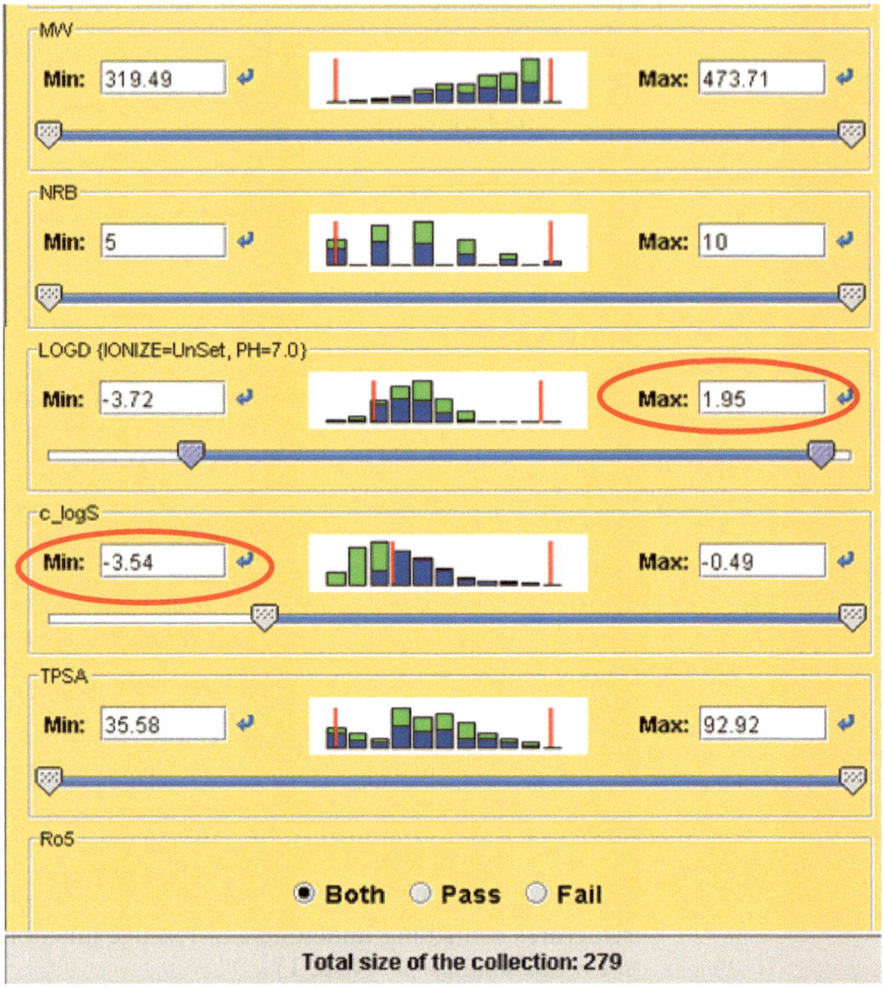

Fig. 10.5. Virtual product property profiling within PGVL Hub. The upper threshold for cLogD and the lower threshold for c_LogS were determined from Spotfire analysis of 2-aminoacetamide lead series. The upper threshold values for MW, number of rotatable bonds (NRB), and polar surface area (TPSA) were user-specified parameters. The rest of the thresholds (e.g., lower threshold for calculated LogD at pH 7.4 or the upper threshold for calculated Log of solubility) are either the lowest or the highest property values of the virtual products.

In this design, the cutoff values include MW <480, NRB ≤10, TPSA <95, cLogD <2.0, and cLogS >−3.5 (for the latter two thresholds, *see* **Note 4**).

7. Export the resulting 279 "predicted property compliant" virtual products as a structure file in SDF format for docking simulations. Our objective was to narrow down the products to 88 compounds to fill a single screening plate at Pfizer.

3.2. Energy Function for Protein–Ligand Interaction

The energy function (24) used to predict the structure and energy of the protein–ligand complex contains an intermolecular term for the interaction between the ligand and the protein and an intramolecular term for the internal energy of the ligand. The

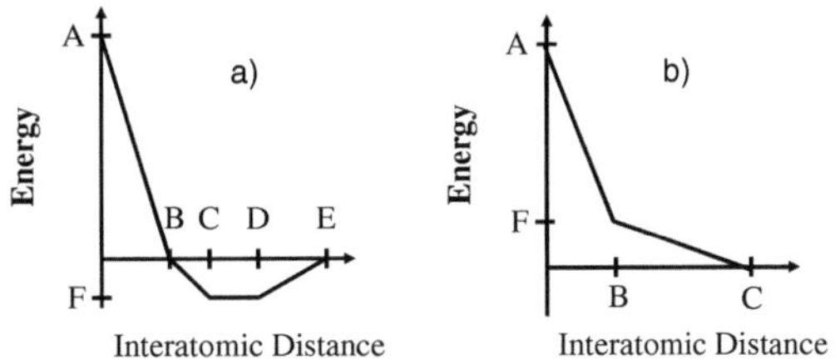

Fig. 10.6. (**a**) Functional form of the hydrogen bond interaction energy ($A = 15.0$, $B = 2.3$, $C = 2.6$, $D = 3.1$, $E = 3.4$, $F = -4.0$) and nonpolar dispersion ($A = 15.0$, $B = 0.93\sigma$, $C = \sigma$, $D = 1.25\sigma$, $E = 1.5\sigma$, $F = -0.4$), where σ is the sum of the atomic radii of the protein and the ligand atoms. (**b**) Functional form of the repulsive interaction ($A = 15.0$, $B = 3.2$, $C = 6.0$, $F = 1.5$). A and F are in kcal/mol and B–E are in Angstroms.

intramolecular term includes the torsional and the van der Waals functions of the DREIDING force field (40) and is useful in differentiating between low- and high-energy ligand geometries and in preventing internal ligand collapse (overlap between ligand atoms). The intermolecular term is a simplified intermolecular potential, which is a pairwise sum of piecewise linear potentials over all ligand and protein heavy (nonhydrogen) atoms (**Fig. 10.6**). Both the hydrogen-bonding and repulsive terms are modulated by a scaling factor based on the relative orientation of the protein and the ligand atoms (**Fig. 10.7**). The piecewise linear potentials for the hydrogen bond and steric interactions have the same functional form but different parameters. This functional form has the advantage of having a finite value compared to the very high energy value in Lennard–Jones potential, when the interatomic distance approaches zero, thereby allowing the ligand to come in close contact with the protein during the early

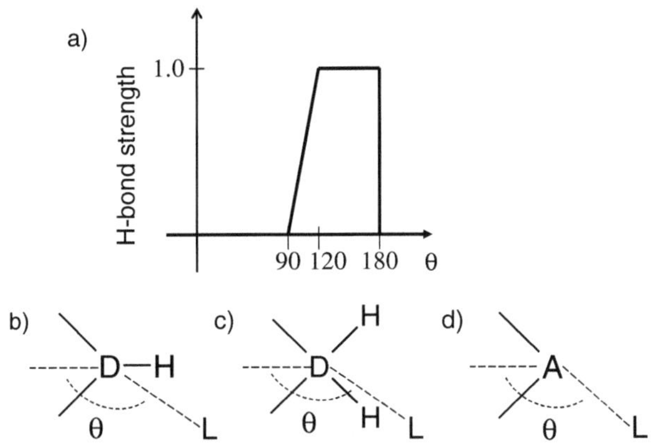

Fig. 10.7. (**a**) Hydrogen bond strength is a function of the angle, θ, determined by the relative orientation of the protein and ligand atoms. (**b**) A protein donor atom D bound to one hydrogen atom H makes an angle θ with the ligand atom L. (**c**) A protein donor atom D bound to the two hydrogen atoms H makes an angle θ with the ligand atom L. (**d**) A protein acceptor atom A makes an angle θ with the ligand atom L.

stages of docking simulations. The parameters used in the piecewise linear potentials depend on the type of interaction and the size of the atom. There are three types of interaction that arise from four different protein and ligand atom types, as follows: (a) hydrogen bond interactions between donors and acceptors, (b) repulsive interactions between pairs of donors or acceptors, and (c) steric interactions between nonpolar atoms or one nonpolar and another atom type (**Table 10.1**). Every pair of interacting atoms is assigned one of these three types of interaction. Atoms are also assigned the atomic radii of 1.4, 1.8, and 2.2 Å corresponding to small (F, metal ions), medium (C, O, N), and large (S, P, Cl, Br) atoms, respectively (36). These parameters were derived from optimized interatomic distances observed in high-quality crystal structures.

Table 10.1
Three types of interaction between ligand and protein heavy atoms arising from different atom types. Primary and secondary amines are defined to be donors while oxygen and nitrogen atoms with no bound hydrogens are defined to be acceptors. Crystallographic water molecules and hydroxyl groups are defined to be both donor and acceptor. Carbon and sulfur atoms are defined to be nonpolar

Ligand	Protein			
	Donor	Acceptor	Both	Nonpolar
Donor	Repulsive	H-bond	H-bond	Steric
Acceptor	H-bond	Repulsive	H-bond	Steric
Both	H-bond	H-bond	H-bond	Steric
Nonpolar	Steric	Steric	Steric	Steric

3.3. Evolutionary Search for Ligand Exploration

Evolutionary programming (28, 29), based on a natural selection process whereby a population of solutions competes for survival, has been adopted as a search technique for finding the optimal binding conformation of the ligand within the protein active site. In this optimization, the population consists of floating point vectors encoding dihedral angles about rotatable bonds, with each vector representing a potential ligand conformation. These dihedral angles are initialized to random values and are allowed to vary during the optimization process. The energy barrier (25) to rotation about a given bond, as defined by the DREIDING force field (40), determines whether this bond is allowed to rotate during optimization (**Table 10.2**). The search process consists of a fixed number of generation cycles. In each cycle, members of a population of ligand conformations are scored using the above

Table 10.2
Rotatable bond types with common threshold energy values

Threshold energy value (kcal/mol)	Rotatable bond type
1.0	sp^2–sp^3 bond only
2.0	Add sp^3–sp^3 bonds
5.0	Add sp^2–sp^2 single bonds
10.0	Add exocyclic aromatic resonant single bonds
25.0	Add resonant bonds
50.0	Add double bonds

energy function. A subset of the population is selected to become parents for the next generation, with the remainder of the population discarded. These parents are then used to produce offspring, thereby restoring the population to its original size. Selection of parents is based on a stochastic competition wherein the energy of each member of the population is compared with the energies of a fixed number of randomly selected subset of the population. A win is assigned to the member with the lowest energy and the number of wins for each member is used to determine whether it survives into the next generation. All surviving members produce offspring by Gaussian mutations of the dihedral angles of the parent vectors. Mutation sizes are allowed to vary as the simulation progresses (24). In the final generation cycle, the best scoring member is minimized using a conjugate gradient method (41). This minimized structure corresponds to the predicted ligand conformation in the active site.

3.4. Docking Simulations of Virtual Library

In the present work, docking of our virtual library was conducted using the available in-house gp11β-HSD1 protein crystal structure. This protein structure was derived from its cocrystal complex (**Fig. 10.4a**) with an adamantyl amide analog that exhibits enzyme activity in hu11β-HSD1 (Ki(app) = 3.6 nM). It was hypothesized that both gp11β-HSD1 and hu11β-HSD1 interact in the same way with the ligand, forming hydrogen bonds with the conserved Tyr-183 and Ser-170 residues and, that, both share a similar hydrophobic pocket for accommodating the adamantyl group. Moreover, the bound ligand is the Reactant A monomer which serves as the anchoring template in our library design. Thus, we used this compound as our reference ligand for defining the active site. Input to docking consisted of a protein PDB file with the gp11β-HSD1 protein crystal structure, a ligand SDF file containing the virtual enumerated products to be docked, a "defining ligand" PDB file with the gp11β-HSD1 bound ligand for defining the active site in the protein structure, and a "core structure" PDB file (**Fig. 10.4b**) derived from the bound ligand. Docking simulations were performed using a docking script which contains the following steps:

1. Prepare the ligand SDF file for docking by titrating at pH 7.
2. Run AGDOCK using the titrated ligand SDF file and the protein PDB file with the following options:
   (a) dl – will use the first entry in the defining ligand file for defining the search area within the active site of the protein where the ligand will be docked; the search area is defined as the minimum bounding rectangle of the defining ligand extended with the cushion site definition
   (b) core – will use the core structure PDB file to specify the coordinates of the core structure that will be used in aligning the virtual products during partially fixed or fixed anchor docking (*see* **Note 5**)
   (c) cushion – 2 Å to be added as cushion to extend the minimum bounding box defined by the defining ligand (*see* **Note 6**)
   (d) maxbarrier – this value is used to indicate what bonds will be considered rotatable during the ligand conformational search; a value of 25 will allow rotation of conjugated single bonds in addition to sp^2–sp^3, sp^3–sp^3, non-conjugated sp^2–sp^2 single bonds, and exocyclic aromatic single bonds (**Table 10.2**)
3. Run PLCALC to calculate the receptor-ligand interaction (HT) scores (42) for each docked ligand pose in the above output file and output the scored dock poses to an SDF file.
4. Evaluate dock results by invoking in-house utility tools that execute the following:
   (a) Retrieval of the best ligand conformations from the output file with scored dock poses based upon specified criteria. In this work, the five best conformations with the lowest HT scores were specified for retrieval
   (b) Sorting the best dock poses by HT scores and storing the sorted ligands to an output SDF file
   (c) Converting the output file into a table with compound ID, HT scores, and other optional parameters
5. View dock poses in a 3D molecular viewing tool and select dock hits based on predicted binding modes and HT scores (*see* **Note 7**).

3.5. Library Synthesis and Purification

In a glove box, the alcohol (320 μL, 80.0 μmol, 1.0 eq, 0.25 M in anhydrous DCE), TEA (33 μL, 240 μmol, 3.0 eq, neat TEA),

DMAP (40 μL, 8.0 μmol, 0.1 eq, 0.2 M in anhydrous DCE), and methanesulfonyl chloride (320 μL, 160 μmol, 2.0 eq, 0.5 M in anhydrous DCM) were added to a 10×95 mm test tube. The test tube was sealed with a test tube cap and stirred in glove box for 3 h at ambient temperature. The solvent was evaporated (SpeedVac or GeneVac, vacuum, medium heat, 16 h) and the residue was dissolved in anhydrous DMF (400 μL). TEA (80 μL, 80.0 μmol, 1.0 eq, 1 M in anhydrous DMF) and the amine (480 μL, 240.0 μmol, 3.0 eq, 0.5 M in anhydrous DMF) were added. The test tube was sealed with a test tube cap. The reaction was heated and stirred at 80°C for 5 h. The solvent was evaporated (SpeedVac or GeneVac, vacuum, medium heat, 16 h), and the residue was dissolved in DMSO (1.340 mL). The reaction mixtures were analyzed by LCMS and the products isolated by automated mass-directed HPLC.

All chromatographic separations were at ambient temperature. Analytical-scale separations were achieved using Agilent HP 1100® MSD systems with a Phenomenex Gemini™ C18 column (4.6 × 50 mm ID, 5.0 μm) or Agilent Zorbax® Extend™ C18 column (4.6 × 50 mm ID, 3.5 μm). The mobile phase consisted of water and acetonitrile, each with 0.05% trifluoroacetic acid and was applied as linear gradient 0–100% organic solvent in 3.0 or 1.75 min, depending on the column used. The MSD utilized positive mode APCI with a scan range from 100 to 1,000 amu. The mass-directed preparative HPLC was a Waters Fractionlynx™ system operating at 50 mL/min using the Gemini C18 stationary phase in a 20 × 50 mm ID column. The mobile-phase solvents were the same as the analytical scale with a 1 min hold to allow for 1,200 μL injection of the crude sample. The gradient was 0–100% organic in 5.4 min. The Waters Micromass® ZQ™ single quad MS utilized positive mode electrospray ionization with a 1:10,000 split from the preparative flow to the MS using a methanol carrier fluid.

The library synthesis steps are as follows:

(1) Prepare an 8×11 array of 10×95 mm test tubes in a test tube rack.

(2) Add one 6×3 mm stir bar into each of the test tubes.

(3) Dry the rack of test tubes, the vials, and caps needed to make the stock solutions at 100°C for 16 h (overnight). Predried vials and caps must be used in subsequent steps.

(4) Transfer the rack of test tubes, the vials, and the caps into a glove box until future use.

(5) In the glove box, prepare a 0.25 M stock solution of each alcohol (Reactant A) in anhydrous DCE. *Note: In case of salt, equal amount of equivalents of TEA should be added.*

(6) In the glove box, prepare a 0.5 M stock solution of methanesulfonyl chloride (MW = 114.55) in anhydrous DCE.

(7) In the glove box, prepare a 0.2 M stock solution of DMAP (MW = 122.2) in anhydrous DCE.

(8) Outside the glove box, prepare a 1 M stock solution of TEA in anhydrous DMF for use in *Step 21.*

(9) Outside the glove box, prepare a 0.5 M stock solution of each amine (Reactant B) in anhydrous DMF for use in *Step 22. Note: In case of salt, equal amount of equivalents of TEA should be added.*

(10) In the glove box, add 320 μL (80 μmol, 1.0 eq) of the appropriate alcohol (Reactant A) solutions into the appropriate predried test tubes. *Note: The reaction is sensitive to the order of the addition of reagents.*

(11) In the glove box, add 33 μL (240 μmol, 3.0 eq) of neat TEA into each test tube.

(12) In the glove box, add 40 μL (8 μmol, 0.1 eq) of the DMAP solution into each test tube.

(13) In the glove box, add 320 μL (160 μmol, 2.0 eq) of the methanesulfonyl chloride solution into each test tube.

(14) In the glove box, cover each test tube with a test tube cap.

(15) In the glove box, stir the reactions at ambient temperature for 3 h.

(16) Take the rack of test tubes out of the glove box.

(17) Remove the volatiles and solvents from the reactions until dryness using a GeneVac™ or SpeedVac™ (medium heat, 6 h).

(18) Add 400 μL of anhydrous DMF to each test tube.

(19) Cover the test tubes with Parafilm.

(20) Vortex and sonicate the covered test tubes until the residues are dissolved.

(21) To each test tube, add 80 μL (80 μmol, 1.0 eq) of the TEA/DMF stock solution from *Step 8.*

(22) To each test tube, add 480 μL (240 μmol, 3.0 eq) of the appropriate amines (Reactant B) solution from *Step 9.*

(23) Cap each test tube with a test tube cap.

(24) Transfer the test tubes to a test tube heating block that has been preheated to 80°C.

(25) Stir the reactions in the test tubes at 80°C for 5 h.

(26) Remove the volatiles and solvents from the reactions until dryness using a GeneVacTM or SpeedVacTM (medium heat, 16 h).

(27) Dissolve the residue in each test tube in 1340 μL DMSO (containing 0.01% BHT) to reach a final concentration of 0.0572 M.

(28) Using a liquid handler, transfer the contents of each test tube to its corresponding well in a 2-mL 96-well polypropylene deep-well plate for purification by HPLC.

(29) Using a liquid handler, remove 5 μL of the solution from each well, dilute the aliquot to 1.0 mL with MeOH/water 95:5 in a new deep-well plate for LC/MS analysis.

(30) Seal each plate with an aluminum foil lid.

(31) Deliver each plate to its appropriate destination.

3.6. Library Assay Results

Initial screening for enzymatic inhibition of human 11β-HSD1 was carried out on the 88 raw products of the library. This screening resulted in 100% hit rate with activity ranging from 76 to 103% inhibition at 0.2 μM concentration. Of these 88 crude products, 37 compounds were purified and submitted for enzymatic and cellular assays in human 11β-HSD1. The enzymatic Ki(app) was determined using human 11β-HSD1 and labeled 3H-cortisone substrate in a triethanolamine buffer in the presence of NADPH, glucose-6-phosphate, glucose-6-phosphate dehydrogenase, and $MgCl_2$. The ratio of the resulting radioactive cortisone and cortisol was determined by radio-HPLC and used to calculate the Ki(app) values (43). The cellular (HEK293T-11_HSD1 Cell Reporter) EC_{50} was determined using human kidney cells that had been stably transfected with human11_HSD1 gene, a reporter plasmid containing DNA sequences of glucocorticoid-activated glucocorticoid receptors, and a luciferase reporter gene (Luc), allowing for quantification of 11_HSD1 enzyme modulation (44). Cellular activity and hepatocyte stability results for the 37 purified compounds are shown in **Fig. 10.8**. In **Fig. 10.8a**, calculated lipophilic efficiency (cLipE) was plotted against cellular EC_{50} in order to rapidly identify potent compounds that are highly lipophilic efficient. Lipophilic efficiency (45, 46) is defined as pIC50 (or pKi) minus cLogP (or LogD) and is a measure of how efficient the ligand is in terms of achieving a high binding affinity (Ki) or cellular potency (EC_{50}) without increasing the ligand lipophilicity (cLogP or LogD). Increasing lipophilicity tends to increase binding affinity through greater van der Waals interactions with the protein target, but it also tends to promote binding to unwanted drug targets, leading to toxicity and other side effects. Thus, designing a high-affinity ligand that engages in other types of interactions (e.g., hydrogen bonding), while keeping the LogP or LogD low, has

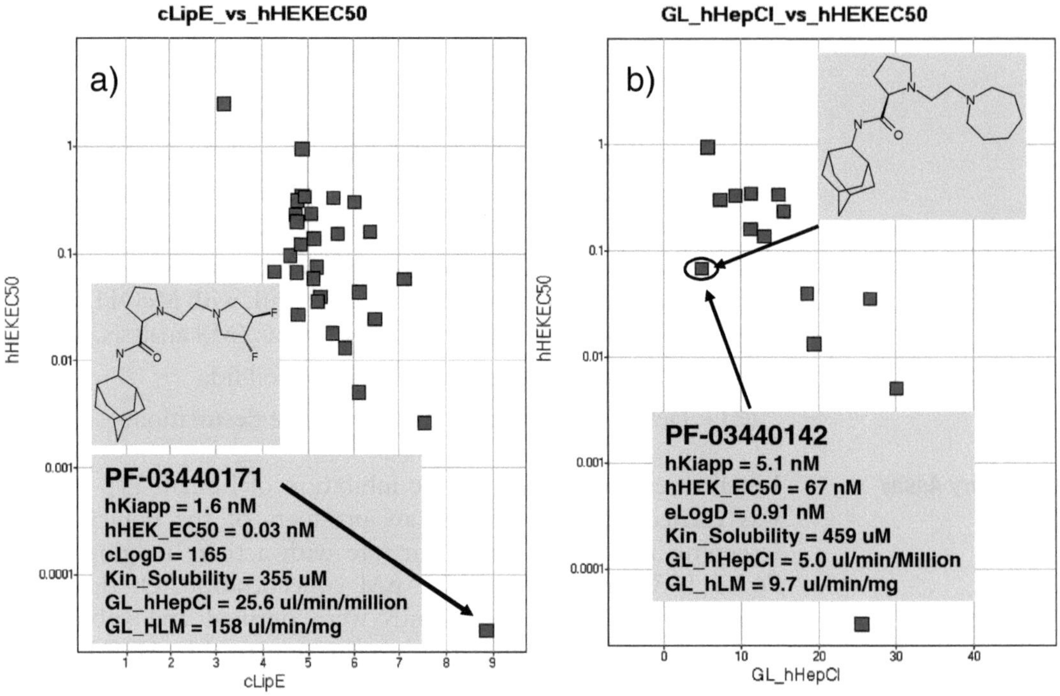

Fig. 10.8. Cellular activity and hepatocyte stability of the 37 purified library compounds. (**a**) cLipE vs. human 11β-HSD1 EC_{50} (hHEKEC50). Chart enables rapid identification of the active and lipophilic efficient compounds (in lower right corner), e.g., PF-03440171 with cLipE = 8.7 and EC_{50} = 0.03 nM. (**b**) Human hepatocyte intrinsic apparent clearance (GL_hHepCl) vs. hHEKEC50. Chart enables rapid identification of active and metabolically stable compounds, such as PF-03440142 with GL_hHepCl = 5 μL/min/million and EC_{50} = 67 nM.

always been the goal for optimization. In general, a LipE range of 5–7 or higher is desired based on the average oral drug with potency in the range of 1–10 nM and cLogP ~2.5 (46). Since the experimental LogD values for the purified products were not available, calculated LogD (cLogD) was used in estimating LipE. Hence, cLipE was calculated as the negative $Log_{10}EC_{50}$ minus cLogD and the values range from 3 to 8.9 with the most cellular potent compound having the highest cLipE value (cLipE = 8.87, EC_{50} = 0.03 nM), albeit with a low stability in HHep (**Fig. 10.8a**). A plot of the human hepatocyte clearance versus cellular activity (**Fig. 10.8b**) shows four compounds with good stability (GL_hHepCl < 10 μL/min/million) and with high to moderate activity. The library was able to achieve its objective by giving rise to PF-03440142, with improved cellular activity (EC_{50} = 67 nM) and retention of favorable solubility (kinetic solubility = 459 μM), while maintaining stability in human liver microsomes (GL_HLM = 9.7 μL/min/mg) and human hepatocytes (GL_hHepCl = 5 μL/min/million).

4. Notes

1. "Structural alerts" (STA) exemplify substructures which have been associated with multiple examples of adverse in vivo events (47). Information on the chemistry, pharmacology, and toxicology for these substructures has been collated to provide a strong rationale for their classification as STA due to their intrinsic reactivity, ability to intercalate to DNA, coordination with a metal, metabolic activation, or transformation to a species capable of covalently binding to biological macromolecules. A common example of STA is the aniline functional group which can get oxidized either at the aniline nitrogen or at the *ortho* and *para* aromatic carbon atoms to form the reactive nitroso and iminoquinone intermediates, respectively, that can lead to increased risk of mutagenicity, in vivo carcinogenicity, and hepatoxicity. Hence, chemists who design libraries must be aware of all these potential liabilities and understand the chemistry and mechanism involved in the metabolism of the design compounds. If the purpose of the library is to probe the receptor pocket to find novel chemotypes, then compounds with STA may be included in the design, provided they can be modified later to eliminate the risks. In general, chemists are advised to avoid compounds with STA in order to minimize the risk of adverse outcome, thereby ensuring the safety and efficacy of drugs.
2. PGVL Hub has access to a plethora of computational models for predicting molecular properties. In this library design, the ACD LogD was used in profiling the virtual products. LogD is an experimental measurement of lipophilicity, where D is a pH-dependent parameter which changes with the degree of ionization. The ACD LogD was selected after careful comparison with alternative LogD predictors, such as the Pallas LogD and the in-house Cubist models based on Pfizer experimental LogD. Chemists are encouraged to test different LogD models to see how well they correlate with the experimental data of their lead series.
3. The legacy aqueous thermodynamic solubility model is an in-house linear regression model consisting of 11 calculated descriptors which include cLogP and polar surface area. This is the model that was used to shape the property profile of the virtual products. This model has now been superseded by a Cubist (48) thermodynamic solubility model (R^2 = 0.93 and RMSE = 0.44, N = 2794) based on a data set of 3,075 compounds with "intrinsic" solubility. This model

calculates the “intrinsic” solubility and provides an estimate of the apparent solubility at a given pH based on the computed “intrinsic” solubility and calculated pK_a using the ACD software. The accuracy of the model is highly dependent on the pK_a prediction given by the ACD software. This model has been validated using Cubist ($R^2 = 0.86$ and RMSE $= 0.61$, $N = 281$).

4. The multiproperty filtering criteria used in profiling the virtual library were derived empirically from Spotfire analysis of biological and physicochemical data for existing compounds belonging to the 2-aminoacetamide lead series (**Fig. 10.9**). These compounds consisted mostly of pyrrolidine-2-carboxylic acid N-(R1)-substituted amides (**Fig. 10.9a**) and morpholine-3-carboxylic acid N-(R1)-substituted amides (**Fig. 10.9b**), where R1 is an adamantyl, a cycloalkyl, a benzyl, an aryl, or a heteroaryl group. For our in-house 11β-HSD1 inhibitors, correlations of biological stability endpoints with physicochemical parameters have been shown to be lead-series specific. Hence, we selected the 2-aminoacetamide lead series, which is closest to our adamantyl amide virtual library, as our training set to guide the selection of parameters to be used for filtering this library. For example, a plot of the experimental LogD values (eLogD) versus human liver microsome stability measured as % remaining (HLM_%Rem@1μM) shows that in order to achieve our laboratory objective of designing compounds with HLM_%Rem@1μM > 70%, an eLogD cutoff value of <2.7 is desired since this includes the highly stable compounds (**Fig. 10.10a**). Since we are using the calculated LogD at pH 7.4 (LogD_pH74) for the virtual products, a plot of the experimental versus calculated LogD translates this cutoff to <2.0 (red horizontal line in **Fig. 10.10b**). For stability in human hepatocytes, we have two types of biological endpoints, % remaining (HHEP_%Rem@1μM) and intrinsic clearance (HHEP_CL(int)_μL/min/M). In order to achieve the desired stability in HHep, i.e., >70% remaining or <3 μL/min/million, a calculated Log of solubility (c_LogS) threshold of >−3 is desired (**Fig. 10.11**). Along the same vein, a cLogS >−3 based on HLM_%Rem@1μM plot is required to achieve >70%, while a cLogS >−4 based on the HLM_CL(int)_μL/min/mg plot is needed to satisfy the laboratory objective of < 30 μL/min/mg (**Fig. 10.12**). In the library design, we used c_LogS >−3.6 as our threshold for filtering the virtual library for compounds with the desired property profile.

5. Fixed anchor docking allows the docking of one part of a ligand while keeping the portion of the ligand that is primarily

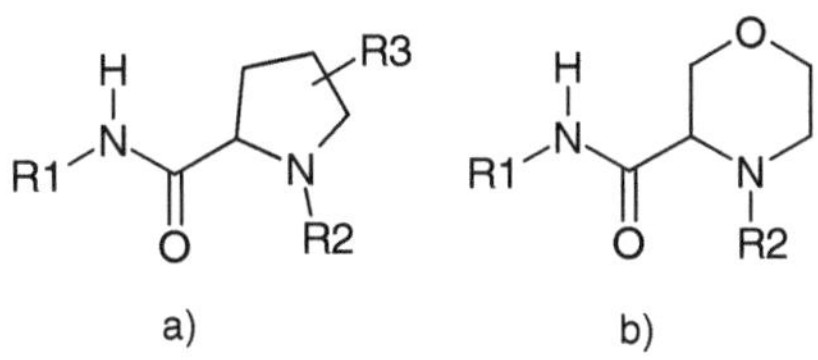

Fig. 10.9. Aminoacetamide lead series used in establishing thresholds for solubility and metabolic stability to guide the library design. R1 can be adamantyl, cycloalkyl, benzyl or substituted benzyl, aryl, or heteroaryl. R2 can be alkyl, substituted alkyl, cycloalkyl, benzyl, substituted benzyl, or acetyl. R3 can be H or OH.

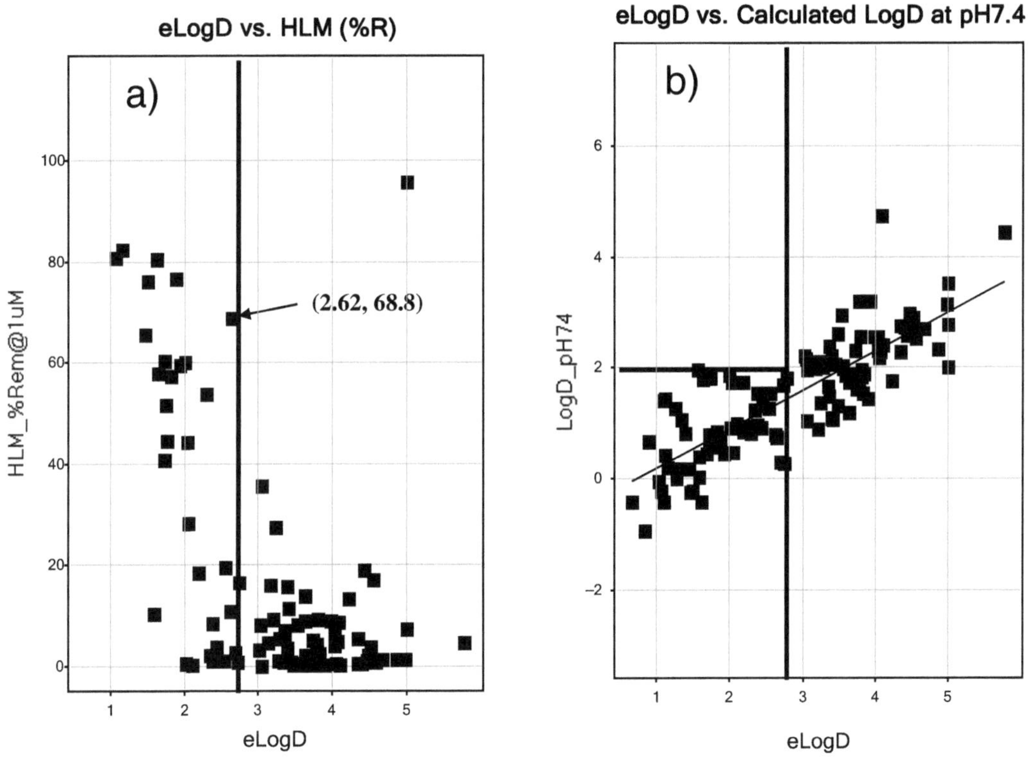

Fig. 10.10. (**a**) Experimental LogD (eLogD) vs. human liver microsome stability (HLM_%Rem@1μM). A threshold value of eLogD < 2.7 is required for >70% stability in HLM. (**b**) Experimental vs. calculated LogD. eLogD < 2.7 translates to cLogD < 2.0 at pH 7.4.

responsible for molecular recognition fixed within the active site (36, 37). In the current work, the adamantyl amide moiety acts as a molecular anchor, with the adamantyl group occupying a specific lipophilic pocket within the enzyme active site and with the amide carbonyl oxygen atom forming hydrogen bond interactions with the conserved Ser-170 and Tyr-183 residues (**Fig. 10.13**). This computational approach will work only if the binding mode of the anchor fragment is not significantly affected by the different sub-

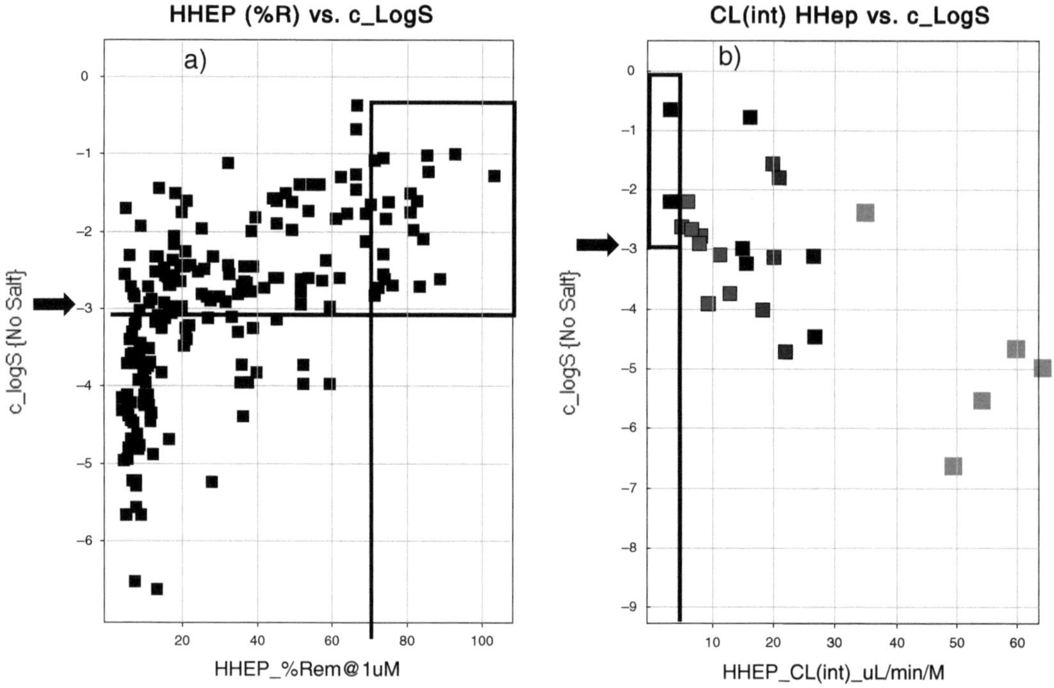

Fig. 10.11. (**a**) Experimental human hepatocyte stability (HHEP_%Rem@1 μM) vs. calculated Log of solubility (c_LogS). (**b**) Experimental human hepatocyte stability expressed as apparent intrinsic clearance (HHEP_CL(int)_μL/min/million) vs. calculated Log of solubility (c_LogS). A threshold value of cLogS > −3.0 is required for stability in human hepatocytes.

stituents in the analogs. One must be careful when selecting a ligand fragment to fix during docking since not all ligand fragments can act as molecular anchors. A molecular anchor is characterized by a specific binding mode with a dominant free energy minimum and a large stability gap, defined as the free energy of the crystal binding mode relative to the free energy of alternative binding modes (26). An advantage of fixed anchor docking is that the large number of degrees of freedom due to ligand flexibility is drastically reduced and that calculation of the free energy of binding for close analogs containing the anchor fragment is significantly facilitated.

6. During docking, the ligand is required to remain in a rectangular box that encompasses the active site. Ligand conformations and orientations are searched via an evolutionary programming algorithm within this rectangular box. A constant energy penalty is added to every ligand atom outside this box. If the virtual library of compounds contain a lot of large substituents (Reactant B), it is advisable to increase this cushion to a larger value in order to accommodate the

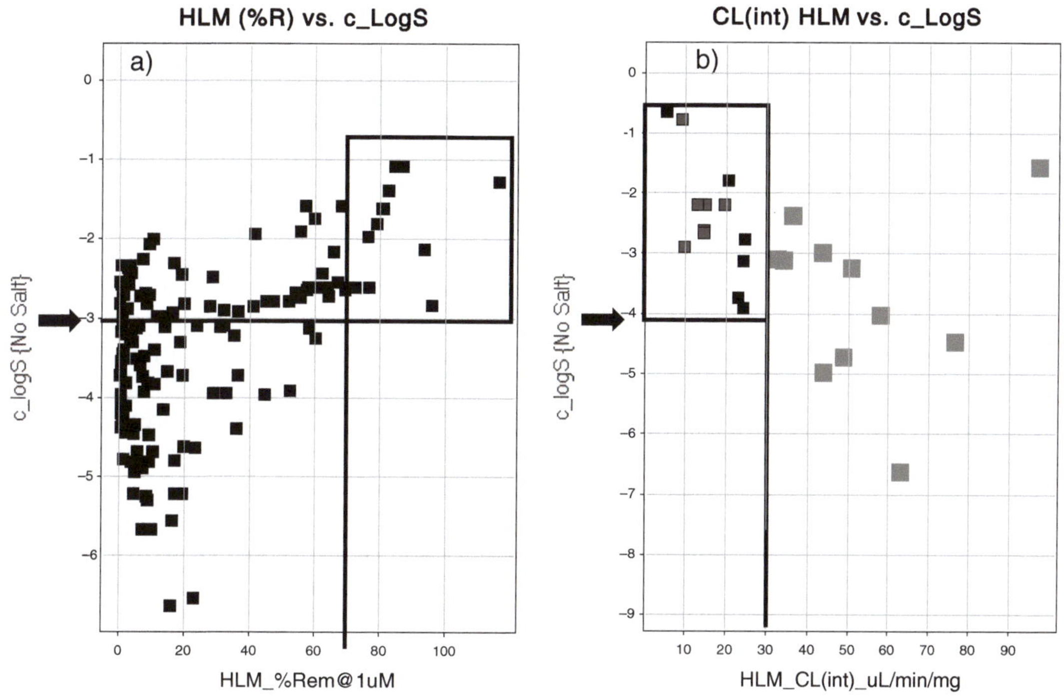

Fig. 10.12. (**a**) Experimental human liver microsome stability (HLM_%Rem@1μM) vs. calculated Log of solubility (c_LogS). (**b**) Experimental human liver microsome stability expressed as apparent intrinsic clearance (HLM_CL(int)_μL/min/mg) vs. calculated Log of solubility (c_LogS). A threshold value of cLogS > −3 and −4.0 is required for stability in HLM based on %R and intrinsic clearance, respectively.

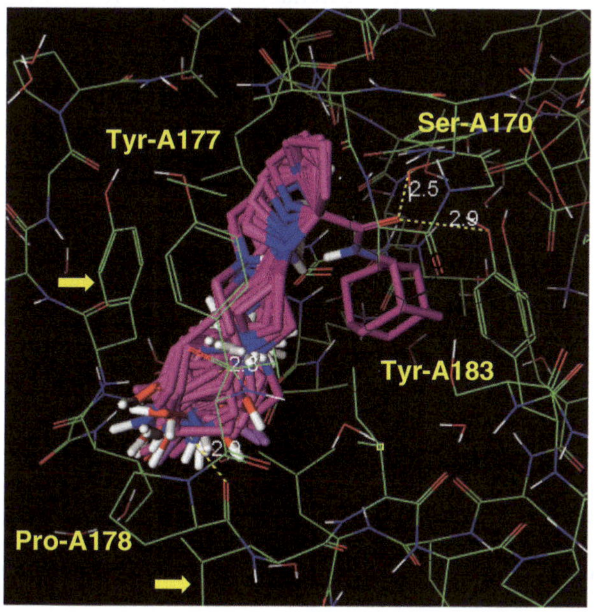

Fig. 10.13. Examples of dock poses from fixed anchor docking of the virtual library in gp11β-HSD1 crystal structure.

various conformations of the larger ligands and minimize the energy penalty.

7. While the simplified piecewise linear potential energy function is able to reproduce the crystallographic bound complexes and predict the structure of the bound ligand in the protein active site, the current high-throughput (HT) scoring function (42) is not sufficient to predict the free energy of binding accurately. Hence, HT scores (42) must be interpreted with caution since these do not necessarily correlate with binding affinities, especially for structurally diverse ligands. In the case of the current library design in which the binding mode of the adamantyl amide anchor is likely to be preserved in the docked analogs, the HT scores represent the free energy differences in the substituents and can be used in weeding out the least active compounds from the virtual library. After visual inspection of the predicted binding modes, 82 virtual compounds with HT scores ranging from −6 to −8 were selected along with six others for the library design.

Acknowledgments

The authors would like to thank Simon Bailey, Martin Edwards, and Michael McAllister for their valuable advice, encouragement, and guidance. Specifically, the authors are grateful to Stanley Kupchinsky for the synthesis of the starting adamantyl amide lead and to the Discovery Computation group at PGRD La Jolla for the development of PGVL and AGDOCK, under the leadership of Atsuo Kuki and Peter Rose, respectively. Thanks are especially due to the following colleagues who developed and performed our project assays, specifically, Jacques Ermolieff (11β-HSD1 enzyme assays); Andrea Fanjul (11β-HSD1 cellular assays); Nora Wallace, Christine Taylor, and Rob Foti (HLM assays); and Veronica Zelesky, Kevin Whalen, and Walter Mitchell (HHEP assays). This work was supported by the 11β-HSD1 project team and the Pfizer Diabetes Therapeutic Area management.

References

1. Charmandari, E., Kino, T., Chrousos, G. P. (2004) Glucocorticoids and their actions: an introduction. *Ann N Y Acad Sci* **1024**, 1–8.

2. Tomlinson, J. W., Walker, E. A., Bujalska, I. J., Draper, N., Lavery, G. G., Cooper, M. S., Hewison, M., Stewart, P. M. (2004) 11β-Hydroxysteroid dehydrogenase type 1:

a tissue-specific regulator of glucocorticoid response. *Endocr Rev* **25**(5), 831–866.
3. Draper, N., Stewart, P. M. (2005) 11β-Hydroxysteroid dehydrogenase and the pre-receptor regulation of corticosteroid hormone action. *J Endocrinol* **186**, 251–271.
4. Walker, E., Stewart, P. M. (2003) 11β-Hydroxysteroid dehydrogenase: unexpected connections. *Trends Endocrinol Metab* **14**, 334–339.
5. Masuzaki, H., Paterson, J., Shinyama, H., Morton, N. M., Mullins, J. J., Seckl J. R., Flier, J. S. (2001) A transgenic model of visceral obesity and the metabolic syndrome. *Science* **294**, 2166–2170.
6. Kotelevtsev, Y., Holmes, M. C., Burchell, A., Houston, P. M., Schmoll, D., Jamieson, P., Best, R., Brown, R., Edwards, C. R. W., Seckl, J. R., Mullins, J. J. (1997) 11β-Hydroxysteroid dehydrogenase type 1 knockout mice show attenuated glucocorticoid-inducible responses and resist hyperglycemia on obesity or stress. *Proc Natl Acad Sci USA* **94**, 14924–14929.
7. Morton, N. M., Holmes, M. C., Fievet C., Staels, B., Tailleux, A., Mullins, J. J., Seckl, J. R. (2001) Improved lipid and lipoprotein profile, hepatic insulin sensitivity, and glucose tolerance in 11β-hydroxysteroid dehydrogenase type 1 null mice. *J Biol Chem* **276**, 41293–41300.
8. Rask, E., Walker, B. R., Soderberg, S., Livingstone, D. E. W., Eliasson, M., Johnson, O., Andrew, R., Olsson, T. (2002) Tissue-specific changes in peripheral cortisol metabolism in obese women: increased adipose 11β-hydroxysteroid dehydrogenase type 1 activity. *J Clin Endocrinol Metab* **87**, 3330–3336.
9. Paulmyer-Lacroix, O., Boullu, S., Oliver, C., Alessi, M. C., Grino, M. (2002) Expression of the mRNA coding for 11β-hydroxysteroid dehydrogenase type 1 in adipose tissue from obese patients: an in situ hybridization study. *J Clin Endocrinol Metab* **87**, 2701–2705.
10. Kannisto, K., Pietilainen, K. H., Ehrenborg, E., Rissanen, A., Kaprio, J., Hamsten, A., Yki-Jarvinen, H. (2004) Overexpression of 11β-hydroxysteroid dehydrogenase-1 in adipose tissue is associated with acquired obesity and features of insulin resistance: studies in young adult monozygotic twins. *J Clin Endocrinol Metab* **89**, 4414–4421.
11. Abdallah, B. M., Beck-Nielsen, H., Gaster, M. (2005) Increased expression of 11β-hydroxysteroid dehydrogenase type 1 in type 2 diabetic myotubes. *Eur J Clin Invest* **35**, 627–634.
12. Valsamakis, G., Anwar, A., Tomlinson, J. W., Shackleton, C. H. L. McTernan, P. G., Chetty, R., Wood, P. J., Banerjee, A. K., Holder, G., Barnett, A. H., Stewart, P. M., Kumar, S. (2004) 11β-hydroxysteroid dehydrogenase type 1 activity in lean and obese males with type 2 diabetes mellitus. *J Clin Endocrinol Metab* **89**, 4755–4761.
13. Walker, B. R., Connacher, A. A., Lindsay R. M., Webb, D. J., Edwards, C. R. (1995) Carbenoxolone increases hepatic insulin sensitivity in man: a novel role for 11-oxosteroid reductase in enhancing glucocorticoid receptor activation. *J Clin Endocrinol Metab* **80**, 3155–3159.
14. Andrews, R. C., Rooyackers, O., Walker, B. R. (2003) Effects of the 11β-hydroxysteroid dehydrogenase inhibitor carbenoxolone on insulin sensitivity in men with type 2 diabetes. *J Clin Endocrinol Metab* **88**, 285–291.
15. Barf, T., Williams, M. (2006) Recent progress in 11b-hydroxysteroid dehydrogenase type 1 (11b-HSD1) inhibitor development. *Drugs Future* **31**(3), 231–243.
16. Barf, T., Vallgarda, J., Emond, R., Haggstrom, C., Kurz, G., Nygren, A., Larwood, V., Mosialou, E., Axelsson, K., Olsson, R., Engblom, L., Edling, N., Ronquist-Nii, Y., Ohman, B., Alberts, P., Abrahmsen, L. (2002) Arylsulfonamidothiazoles as a new class of potential antidiabetic drugs. Discovery of potent and selective inhibitors of the 11β-hydroxysteroid dehydrogenase type 1. *J Med Chem* **45**, 3813–3815.
17. Hult, M., Shafqat, N., Elleby, B., Mitschke, D., Svensson, S., Forsgren, M., Barf, T., Vallgarda, J., Abrahmsen, L., Oppermann, U. (2006) Active site variability of type 1 11b-hydroxysteroid dehydrogenase revealed by selective inhibitors and cross-species comparisons. *Mol Cell Endocrinol* **248**, 26–33.
18. Xiang, J., Ipek, M., Suri, V., Massefski, W., Pan, N., Ge, Y., Tam, M., Xing, Y., Tobin, J. F., Xu, X., Tam, S. (2005) Synthesis and biological evaluation of sulfonamidooxazoles and B-keto sulfones: selective inhibitors of 11β-hydroxysteroid dehydrogenase type I. *Bioorg Med Chem Lett* **15**, 2865–2869.
19. Olson, S., Balkovec, J., Gao, Y. -D, et al. (2004) Selective inhibitors of 11β-hydroxysteroid dehydrogenase type 1. Adamantyl triazoles as pharmacological agents for the treatment of metabolic syndrome. *Keystone Symp* Abst X2–239.
20. Berwaer, M. (2004) Promising new targets. The therapeutic potential of 11β-HSD1 inhibition. *6th Annu Conf Diabetes* (Oct. 18–19, London).

21. Webster, S. P., Ward, P., Binnie, M., Craigie, E., McConnell, K. M. M., Sooy, K., Vinter, A., Seckl, J. R., Walker, B. R. (2007) Discovery and biological evaluation of adamantyl amide 11β-HSD1 inhibitors. *Bioorg Med Chem Lett* **17**, 2838–2843.
22. Becker, D. P., Flynn, D. L., Villamil, C. I. (2004) Bridgehead-methyl analog of SC-53116 as a 5-HT4 agonist. *Bioorg Med Chem Lett* **14**(12), 3073–3075.
23. Reddy, P. G., Baskaran, S. (2004) Epoxide-initiated cationic cyclization of azides: a novel method for the stereoselective construction of 5-hydroxymethyl azabicyclic compounds and application in the stereo- and enantioselective total synthesis of (+)- and (−) -indolizidine 167B and 209D. *J Org Chem* **69**, 3093–3101.
24. Gehlhaar, D. K., Verkhivker, G. M., Rejto, P. A., Sherman, C. J., Fogel, D. B., Fogel, L. J., Freer, S. T. (1995) Molecular recognition of the inhibitor AG-1343 by HIV-1 protease: conformationally flexible docking by evolutionary programming. *Chem Biol* **2**, 317–324.
25 Gehlhaar, D. K., Bouzida, D., Rejto, P. A. (1998) Fully automated and rapid flexible docking of inhibitors covalently bound to serine proteases. *Proceedings of the 7th International Conference on Evolutionary Programming*, MIT Press, Cambridge, MA., pp. 449–461.
26. Rejto, P. A., Verkhivker, G. M. (1998) Molecular anchors with large stability gaps ensure linear binding free energy relationships for hydrophobic substituents. *Pacific Symp Biocomput* **1998**, 362–373.
27. Bouzida, D., Rejto, P. A., Arthurs, S., Colson, A. B., Freer, S. T., Gehlhaar, D. K., Larson, V., Luty, B. A., Rose, P W., Verkhivker, G. M. (1999) Computer simulations of ligand-protein binding with ensembles of protein conformations: a Monte Carlo study of HIV-1 protease binding energy landscapes. *Intl J Quantum Chem* **72**, 73–84.
28. Fogel, D. B. (1995) *Evolutionary Computation: Toward a New Philosophy of Machine Intelligence*, IEEE Press, Piscataway.
29. Fogel, L. J., Owens, A. J., Walsh, M. J. (1966) *Artificial Intelligence Through Simulated Evolution*, Wiley, New York.
30. Ogg, D., Elleby, B., Norstrom, C., Stefansson, K., Abrahmsen, L., Oppermann, U., Svensson, S. (2005) The crystal structure of guinea pig 11β-hydroxysteroid dehydrogenase type 1 provides a model for enzyme-lipid bilayer interactions. *J Biol Chem* **280**, 3789–3794.
31. Hosfield, D. J., Wu, Y., Skene, R. J., Hilgers, M., Jennings, A., Snell, G. P., Aertgeerts, K. (2005) Conformational flexibility in crystal structures of human 11β-hydroxysteroid dehydrogenase type 1 provide insights into glucocorticoid interconversion and enzyme regulation. *J Biol Chem* **280**, 4639–4648.
32. Zhang, J., Osslund, T. D., Plant, M. H., Clogston, C. L., Nybo, R. E., Xiong, F., Delaney, J. M., Jordan, S. R. (2005) Crystal structure of murine 11β-hydroxysteroid dehydrogenase 1: an important therapeutic target for diabetes. *Biochemistry* **44**, 6948–6957.
33. Polinsky, A., Feinstein, R. D., Shi, S., Kuki, A. (1996) LiBrain: software for automated design of exploratory and targeted combinatorial libraries. *Colorado Conf*, Chapter 20, 219–232.
34. Hoorn, W. P., Bell, A. S. (2009) Searching chemical space with the Bayesian idea generator. *J Chem Inf Model* **49**(10), 2211–2220.
35. Lee, P. H., Cucurull-Sanchez, L., Lu, J., Du, Y. J. (2007) Development of in silico models for human liver microsomal stability. *J Comput Aided Mol Des* **21**(12), 665–673.
36. Gehlhaar, D. K., Bouzida, D., Rejto, P. A. (1999) Reduced dimensionality in ligand-protein structure prediction: covalent inhibitors of serine proteases and design of site-directed combinatorial libraries. *Proceedings of the Division of Computers in Chemistry, ACS.* Chapter 19, pp. 292–310.
37. Bouzida, D., Gehlhaar, D. K., Rejto, P. A. (1997) Application of partially fixed docking towards automated design of site-directed combinatorial libraries. *ACS National Meeting*, COMP 156.
38. Bouzida, D., Arthurs, S., Colson, A. B., Freer, S. T., Gehlhaar, D. K., Larson, V., Luty, B. A., Rejto, P. A., Rose, P. W., Verkhivker, G. M. (1999) Thermodynamics and kinetics of ligand-protein binding studied with the weighted histogram analysis method and simulated annealing. *Pacific Symp Biocompu*, pp. 426–437.
39. Weiner, S. J., Kollman, P. A., Case, D. A., Singh, U. C., Ghio, C., Alagona, G., Profeta, S. Jr., Weiner, P. (1984) A new force field for molecular mechanical simulation of nucleic acids and proteins. *J Am Chem Soc* **106**, 765–784.
40. Mayo, S. L., Olafson, B. D., Goddard, W. A. III (1990) DREIDING: a generic force field for molecular simulations. *J Phys Chem* **94**, 8897–8909.
41. Press, W. H., Teukolsky, S. A., Vettering, W T., Flannery, B. P. (1992) *Numerical Recipes in C. The Art of Numerical Com-*

puting, 2nd ed. Cambridge University Press, Cambridge.

42. Marrone, T. J., Luty, B. A., Rose, P. W. (2000) Discovering high-affinity ligands from the computationally predicted structures and affinities of small molecules bound to a target: a virtual screening approach. *Perspect Drug Discovery Design*, **20**, 209–230.
43. Castro, A., Zhu, J. X., Alton, G. R., Rejto, P., Ermolieff, J. (2007) Assay optimization and kinetic profile of the human and the rabbit isoforms of 11b-HSD1. *Biochem Biophys Res Commun* **357**(2),561–566.
44. Bhat, B. G., Hosea, N., Fanjul, A., Herrera, J., Chapman, J., Thalacker, F., Stewart, P. M., Rejto, P. A. (2008) Demonstration of proof of mechanism and pharmacokinetics and pharmacodynamic relationship with 4′-cyanobiphenyl-4-sulfonic acid(6-amino-pyridin-2-yl)amide (PF-915275), an inhibitor of 11b-hydroxysteroid dehydrogenase type 1, in cynomolgus monkeys. *J Pharm Exp Ther* **324**(1), 299–305.
45. Ryckmans, T., Edwards, M. P., Horne, V. A., Correia, A. M., Owen, D. R., Thompson, L. R., Tran, I., Tutt, M. F., Young, T. (2009) Rapid assessment of a novel series of selective CB2 agonists using parallel synthesis protocols: a lipophilic efficiency (LipE) analysis. *Bioorg Med Chem Lett* **19**(15), 4406–4409.
46. Leeson, P. D., Springthorpe, B. (2007) The influence of drug-like concepts on decision-making in medicinal chemistry. *Nat Rev Drug Disc* **6**(11), 881–890.
47. Blagg, J. (2006) Structure-activity relationships for in vitro and in vivo toxicity. *Ann Reps Med Chem* **41**, 353–368.
48. Quinlan, J. R. (1992) Learning with continuous classes. *In Proc. AI '92*, Adams, Sterling, Eds., 343–348.

Section III

Fragment-Based Library Design

Chapter 11

Design of Screening Collections for Successful Fragment-Based Lead Discovery

James Na and Qiyue Hu

Abstract

A successful fragment-based lead discovery (FBLD) campaign largely depends on the content of the fragment collection being screened. To design a successful fragment collection, several factors must be considered, including collection size, property filters, hit follow-up considerations, and screening methods. In this chapter, we will discuss each factor and how it was applied to the design and assembly of one or more fragment collections in a major pharmaceutical company setting. We will also present examples and statistics of screening results from such collections and how subsequent collections can be improved. Lastly, we will provide a summary comparison of selected fragment collections from literature.

Key words: Fragment-based lead discovery, screening collection, library design, computational filtering, NMR screening

1. Introduction

In the past decade, fragment-based drug discovery (FBDD), or fragment-based lead discovery (FBLD), has become an exciting way for the pharmaceutical industry to discover new medicines (1–5). In addition to biochemical assays, fragment screening takes advantage of several other screening technologies, including NMR (nuclear magnetic resonance), MS (mass spectroscopy), SPR (surface plasmon resonance), X-ray crystallography, and various forms of calorimetry. Several clinical candidates can trace their origin to FBLD from different screening methods; a recent review by de Kloe et al. provided several interesting examples (6).

While most pharmaceutical and biotech companies utilize high-throughput screening (HTS) as their primary assay, FBLD offers numerous advantages. Compared with HTS, there is significantly fewer compounds to be screened. There are typically thousands of compounds for a fragment screen versus hundreds

J.Z. Zhou (ed.), *Chemical Library Design*, Methods in Molecular Biology 685,
DOI 10.1007/978-1-60761-931-4_11, © Springer Science+Business Media, LLC 2011

of thousands or more for an HTS campaign. This reduced set of compounds results in time and resources savings for an FBLD screen as compared to an HTS screen. The smaller compound size also means a relatively small set of fragments can cover a larger chemical space than a typical HTS collection (7). Moreover, fragment screens generally result in much higher hit rates than HTS campaigns, and often the hits are novel with respect to the HTS-derived chemical series. Lastly, a distinct advantage of FBLD is that often orthogonal screening methods are used as confirmation, e.g., NMR screen followed by SPR or MS, so chances of false positives are diminished.

The drawbacks of FBLD are that not all targets are amenable to FBLD, and that fragment hits are usually much weaker binders than typical HTS hits, generally in the millimolar to high micromolar range. There are also concerns about whether the hit compounds are binding at the desired pocket or at a random hotspot on the protein, although this can be resolved by competitive binding experiments or by crystallography.

Because fragment hits are smaller in size compared with a more "drug-like" HTS hit, there is much more room for the medicinal chemists to shape them into more novel leads and eventually lead series. This task is made easier if the fragment hits contain chemistry vectors for elaboration, which can be built in when assembling a fragment screening collection.

A successful FBLD campaign requires a fragment library possessing several important characteristics, including proper collection size, good physicochemical properties, chemical diversity, and chemistry follow-up considerations. There are a number of publications describing the process of building a fragment collection, from a general collection to collection tailored for a specific screening method (8–12). In this chapter, we will discuss some of the factors to consider when assembling a fragment collection. In addition, we will describe the process of how two fragment collections were assembled, and analyze the screening results from multiple fragment screens.

2. Factors to Consider in Creating a Fragment Collection

2.1. Collection Size

Perhaps the biggest distinction between an HTS and a fragment collection is the size of each collection. A typical HTS collection can contain 10^5–10^7 compounds, whereas a fragment collection can range in size from 10^2 to 10^4 compounds. While recent advances have greatly increased the screening capabilities of HTS, the cost factor can still be argued to favor fragment screening in terms of assembling the collection and the compounds and reagent resources consumed per screening campaign. The size of a fragment collection can be created based on the targets being pursued, e.g., a set of fragments most likely to be hits against kinases. In such cases the collection can be relatively small, numbering in the hundred of compounds. Alternatively, the collection size can also be dictated by the screening method. For example, if the screening method is a high-concentration screening (HCS) using biochemical assay with good throughput, then the size of the collection can be relatively large with numbers in the thousands or even 10–20 K compounds. In general, if the desire is to build a generic collection, then the number of compounds can typically number from 5 to 20 K, whereas a screening method or a biased collection will be smaller by one order of magnitude in size.

2.2. Physicochemical Properties

Solubility is an important factor to consider in selecting fragments, especially when the assay of choice is high-concentration screening. This is often the case since fragments are typically weak binders with low millimolar to high micromolar activity. In most instances the solvent used tends to be more polar, such as DMSO or water. Therefore, the fragments have to be water soluble, and compounds having ionizing groups or polar functions favor solubility. There are various methods to calculate the solubility of a compound, see a recent review for details (13). These calculated values can be used to guide the selection of compounds for a given collection.

Besides solubility, other physicochemical factors to be considered in building a fragment collection are molecular weight, number of heavy atoms, rotatable bonds, lipophilicity, and polar surface area. All or some of these factors can be accounted for when building a collection, although molecular weight is one factor that is almost always considered. In general the MW range for a fragment collection should fall within 120–300, with the median MW at around 200–250. Compounds with MW much less than 150 are undesirable due to higher risk of unspecific or undetectable binding, while compounds larger than 300 are becoming more "full-size" molecules rather than fragments. Attenuating the MW range would also affect the number of heavy atoms, a closely related molecular property. For the number of rotatable bonds, typically a fragment with MW around 250 would have 0–3 rotatable bonds, which in general is a desirable range for a fragment collection. Another important factor to consider is lipophilicity, usually measured as logP, which can have a big influence on binding affinity. The generally accepted range for logP is 0–3 (14).

2.3. Hit Follow-Up Consideration

One of the less common factors considered when building a fragment collection is synthetic attractiveness of the fragments, or put differently, ease of hit follow-up for the hit-to-lead process. In general, chemists like to have hits which present opportunities for making analogs, but often fragments lack the synthetic handle(s) that a chemist desires. It would be relatively easy to assemble a fragment collection from a pool of reagent-type compounds which contain one or multiple reactive centers. However, some chemistry savvy must be exercised to avoid highly reactive functional groups such as sulfonyl halides or isocyanates, which can react with the protein side chains. Because fragments are often screened as a mixture, care also must be taken to avoid compounds which may react with each other when mixed together. Another factor to consider is that most reactive functional groups can also elicit binding interaction, and this interaction can be altered or destroyed altogether when reaction occurs at the site. For example, the bromine of an alkyl bromide can elicit a binding interaction with the protein target which can result in the compound being a screening hit. However, if the bromine is used as a chemistry vector for elaboration, the bromine is then displaced upon reaction which eliminates the bromine–protein interaction, which may be an important part of the overall binding interaction. Hence, the selection of compounds which contain reactive functional groups to be included in a screening collection must be done carefully, preferably with input from medicinal/synthetic chemists.

Less reactive synthons which also facilitate the hit follow-up process contain "chemistry handles" or functional groups on a molecule which can be easily converted to grow or shape the molecule. One of the more useful functional groups for this purpose are halo-aromatic compounds, in which the halogen (or even the aromatic hydrogens) can often be the chemistry vector that allows the chemist to explore other parts of the binding pocket. Primary amines and carboxylic acids are other functional groups that can be considered useful chemistry handles to be included in a fragment collection. When screened against a protein target, the binding interaction from these functional groups can mostly be preserved even after they are used as chemistry vectors for elaboration. **Figure 11.1** lists commonly used functional groups which serve as chemistry handles in a fragment.

Sometimes, reactive functional groups are protected. For instance, a novel primary amine can be "protected" by reacting with a small acid (e.g., acetic acid), and the resulting amide would still retain some of the characteristics of the original amine where both the amino R-group and the resulting amide reaction product can elicit binding interaction with the protein. And if this pro-

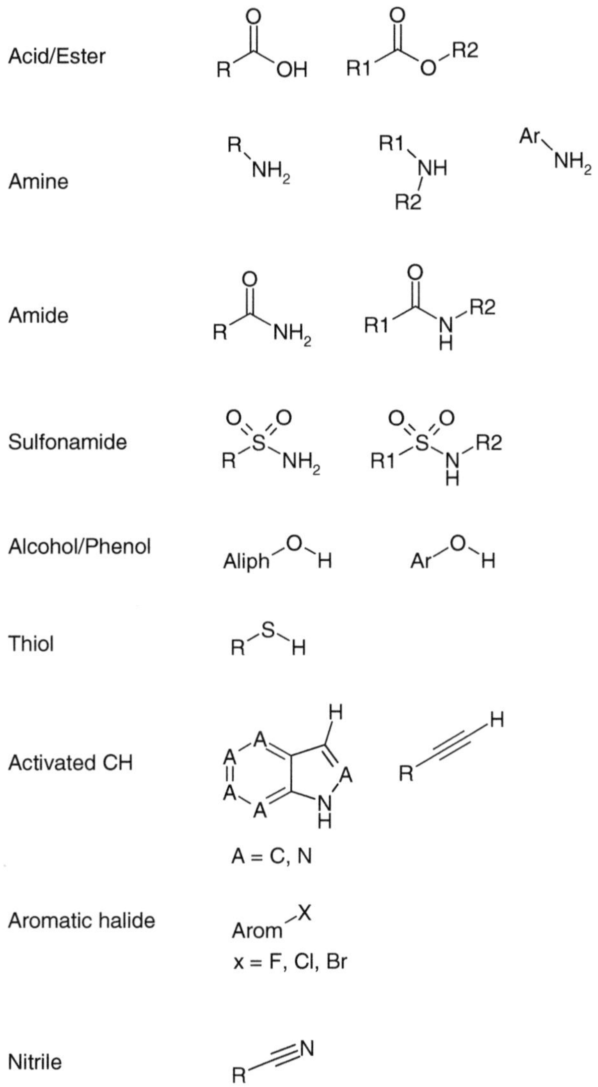

Fig. 11.1. Functional groups that are useful as chemistry handles in a fragment.

tected compound becomes a screening hit, the amide moiety can becomes a useful chemistry vector while preserving the possible binding interactions from the amide itself. Note that the choice of "protecting" groups for reactive monomers as fragments must be carefully selected so as to limit MW of the resulting product to stay within the desirable MW range for fragments.

Of course protecting the reactive functionality of a compound alters its binding characteristic. Therefore, selecting the reactive monomer types to be included in a fragment collection and selecting its protecting group must be carefully considered and chosen wisely.

3. Building a Fragment Collection

There are numerous ways to build a fragment collection. Consideration of factors described in the earlier sections can be used to build a generic collection or to build a collection tailored to a specific screening method or a particular target. One of the more popular and rational approaches to assemble a fragment collection is to build the collection based on the screening technique. Among the more popular fragment screening methods are NMR techniques, beginning with the "SAR by NMR" method pioneered by Fesik et al. at Abbott (5). Other methods using NMR include saturation transfer difference (STD) (15, 16) and water-LOGSY (17). There have been several efforts to assemble fragment collection based on NMR screening techniques (11, 18).

In the following sections, we will describe in detail efforts to build fragment collections and the processes involved in their creation. Two such efforts were performed at a major pharmaceutical company (Pfizer) while a third took place at a biotech company (Vernalis).

3.1. Pfizer Fragment Screening Collections

At the Pfizer research site in La Jolla, the preferred primary screening method for fragments is the STD NMR technique, although other research sites have employed other screening methods. Prior to 2006, Pfizer had a legacy fragment collection of ~5 K compounds, but this collection had two major drawbacks: many of the fragments lacked chemistry handles to facilitate hit follow-up efforts and almost all of the fragments were purchased as screening compounds and therefore were of insufficient quantity for chemistry efforts. Therefore, it was decided to build a fragment collection for the La Jolla NMR screening campaigns. The goal was to create a collection optimized for NMR screening while being chemically attractive for hit follow-up efforts.

The approach taken to achieve these goals was to first select a set of novel reagents, then react these compounds with simple reagents to "cap" the reactive functionalities. Virtual products for the selected compounds were created and are then passed through an in silico filtering process. Finally, the filtered libraries were synthesized via combinatorial libraries.

Selection of the reagents was based on the Pfizer internal compound collection which allowed speedy acquisition of any selected compound. A set of primary amines, secondary amines, and carboxylic acids which were *not* commercially available were chosen for consideration. These acids and amines were designed by medicinal chemists via a Pfizer internal screening file enrichment effort to be novel and diverse, and more importantly, were not part of any existing Pfizer fragment collection. The MW cutoff was set at an upper limit of 200, with most of the amines having a MW range of 100–150. All compounds in consideration had at least 5 g quantity available to ensure that future follow-up activities were enabled.

The combinatorial reactions chosen for the novel amines were amide bond formation and sulfonamide formation. The novel carboxylic acids were derivatized to simple amides. For the amine reactions, we chose two simple carboxylic acids (propionic acid and benzoic acid) and two simple sulfonyl chlorides (methylsulfonyl chloride and benzenesulfonyl chloride) as the "capping groups." Propyl amine and benzylamine were chosen as the capping groups to react with the novel carboxylic acids. Because only one reactant will be variable, these combinatorial libraries were essentially $1 \times N$ libraries, where the one reactant was a simple reactant and the N component is the novel amines or acids.

Next, we used an in-house library design software (*see* details in Chapter 15) to enumerate the virtual libraries and then calculated various physical properties. Products were removed from consideration if MW is > 300, number of rotatable bonds > 3, and ClogP > 3. For solubility, two in-house model calculations were applied as filters: turbidimetric $\geq$ 10 mg/mL and thermodynamic solubility >100 μM. The resulting cherry-picked library was then reviewed by NMR spectroscopists to remove compounds with possible artifacts, likely to be insoluble, or likely to be false positive. These included some conjugated systems and compounds with likelihood of indistinct NMR spectra.

Approximately 1,200 amines and 300 carboxylic acids were selected for inclusion in the fragment libraries, from which approximately 20 fragment libraries were synthesized. These libraries yielded ~2,000 products with sufficient purity (>95%) and quantity (1.2 mL of 30 μM solution), and the product structures were confirmed via 1D NMR. This fragment collection became known as the "NMR Combicores" to denote their purpose and their combichem origin. It was distributed across several major Pfizer research sites and used in multiple fragment screens.

One of the lessons learned from previous fragment screening collections is that fragments which are enabled for chemical expansion was a key factor in engaging chemists in performing hit optimization. This was addressed in the Combicores collection described above via "capped" functional groups. However, Combicores is a specialized collection since it was designed specifically for NMR screening. To build a more generic fragment collection to accommodate protein target screening requirements such as reagent stability, sensitivity of screening methods, and druggability (19) of binding sites, the Pfizer Global Fragment Initiative (GFI) (20) was initiated with the goal of assembling a fragment collection suitable for several screening methods, including

NMR techniques, SPR, high-concentration bioassays, and fragment crystallography. The assembly process for the collection involved several computational filters, chemical complexity analysis, diversity analysis, and manual review by chemists to ensure chemical attractiveness for follow-up. Details of the complete process will be published elsewhere (20). The analysis of the screening results presented early in the 2009 fall ACS meeting showed that fragment screening of GFI offered consistent high hit rate across protein classes (21).

Figure 11.2 shows a typical screening and hit-to-lead cascade utilized in an FBLD campaign at Pfizer. In the first stage, we perform a primary screen (STD) along with a confirmation screen (HCS, MS, or SPR). For a fragment to be considered a primary NMR hit, the STD values must be >10 but less than 40, and MS confirms that at least one copy of the fragment is bound with the protein (binding = YES). In the biophysical confirmation step, we conduct competitive binding studies with MS or a biochemical assay to see if the compound displaces known active site binders in order to confirm that the fragment is bound at the active site. For this purpose we also attempt crystallography on the more active hits when the protein target is amenable. The biochemical assay results also allow us to calculate ligand efficiencies (LE) (22) of the fragment hits. In the second stage, the hits with LE ≥ 0.3 and are chemically attractive are selected to be progressed for hit follow-up activities. These activities include database mining for similar analogs which are submitted for biochemical screening, synthesizing of analogs by chemists, designing core-based fragments to enable further elaboration of the hits, and designing structural-based targeted library based on top selected hits. This is an interactive and iterative process involving a project team

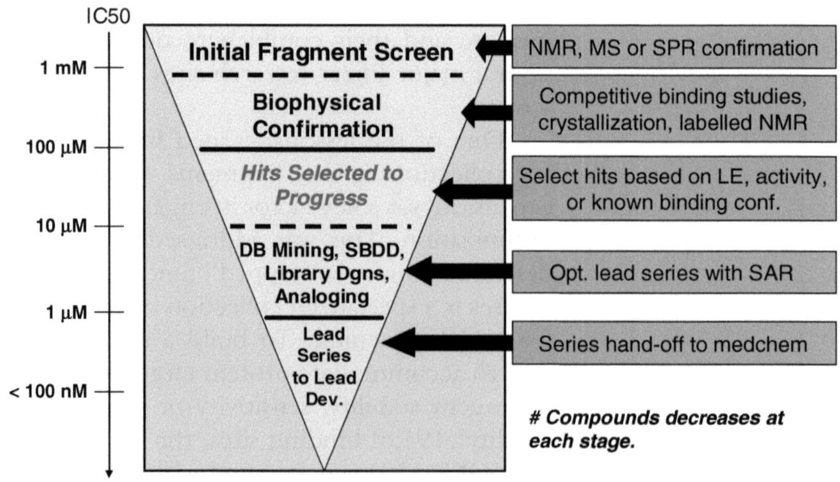

Fig. 11.2. Typical FBLD screening cascade.

consisting chemists, spectroscopists, biologists, and computational chemists. Once lead series are identified with good activity and SAR, they are passed to the lead optimization stage.

3.2. Vernalis NMR Screening Collection

Vernalis Ltd., a UK-based biotech company, has a drug discovery platform based on fragment screening using NMR techniques. The Vernalis FBLD strategy is called SeeDs (Selection of Experimentally Exploitable Drug Startpoints) (18) and an integral part of this strategy involves the creation of their fragment libraries. The following section will describe this effort and how it was applied to iteratively create four separate fragment libraries.

The compound collections used to select the desired fragments were the 2001 version of ACD (Available Chemicals Directory) which contains ~267 K compounds, and a database of ~1.6 M compounds from 23 chemical vendors. Removing duplicates from both databases yields 1.79 M compounds.

Since solubility is an important factor to consider when selecting fragment for NMR screening, the solubility calculation was done using a cross-validated PLS regression algorithm fitting 49 2D descriptors trained with a 3,041 molecules training set (23). This model was shown to be predictive within 1 log unit for a small test set, which is on par with experimental error.

Hit follow-up consideration was accounted for via two filters to remove undesirable functional groups and to include molecules with a desirable functional group that would act as chemistry handles to enhance compound elaboration. These filters were derived from extended discussion with medicinal chemists. A total of 12 undesirable functional group sets were created and collectively used as a negative filter:

- Four aliphatic carbons except if also contains X–C–C–C–X, X–C–C–X, X–C–X with X = O or N
- Any atom different from H, C, N, O, F, Cl, S
- –SH, S–S, O–O, S–Cl, N–halogen
- Sugars
- Conjugated system: R=C–C=O, with R different from O, N, or S or aromatic ring
- (C=O)–halogen, O–(C=O)–halogen, SO_2–halogen,
- N=C=O, N=C=S, N–C(=S)–N
- Acyclic C(=O)–S, acyclic C(=S)–O, acyclic N=C=N
- Anhydride, aziridine, epoxide, *ortho* ester, nitroso
- Quaternary amines, methylene, isonitrile
- Acetals, thioacetal, N–C–O acetals
- Nitro group, >1 chlorine atom

For the desired functionalities, a set of 11 functional groups was used to filter against the compound databases, and all molecules that did not contain at least one of these desired groups were removed. These functional groups are: R–CO_2Me, R–CO_2H, R–NHMe/R–NH$(Me)_2$, R–NH_2, R–CONHMe/R–CON$(ME)_2$, R–$CONH_2$, R–SO_2NHMe/R–SO_2N$(Me)_2$, R–SO_2NH_2, R–OMe, R–OH, and R–SMe. In addition, all compounds also must contain at least one ring system or be removed from consideration. This filtering process was done using 2D SMILES strings.

Molecular complexity is defined by the number of pharmacophoric triangles, where compounds with more triangles are more deemed complex. This is done by first identifying the pharmacophoric elements contained within each molecule, and then a triangle is identified by three features and the shortest bond path between each pair of features (**Fig. 11.3**). There are eight pharmacophoric features used: H-bond donor, H-bond acceptor, polar, hydrophobic, pi donor, pi acceptor, pi polar, and pi hydrophobic. The Molecular Operating Environment (MOE) (24) was used to calculate the pharmacophoric features as well as the pharmacophoric triangles. Each compound is then assigned a set of integers representing the pharmacophoric triangles it contains, which becomes its fingerprint. For a given collection of compounds, these fingerprints can be used to identify which features are present and which ones are missing, and the collective fingerprints becomes a measure of the diversity for the compound collection.

The last filtering step involves experimental quality control, which validates whether a given compound is soluble to 2 mM in buffer solution, has a consistent NMR spectrum with its structure, has 95% or greater purity, and is both stable and soluble for 24 h in buffer solution. In addition, a water-LOGSY spectrum is taken and compounds with positive results are considered to have

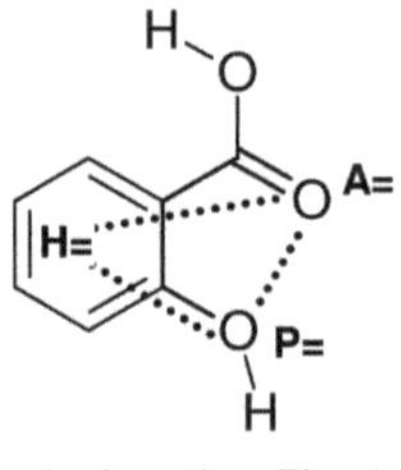

Fig. 11.3. Pharmacophoric triangle detection. The dotted lines define a triangle comprising three features: piHydrophobic (H=, centroid of benzene), piAcceptor (A=, oxygen of carboxylic acid), and piPolar (P=, oxygen of hydroxyl), and the shortest bond path between each pair of features is 2 (A= to H=), 1 (P= to H=), and 4 (A= to P=). Reprinted ("adapted" or "in part") with permission from *Journal of Chemical Information and Modeling*. Copyright 2008 American Chemical Society.

self-association, which can lead to false results in the NMR screen and are thus removed from consideration.

Four fragment libraries were generated with different combination of the compound databases and filtering processes. The first library (SeeDs-1) was designed from a relatively small database of ~87 K compounds comprising compounds available from the Aldrich and Maybridge companies. These compounds were first passed through a MW criterion (110 ≤ MW ≤ 250; 350 for sulfonamides) and then filtered to remove compounds containing a metal, five continuous carbon methylene units, or a reactive functional group. The resulting 7,545 compounds were visually inspected by a medicinal chemist who selected a set mostly based on chemistry follow-up attractiveness. Note that this visual filtering process was captured and became the undesired and desired functional groups filtering described above. Of the 1,078 fragments that passed visual filtering and ordered, 723 passed the QC filtering. The experimental solubility results for these 723 compounds were then used as a further test set for the aqueous solubility prediction model, which showed an 88% correlation for predicting solubility for both the soluble (636 out of 723 correctly predicted to be soluble) and insoluble (84 out of 95 correctly predicted to be insoluble) compounds.

SeeDs-2 library was generated from their in-house database called rCat of 1,622,763 unique chemical compounds assembled from 23 suppliers (25). The filtering cascade began with MW (same as SeeDs-1), then the functional groups and solubility filters which resulted in ~43 K unique compounds (no overlap with SeeDs-1). These were then clustered by 2D, 3-point pharmacophoric features to provide ~3 K clusters, and the centroids of each cluster was submitted for chemist review. Of the 395 selected compounds that were ordered, 357 passed QC to become the SeeDs-2 fragment library.

SeeDs-3 library was designed as a kinase-specific fragment collection. The filtering process began with selecting compounds which had the potential to bind to the ATP-binding site of protein kinases. This was achieved via pharmacophore queries to match the donor–acceptor–donor motif present in the ATP-binding site, which was used as a first pass filter. The compounds matching these queries were then filtered for MW and predicted solubility, as well as the wanted and unwanted functional groups filtering. In the end, only 204 compounds were selected for purchasing, and 174 passed QC filtering.

The final library was designed with the purpose of adding incremental diversity to the first three fragment libraries. The main filtering criterion was novel pharmacophoric triangles not found in the first three libraries. After clustering and visual inspection from a panel of medicinal chemists, only 65 compounds were purchased and 61 compounds passed QC.

Combining all four SeeDs libraries resulted in 1,315 fragments for the collection. Various properties were calculated, analyzed, and compared with a drug-like reference set created from the WDI and a binding reference set created from PDB. These results can be found in the key reference (18).

4. Screening Results

There are various methods to conduct a fragment screening campaign. The most commonly utilized methods include various NMR techniques, mass spectrometry, SPR, and biochemical screens. X-ray crystallization is a preferred method since it provides a binding conformation, but can only be used when the target protein is well behaved. Various calorimetry techniques have also been used for fragment screening, but these have been less commonly utilized. The merits of each method have been discussed in the literature (26) and will not be outlined here.

An analysis of screening hits based on 12 NMR screens (**Table 11.1**) for a range of protein targets conducted over an 8-year period at Vernalis was performed (27). Three main aspects of the analysis were (1) the relationship of the fragment hit rates to the druggability of the target; (2) comparison of hits, nonhits to the entire fragment library; and (3) the specificity and ligand complexity of the fragment hits.

Composition of the Vernalis fragment library evolved over the course of 4 years through changes in what was synthesized in-house, available commercially, and removed from the collection through quality control process. Although the number of compounds remains roughly the same, the content has changed dramatically, which makes the analysis quite challenging and interesting as well.

4.1. Fragment Screening Campaigns

As mentioned above, all data in the analysis are from fragment screens using NMR spectroscopy to detect fragment binding (28). Three NMR spectra (STD, water-LOGSY, and CPMG (29)) were recorded separately for the ligand, ligand + protein, and ligand + protein + known binding ligand (for competitive binding). This approach can identify hits which bind in the same site as the known binding ligand used in the screens. The resulting spectra are then inspected and a hit is defined as a fragment which binds to and can be displaced by the known binder from the protein. Based on the screening results from the three NMR experiments, hits are classified in three categories: Class 1 hit is defined as a fragment which shows evidence for binding in all three NMR experiments, Class 2 hit shows changes in two experiments, and a Class 3 hit in only one experiment.

4.2. Fragment Hit Rates and Druggability Index

One of the interesting observations is that the experimentally observed hit rate for screening fragments can be related to a computationally defined druggability index for the target, which provided an interesting side usage of fragment screening.

Due to the evolutionary nature of the Vernalis fragment collections and the fact that various screens were performed over a period of several years, it is difficult and perhaps unreasonable to directly compare the hit rates across multiple screens. However, it is still helpful to notice that reasonable hit rates (compared to HTS) are obtained across a diverse group of targets (**Table 11.1**).

A very intriguing aspect of the analysis is assessing and ranking the target druggability based on the NMR screenings. This approach was first reported by Abbott in 2005 as a strategy to quickly evaluate protein druggability by screening chemical libraries with 2D heteronuclear-NMR (30). They observed that NMR hit rates were shown to be correlated with a number of surface properties calculated from the binding site. Inspired by the Abbott findings, Vernalis took a similar approach using the druggability score (DScore) calculated by SiteMap (31) from Schrodinger. What they found was that they were able to reach a similar conclusion correlating fragment binding hit rate by 1D NMR with protein druggability. Three aspects of the binding pocket are considered as major contributions to DScore: pocket size, degree of enclosure, and hydrophilicity. The results shown in **Fig. 11.4** indicated that if using a hit rate of 2% as a cutoff, all targets which yielded high hit rates (<2%) have a DScore greater than 0.8, with the only outlier being HSP70. Their internal data revealed significant plasticity of the HSP70 ATP-binding site upon binding to different ligands, which currently cannot be captured by the SiteMap calculation. Excluding HSP70, SiteMap DScore appears to be a good indicator for the NMR hit rate one should expect for a target. As the authors pointed out, few data points are available in the DScore range between 0.6 and 0.8, but they are optimistic this gap will be filled from future screens so that a more complete evaluation of DScore can be achieved.

4.3. Comparison of Hits, Nonhits, and the Entire Fragment Library

The importance of physicochemical properties and structural diversity to the assembly of a successful fragment collection has been described in earlier sections. In this section, we will focus on an interesting analysis done on the distribution of the physicochemical properties and 3D pharmacophore triplet in three groups of molecules: hits, nonhits, and the whole library. Hits are defined as fragments which have been identified as

Table 11.1
SeeDs screening hit rates for 12 protein targets. With kind permission from Springer Science+Business Media: *Journal of Computer-Aided Molecular Design*, 23, 2009, 603, I-Jen Chen and Roderick E. Hubbard, Table 4

Protein	Number of hits			Library size	Class 1 hit rate		High-affinity ligands[d]
	Total[a]	Class 1[b]	Class 1 series[c]		%	Category	
AK	15	11	10	308	3.6	High	Yes
CDK2	109	40	35	1,250	3.1	High	Yes
DNA gyrase	54	44	39	855	4.9	High	Yes
FAAH	81	63	51	868	7.3	High	Yes
HSP70	38	6	5	1,351	0.4	Low	No
HSP90	82	60	42	1,351	4.4	High	Yes
JNK3	101	55	53	1,351	4.0	High	Yes
PDPK1	119	58	54	1,260	4.5	High	Yes
PIN-1	13	5	4	1,351	0.4	Low	No
PPI-1	40	34	23	1,064	3.2	High	Yes
PPI-2	52	24	20	1,068	2.2	High	Yes
PPI-3	39	10	9	1,351	0.7	Low	No

[a]Total number of fragments identified by at least one NMR experiment to interfere with the binding of known competitor compound.
[b]Number of fragments identified by all three NMR experiments (STD, water-LOGSY, and CPMG) to interfere with the binding of known competitor compound.
[c]Total number of unique chemical series suggested by the clustering results of Class 1 fragment hits with a Tanimoto coefficient of 0.70 and MACCS keys.
[d]Reported affinities <300 nM. Please refer to the paper for references.

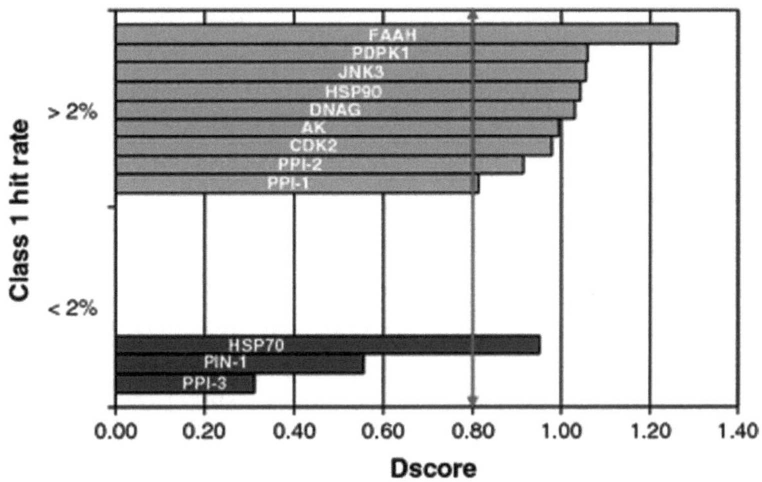

Fig. 11.4. Targets with observed high (>2%, *light bars*) and low (<2%, *darker bars*) Class 1 hit rates compared to the druggability score (Dscore) calculated by SiteMap. The red arrow indicates the minimum Dscore for targets yielding high hit rates for the current data set. With kind permission from Springer Science+Business Media: *Journal of Computer-Aided Molecular Design*, 23, 2009, 603, I-Jen Chen and Roderick E. Hubbard, Fig. 4.

competitive binders by at least one NMR experiment in any one of the 12 screens. The nonhits are the compounds which are not recognized as hits by any of the 12 screens. The sum of hits (29%) and nonhits (71%) equal the entire library.

As seen in **Fig. 11.5**, among the five properties plotted, the distribution of MW, NRot (number of rotatable bonds), and number of pharmacophore triangles show no clear differences among hits and nonhits, while SlogP which represents ligand lipophilicity and number of rings showed clear separation between the hits and nonhits. The more hydrophobic nature of the hits in comparison to the nonhits is in good agreement with a general observation that binding is largely driven by hydrophobicity (32). **Figure 11.5E** shows that there are more two-member rings in the hits than in the nonhits.

4.4. Specificity and Ligand Complexity of the Fragment Hits

The screening results also clearly indicate that small fragments can be specific binders, even for proteins within the same family.

Given their relatively small size, there are natural concerns regarding nonspecific binding of fragments. Based on the Vernalis study, 62% of the fragments were competitive binders with just one target and another 24% were hits for just two targets. This study shows that most fragments are in actuality quite target specific. The pie chart in **Fig. 11.6** focuses on the hits from three kinase screenings, PDPK1, CDK2, and JNK3. It shows that at least 52% of the fragment hits are unique to one kinase, and only 11% of the hits are shared among all three of proteins.

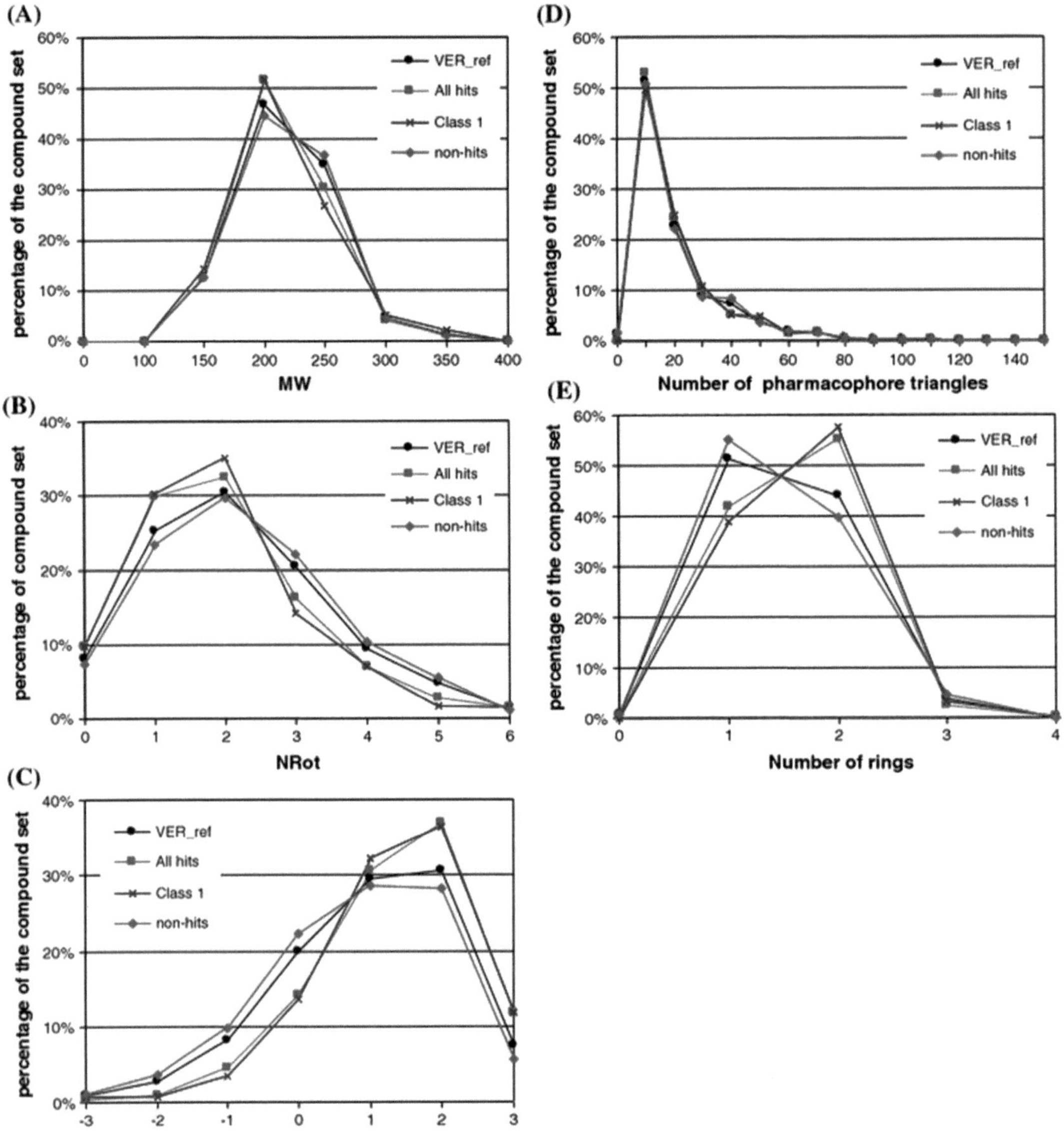

Fig. 11.5. Distribution plots of (**a**) molecular weight (MW), (**b**) number of rotatable bonds (NRot), (**c**) SlogP, (**d**) number of pharmacophore (ph4) triangles, and (**e**) number of rings for the whole library (VER_ref), all hits (Class 1–3), Class 1 hits, and nonhits. With kind permission from Springer Science+Business Media: *Journal of Computer-Aided Molecular Design*, 23, 2009, 603, I-Jen Chen and Roderick E. Hubbard, Fig. 6.

Ligand complexity can be represented by the number of pharmacophore triangles in fragment structures. **Figure 11.7** plots the averaged pharmacophore complexity of both the hits and the Class 1 hits (all three NMR spectra confirm binding) for each target. It would appear that the level of complexity required for a fragment to be detected in binding varies from target to target. HSP70 appears to be the most demanding target as it requires the

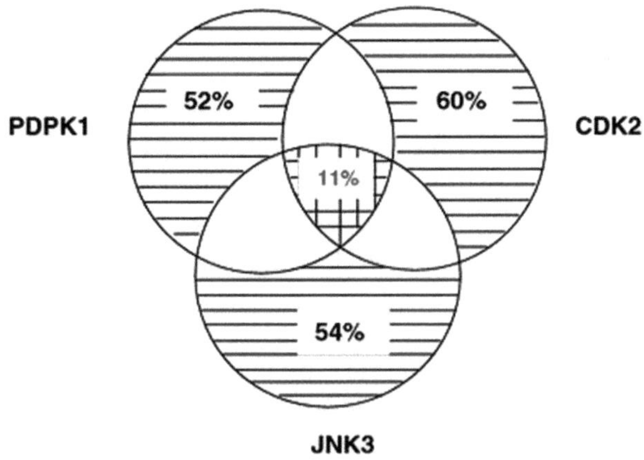

Fig. 11.6. Overlap of kinase fragment hits. The horizontal lines indicate the portion of unique fragment hits to each kinase. The crossed area (11%) is the portion of common hits to all three kinases. With kind permission from Springer Science+Business Media: *Journal of Computer-Aided Molecular Design*, 23, 2009, 603, I-Jen Chen and Roderick E. Hubbard, Fig. 9.

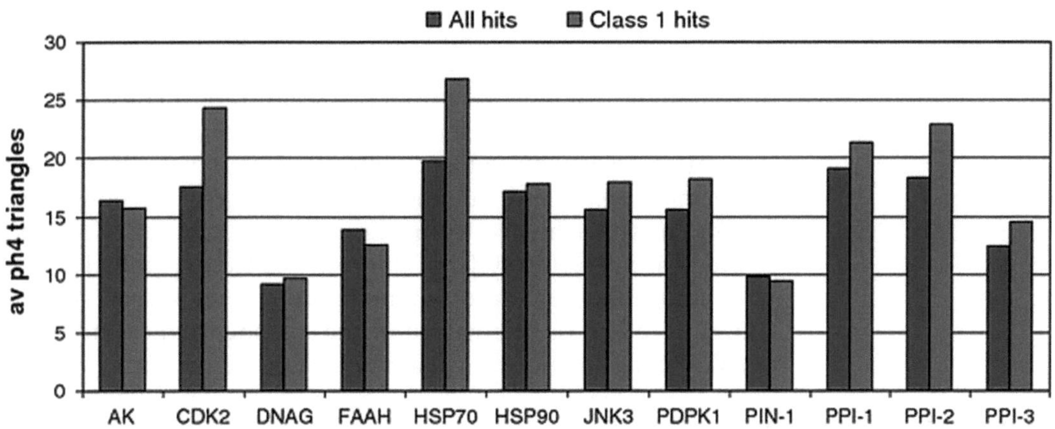

Fig. 11.7. Pharmacophore complexity observed for all fragment hits and Class 1 hits for 12 protein targets. With kind permission from Springer Science+Business Media: *Journal of Computer-Aided Molecular Design*, 23, 2009, 603, I-Jen Chen and Roderick E. Hubbard, Fig. 8.

most complex fragments (20 and 27 triangles for all hits and Class 1 hits) among all targets studied. Perhaps HSP70's hit rate was among the lowest because fewer fragments have the complexity required for HSP70 binding. On the contrary, both DNA gyrase and FAAH showed low average ligand complexity, and they are indeed top two targets among the 12 screens with highest Class 1 hit rates, 4.9 and 7.3%, respectively.

5. Overview of Published Fragment Libraries

Over the past several years, descriptions of fragment collections have been published in journals as well as book chapters. In two recent reviews, a chapter from Evotec (33) compared several collections based their origins, while a journal article from Leiden University (26) summarized fragment collections based on the intended screening methods. We would like to present an overview by blending information from both reviews together to provide a more complete and updated picture (*see* **Table 11.2** for details).

5.1. Origins and Library Size

Table 11.2 is divided into three sections: the first section is a group of large or mid-sized pharma/biotech; the second section contains all small biotechs specializing in FBDD; and the last section contains the companies offering fragment screening as a service. This table demonstrates that FBDD has been adopted as a drug discovery platform throughout the drug discovery industry regardless of company size or screening technology, and also illustrates the fact that FBDD has now become a CRO service. It is also interesting to see that within each group, the fragment collection sizes vary from several hundreds to tens of thousands of compounds. While there does not seem to be a consensus for the collection size, most fragment libraries are assembled based on some fundamental guiding principles which include sampling of chemical space and alignment with a chosen screening technology. It is worthwhile to mention that both AstraZeneca and Evotec have a relatively large fragment collection (>20 K), but both companies stated they intend to screen only subsets of their collection based on the nature of the intended target and some practical considerations, such as the cost of the reagent.

5.2. Physical Properties

Most of the fragment libraries are designed with physical properties within the rule-of-3 constraints (14), which were originally proposed for molecules used for high-throughput fragment crystallography. For solubility, since there are no reliable methods to predict aqueous solubility, clogP is often used as a guide. In the case of AstraZeneca, a clogP value above 2.2 warrants enough risk for a neutral compound that its solubility will be experimentally determined (8).

5.3. Screening Technologies

In recent years there have been encouraging advances in screening technologies suitable for fragment screening. Even for an established method such as NMR, new techniques have been developed and put into practice at various companies that have expanded the scope of targets for FBDD. For example,

Table 11.2
Overview of some key physical properties for selected fragment libraries and their associated screening methods

Company	Size	MW	ClogP	H-bond donors	H-bond acceptors	Number of rotatable bonds	Polar surface area (Å)	Screening technologies	References
Abbott	~10,000	220	1.5	NA	NA	NA	NA	SAR by NMR	(26, 33)
Pfizer[a]	2,792	205	1.0	1	3	2	56.9	Multiple[b]	(20, 21)
Roche	~2,000	≤300	≤3	≤3	NA	≤3	≤60	SPR (target immobilized)	(26, 33)
AstraZeneca	1,200	NA	1.5	NA	NA	NA	NA	NMR	(8)
AstraZeneca	20,342	NA	1.6	NA	NA	NA	NA	Multiple[b]	(8)
Vertex	132	194	−2.2 to 5.5	NA	NA	NA	NA	NMR	(11, 12)
Vernalis	~1,400	190	0.3	1.9	3.2	2.3	NA	NMR	(18, 27)
Astex	850	≤300	≤3	≤3	≤3	≤3	≤60	X-ray	(34)
Plexxikon	~20,000	127 to 350	NA	NA	NA	NA	NA	X-ray	(35)
SGX[c]	1,000	174	≤2	≤3	≤3	≤3	NA	X-ray	(36)
Sunesis	7,000	<200	NA	NA	NA	NA	NA	Tether	(37)
Graffinity	10,000	≤300	≤3	≤3	NA	NA	NA	SPR (ligand immobilized)	(26, 33)
Evotec	20,063	276	2.2	≤4	≤7	≤5	NA	NMR	(33)
Evotec	21,869	247	1.6	1 to 3	≤4	≤5	70	Biochemical assay	(33)
ZoBio/Pyxis Discovery	2,000	218	1.2	1.3	2.7	1.9	52.6	NMR (target immobilized)	(26)

[a] All single values in this table for properties are mean, except the values for the Pfizer collection which are median.
[b] Multiple means more than one screening technology, including NMR, SPR, biochemical assay, and X-ray.
[c] The property values reported for SGX applies to ~90% of the molecules for the SGX collection.

membrane-bound protein can be made suitable for fragment screening using the target-immobilized NMR screening (TINs) (26). For an excellent comparison of all fragment screening technologies, please refer to the review from Siegal et al. (26).

6. Conclusion

The composition of a fragment collection can have a profound effect on the success of an FBLD campaign. Consideration of the screening method and ease of chemistry follow-up are two of the more important factors in creating a fragment collection. It has been shown that by using a combination of computational analysis and human expertise, a fragment collection can be created to accommodate a single method or several screening methods without being target or protein family specific. Further, a carefully designed fragment collection can result in high hit rates across a variety of targets to produce hits with novelty and good ligand efficiency, thereby accelerating the lead discovery process.

Acknowledgments

We would like to thank Drs. Ben Burke and Zhongxiang (Joe) Zhou for their valuable comments and insights throughout the preparation of this manuscript.

References

1. Congreve, M., Chessari, C., Tisi, D., Woodhead, A. (2008) Recent advances in fragment-based drug discovery. *J Med Chem* **51**, 3661–3680.
2. Hesterkamp, T., Whittaker, M. (2008) Fragment-based activity space: smaller is better. *Curr Opin Chem Biol* **12**, 260–268.
3. Hajduk, P. J., Greer, J. (2007) A decade of fragment-based drug design: strategic advances and lessons learned. *Nat Rev Drug Discov* **6**, 211–219.
4. Albert, J. S., Blomberg, N., Breeze, A. L., Brown, A. J. H., Burrows, J. N., Edwards, P. D., Folmer, R. H. A., Geschwindner, S., Griffen, E. J., Kenny, P. W., Nowak, T., Olsson, L. -L., Sanganee, H., Shapiro, A. B. (2007) An integrated approach to fragment-based lead generation philosophy, strategy and case studies from AstraZeneca's drug discovery programmes. *Curr Top Med Chem* **7**, 1600–1629.
5. Shuker, S. B., Hajduk, P. J., Meadows, R. P., Fesik, S. W. (1996) Discovering high-affinity ligands for proteins: SAR by NMR. *Science* **274**, 1531–1534.
6. de Kloe, G. E., Bailey, D., Leurs, R., de Esch, I. J. P. (2009) Transforming fragments into candidates: small becomes big in medicinal chemistry. *Drug Discovery Today* **14**, 630–646.
7. Hann, M. M., Oprea, T. I. (2004) Pursuing the lead-likeness concept in pharma-

ceutical research. *Curr Opin Chem Biol*, **8**, 255–263
8. Blomberg, N., Cosgrove, D. A., Kenny, P. W., Kolmodin, K. (2009) Design of compound libraries for fragment screening. *J Comput Aided Mol Des* **23**, 513–525.
9. Barker, J., Courtney, S., Hesterkamp, T., Ullmann, D., Whittaker, M. (2005) Fragment screening by biochemical assay. *Exp Opin Drug Discov* **1**, 225–236.
10. Schuffenhauer, A., Ruedisser, S., Marzinzik, A., Jahnke, W., Selzer, P., Jacoby, E. (2005) Library design for fragment based screening. *Curr Top Med Chem* **5**, 751–762.
11. Fejzo, J., Lepre, C. A., Peng, J. W., Bemis, G. W., Ajay, Murcko, M. A., Moore, J. M. (1999) The SHAPES strategy: an NMR-based approach for lead generation in drug discovery. *Chem Biol* **6**, 755–769.
12. Lepre, C. A. (2001) Library design for NMR-based screening. *Drug Discov Today* **6**, 133–140.
13. Taskinen, J. Norinder, U. (2006) In silico predictions of solubility, *Comprehen Med Chem II,* edited by Taylor, J. B., Triggle, D. J. **5**, 627–648.
14. Congreve, M., Carr, R., Murray, C., Jhoti, H. (2003) A 'rule of three' for fragment-based lead discovery. *Drug Discov Today* **8**, 876–877.
15. Mayer, M., Meyer, B. (1999) Characterization of ligand binding by saturation transfer difference NMR spectroscopy. *Angew Chem Int Ed* **38**, 1784–1788.
16. Wang, Y., Liu, D., Wyss, D. F. (2004) Competition STD NMR for the detection of high-affinity ligands and NMR-based screening. *Magn Reson Chem* **42**, 485–489.
17. Dalvit, C., Pevarello, P., Tato, M., Vulpetti, A., Sundstrom, M. (2000) Identification of compounds with binding affinity to proteins via magnetization transfer from bulk water. *J Biomol NMR* **18**, 65–68.
18. Baurin, N., Aboul-Ela, F., Barril, X., Davis, B., Drysdale, M., Dymock, B., Finch, H., Fromont, C., Richardson, C., Simmonite, H., Hubbard, R. E. (2004) Design and characterization of libraries of molecular fragments for use in NMR screening against protein targets. *J Chem Inf Comput Sci* **44**, 2157–2166.
19. Hajduk, P. J., Huth, J., Tse, C. (2005) Predicting protein druggability. *Drug Discov Today* **10**, 1675–1682.
20. Lau, W. F., Hepworth, D., Magee, T. V., Du, J., Bakken, G. A., Miller, M. D., Hendsch, Z. S., Thanabal, V., Kolodziej, S. A., Xing, L., Hu, Q., Narasimhan, L. S., Love, R., Charlton, M. E., Hughes, S., Van Hoorn, W., Mills, J. E., Withka, J. M. (2010) Design of a multi-purpose fragment screening library using molecular complexity and orthogonal diversity metrics. *J Comput-Aided Mol Des. Manuscript in preparation.*
21. Hu, Q., Yan, J., Withka, J. M., Sahasrabudhe, P., Moore, C., Na, J., Narasimhan, L. S. (2009) Computational analysis on NMR screenings of the Pfizer Fragment Initiative collection. *238th ACS National Meeting, Washington, DC, United States.*
22. Hopkins, A. L., Groom, C. R., Alex, A. (2004) A useful metric for lead selection. *Drug Disc Today* **9**, 430–431.
23. Huuskonen, J., Rantanen, J., Livingstone, D. (2000) Prediction of aqueous solubility for a diverse set of organic compounds based on atom-type electrotopological state indices. *Eur J Med Chem* **35**, 1081–1088.
24. Chemical Computing Group Inc., Montreal, H3A 2R7 Canada.
25. Baurin, N., Baker, R., Richardson, C., Chen, I., Foloppe, N., Potter, A., Jordan, A., Roughley, S., Parratt, M., Greaney, P., Morley, D., Hubbard, R. E. (2004) Drug-like annotation and duplicate analysis of a 23-supplier chemical database totalling 2.7 million compounds. *J Chem Inf Comput Sci* **44**, 643–651.
26. Siegal, G., Ab, E., Schultz, J. (2007) Integration of fragment screening and library design. *Drug Discov Today* **12**, 1032–1039
27. Chen, I., Hubbard, R. E. (2009) Lessons for fragment library design: analysis of output from multiple screening campaigns. *J Comput Aided Mol Des* **23**, 603–620.
28. Hubbard, R. E., Davis, B., Chen, I., Drysdale, M. (2007) The SeeDs approach: integrating fragments into drug discovery. *Curr Top Med Chem* 7, 1568–1581.
29. Meiboom, S., Gill, D. (1958) Modified spin-echo method for measuring nuclear relaxation times. *Rev Sci Instrum* **29**, 688–691.
30. Hajduk, P. J., Huth, J. R., Fesik, S. (2005) Druggability indices for protein targets. *J Med Chem* **48**, 2518–2525.
31. Halgren, T. A. (2009) Identifying and characterizing binding sites and assessing druggability. *J Chem Inf Model* **49**, 377–389.
32. Ruppert, J., Welch, W., Jain, A. N. (1997) Automatic identification and representation of protein binding sites for molecular docking. *Prot Sci* **6**, 524–533.
33. Brewer, M., Ichihara, O., Kirchhoff, C., Schade, M., Whittaker, M. (2008) Assembling a fragment library. *Fragment-Based Drug Discovery: A Practical Approach,* in (Zartler, E., Shapiro, M. J. eds.), pp. 39–62.

34. Hartshorn, M. J., Murray, C. W., Cleasby, A., Frederickson, M., Tickle, I. J., Jhoti, H. (2005) Fragment-based lead discovery using X-ray crystallography. *J Med Chem* **48**, 403–413.
35. Card, G.L., Blasdel, L., England, B. P., Zhang, C., Suzuki, Y., Gillette, S., Fong, D., Ibrahim, P. N., Artis, D. R., Bollag, G., Milburn, M. V., Kim, S., Schlessinger, J., Zhang, K. Y. J. (2005) A family of phosphodiesterase inhibitors discovered by cocrystallography and scaffold-based drug design. *Nat Biotechnol* **23**, 201–207.
36. Blaney, J., Nienaber, V., Burley, S. K. (2006) Fragment-based lead discovery and optimization using X-ray crystallography, computational chemistry, and highthroughput organic synthesis, *Fragment-Based Approaches in Drug Discovery,* in (Jahnke, W., Erlanson, D. A., Mannhold, R., Kubinyi, H., and Folkers, G., eds.), pp. 215–248.
37. Erlanson, D.A., Ballinger, M. D., Wells, J. A. (2006) Tethering, *Fragment-Based Approaches in Drug Discovery*, in (Jahnke, W., Erlanson, D. A., Mannhold, R., Kubinyi, H., and Folkers, G., eds.), pp. 285–312.

Chapter 12

Fragment-Based Drug Design

Eric Feyfant, Jason B. Cross, Kevin Paris, and Désirée H.H. Tsao

Abstract

Fragment-based drug design (FBDD), which is comprised of both fragment screening and the use of fragment hits to design leads, began more than 15 years ago and has been steadily gaining in popularity and utility. Its origin lies on the fact that the coverage of chemical space and the binding efficiency of hits are directly related to the size of the compounds screened. Nevertheless, FBDD still faces challenges, among them developing fragment screening libraries that ensure optimal coverage of chemical space, physical properties and chemical tractability. Fragment screening also requires sensitive assays, often biophysical in nature, to detect weak binders. In this chapter we will introduce the technologies used to address these challenges and outline the experimental advantages that make FBDD one of the most popular new hit-to-lead process.

Key words: Fragment-based drug design, fragment screening, ligand efficiency, NMR, X-ray crystallography.

1. Introduction

1.1. General Views

In recent decades, high-throughput screening (HTS) has become the most established method in the pharmaceutical industry for identifying potential lead compounds. Despite extensive effort in designing better and larger libraries for screening, the attrition rate of compounds entering clinical trials has continued to increase. The industry has attempted to address this issue by focusing on the improvement of compound properties, using schemes such as the Lipinski "Rule of 5" for oral drugs (1). A recent study (2) has shown that the decisive factor in designing successful clinical candidates is the quality of the initial HTS hit. Since the primary criterion for HTS hit selection has been potency, often at the expense of other important physicochemical

J.Z. Zhou (ed.), *Chemical Library Design*, Methods in Molecular Biology 685,
DOI 10.1007/978-1-60761-931-4_12, © Springer Science+Business Media, LLC 2011

properties, these initial leads are less likely to evolve to successful candidates.

To improve the lead identification process, the pharmaceutical industry has invested in new approaches. Fragment-based drug design (FBDD) is a relatively new technology that has shown remarkable potential in a short period of time. The foundation of FBDD was described by Jencks (3) and supported by Nakamura and Abeles (4), who showed that drug-like molecules can be regarded as the combination of several binding epitopes or fragments. FBDD presents two main advantages over HTS screening methods. The first is that fragment libraries can cover more of chemical space than HTS screening libraries. Even very large HTS screening decks, with over a million compounds, can explore only a vanishingly small fraction of the estimated 10^{60} compounds (5) with up to 30 heavy atoms. Since the number of possible compounds increases exponentially with molecular size, a library of only a few thousand fragments with fewer than 17 heavy atoms is capable of covering a larger fraction of chemical space. The second advantage lies with the concept of ligand efficiency, i.e., the average contribution of each atom of the molecule to the binding affinity. This concept was introduced by Kuntz et al. in 1999 (6) and later proposed as a criterion for hit selection by Hopkins (7). Interestingly, this model suggests that the level of energy contribution of each ligand atoms to the binding energy is inversely proportional to the molecular weight. Even for fragment binders whose affinity ranges from high micromolar to millimolar, whereas HTS hits range from high nanomolar to low micromolar, fragments are more efficient binders. On the other hand, weak binders are more difficult to detect by common high-throughput techniques like displacement assays. This chapter will illuminate the principles used to design fragment libraries for drug discovery and also describe two different screening methods, NMR and X-ray crystallography.

1.2. Designing Fragment Screening Libraries

Fragment libraries differ from drug-like and lead-like libraries primarily by having members with a significantly lower molecular weight (MW), typically in the 140–300 Da range. However, as fragment screening programs have matured over recent years, other key factors that help improve the success rates of these projects have been identified.

There has been much effort in identifying physical properties of ideal fragment libraries. Since fragments tend to be smaller than most drugs, clinical candidates, leads, or high-throughput screening (HTS) hits, they are able to make fewer interactions on average and tend to have lower affinities for their protein targets. Affinities are often in the high micromolar or low millimolar range, necessitating solubility at least to that degree and potentially higher depending on the assay protocol. Congreve and coworkers (8) introduced the "Rule of 3," which showed that diverse hits from fragment libraries tend to have the following physical properties: MW < 300 Da, clogP $\leq$ 3, H-donors $\leq$ 3, H-acceptors $\leq$ 3, polar surface area (PSA) $\leq$ 60 Å^2, and rotatable bonds $\leq$ 3. These properties not only partially address fragment solubility, but also help to ensure that compounds resulting from the elaboration of fragment hits have a higher likelihood obeying the "Rule of 5." (1)

In addition to physical properties, the chemical structure of fragments can play a role in their success as screening hits. Hajduk and coworkers showed that certain "privileged" scaffolds tend to show up repeatedly in successful fragment screening campaigns (9). Bemis and Murcko (10) analyzed known drugs in an effort to identify common features and scaffolds, which could be used to bias fragment libraries toward drug-like structures. An optimal molecular complexity was also discussed by Hann and coworkers (11), which would ensure that fragments have sufficient chemical features to keep from being overly promiscuous while at the same time not making them overly specific by introducing too many features. Having a good balance of chemical features also ensures that fragment hits will have a sufficient number of chemical handles to allow for synthetic chemistry follow-up and elaboration. Schuffenhauer and coworkers (12) took this idea one step further by suggesting that fragments should have chemical features that mask reactive functional groups, thereby simplifying synthesis of analogs. Having a diverse set of chemical structures in a fragment library is not only ideal for improving the odds of finding interesting hits that bind to the target, but can also assist in deconvolution of fragment structures if they are screened as mixtures. The Novartis approach for synthetic accessibility is very attractive and can be managed successfully for a smaller library. At Wyeth, we created a larger library of 10,000 compounds, since surface plasmon resonance would be the primary screening technique, which is able to support a higher throughput than NMR or X-ray crystallography. Due to the elaborate work in synthesizing a parent library and a screening library of such size, a more practical but less exact approach was taken, by choosing fragments from our corporate library and predicting synthetic accessibility as a function of number and diversity of substituent of the fragment core. The fragment core is always a ring system and was considered synthetically tractable if at least two distinct analogs existed in our compound catalogs (internal and external chemical collections).

Although many groups that use fragment screening develop their own internal libraries, many commercial vendors now offer fragment screening collections that are "Rule of 3" compliant and optimized for chemical diversity (13). Some of these libraries are even targeted at specific screening methodologies, such as brominated fragment libraries for use in X-ray crystallographic screening and fluorinated fragment libraries for use in NMR screening.

1.3. Screening Methods

The main challenge for any fragment screening method is the detection of weak binders. The methods commonly used to detect fragments can be broadly broken down into biophysical and functional assays.

Several biophysical methods are commonly employed in fragment-based screening, including NMR, X-ray crystallography, mass spectrometry (MS), and surface plasmon resonance (SPR). NMR is a popular screening method since it can detect weak binders and is also a flexible technique. Large fragment libraries can be screened using ligand-observe NMR experiments, such as saturation transfer difference (14) (STD) and WaterLOGSY (15). Structural information detailing the fragment's binding mode can be obtained through protein-observe experiments such as heteronuclear single quantum coherence (HSQC) (42). X-ray crystallography is used as a primary screening tool less often than NMR because it often requires higher protein quantity, has a slower throughput, and needs robust crystallization conditions. Crystallography is commonly used for the follow-up of fragments identified using other methods (16), since the structural information gained can be used to direct fragment elaboration. MS can also be used as a detection method for fragment binding (17), as it forms the basis of the tethering strategy used by Sunesis (18). In this case, a native or mutated cysteine residue adjacent to the binding site in the protein target is exposed to a library of disulfide fragments. A covalent bond between a fragment and the protein will then be formed if the fragment has sufficient affinity for the binding site. SPR is gaining popularity not only as a primary screening tool but also as an orthogonal confirmation for fragments identified by other assays (19). This method requires that either the protein or the fragment is immobilized onto the chip. Immobilization of the protein provides an excellent confirmation of fragment binding in cases where the protein is poorly behaved in the primary screening assay (i.e., aggregation). Immobilization of the fragment library leads to high-throughput fragment screening with a robust signal, but the protein must be soluble in the assay conditions.

High-concentration screening (HCS) assays tend to be used less frequently than biophysical methods, but can be employed to good effect due to their high-throughput capacity. These assays require biochemical function of the protein, up front knowledge of biochemical activity, and it can be difficult to reliably detect weak binders in the millimolar range. Fluorescence correlation techniques have also been employed successfully (20) and have the advantage that protein modulation (i.e., stabilization) can be detected, but this method requires large quantities of soluble protein.

Computational chemistry techniques can also be employed in fragment screening to great effect. Virtual screening using methods such as pharmacophore matching and molecular docking or determination of hot spots within the protein active site using computational tools such as HSITE (21), HIPPO (22), or MCSS (23, 24) can be used to bias fragment libraries for a particular protein target, as long as information detailing the protein structure or chemical structures of several ligands are known. Some fragment identification methods are based on the use of molecular dynamics, in particular grand canonical ensemble calculations (25). All computational methods require the use of an experimental screening assay to validate binding.

In this overview of fragment-based drug design, we will expand on the application of NMR and X-ray crystallography as the tools for biophysical screening. Common protocols will be presented, as well as the strengths and limitations of each approach.

1.4. From Fragment to Lead

Although fragments can be considered efficient binders, given their size, their binding potencies are still order(s) of magnitude too "weak" for them to be considered true leads. There are two ways of optimizing these ligands: either linking them when two fragments are bound in distinct, but not too distant, parts of the binding pocket or growing them using combinatorial chemistry with (or without) the support of computational methods. A thorough review of the use of computational approaches to the fragment-based *de novo* design problem can be found in reference (20).

In the growing approach an initial fragment is grown in an attempt to add interactions between the receptor and the ligand. There are numerous examples of software applying the growing approach, e.g., SPROUT (26, 27), LEGEND (28), LUDI (29), GROWMOL (30), SkelGen (31, 32), SMoG (33), and LigBuilder (34). In the linking approach, fragment building blocks are positioned in the target binding site as described above and then computationally connected to each other by linkers to yield a complete molecule that satisfies all of the key interactions. The tacit assumption here is that the binding affinity of fragments is additive, the loss in rigid body entropy on binding of all components of the molecule is small (35), and, moreover, that the affinity contribution from the linker is negligible or favorable. CONFIRM (36), LUDI (29), HOOK (37), PRO_LIGAND

(38), CAVEAT (39, 40) are examples of software using the linking approach. ReCore (41) and MOE Scaffold Replacement are capable of performing both fragment linking and building.

2. Materials

2.1. STD NMR Screening

1. Fragment stock solutions in DMSO, usually 80 mM or higher. Fragments should have good solubility in biological buffer, at least 200 μM.
2. Biological buffer where protein remains stable for a few hours. Typically neutral buffer such as HEPES, Tris with salt (10–200 mM). Protein target needs to be stable in the presence of DMSO (1–10%), usually ~5%.
3. NMR spectrometer operating at 500 MHz or higher in frequency, use of cryoprobe preferable for higher sensitivity.
4. 5, 3, or 1.7 mm id NMR tubes (Wilmad).
5. High purity (>95%) protein target, at concentrations between 0.2 and 10 μM.
6. Known inhibitor to check binding specificity (ex: ATP or staurosporine for kinases).

2.2. X-Ray Screening

1. Protein crystals suitable for soaking. Preferable if known inhibitor(s) have already verified the suitability of the crystals for soaking. If co-crystallization with the fragments is to be used instead of soaking, sufficient protein to grid around a robust crystallization condition for all fragments of interest is required.
2. Fragment stock solutions in DMSO, usually 100 mM or higher (*see* **Note 7**).
3. (Optional) Liquid handling robot to dispense the solutions for co-crystallization.
4. (Optional) Crystallization robot to dispense protein and fragments for co-crystallization. Ability to do small volume drops is very desirable.
5. Trays and cover slips appropriate to the technique used.
6. An X-ray generator with high flux (home source) or access to a synchrotron beam line.
7. (Optional) Robotic sample changer to facilitate exchange of samples for diffraction analysis (*see* **Note 8**).
8. Computers and software to analyze the X-ray diffraction data.

3. Methods

3.1. STD NMR Screening

1. Check protein activity and integrity in the working buffer and the final DMSO amounts (1–10% v/v), at 0.2–10 μM concentration.
2. Prepare NMR samples of the fragments at the desired concentration for solubility, integrity, and reference spectrum, usually at 200–500 μM.
3. Mixtures of six to eight fragments are prepared at 200–500 μM each, with the final % DMSO calculated to be ~5%. Protein for a final working concentration ~5 μM and an internal inert standard, such as TSP, are added. Final volume will be 500 μL (if working with a 5 mm NMR tube), 180 μL (3 mm NMR tube), or 30 μL (1.7 mm tube).
4. Sample is loaded in the NMR spectrometer, parameters are optimized for the STD experiment, and data are acquired.
5. If binding is observed with positive STD signal, the binder is identified by comparison of the STD signal and the reference data from step 2.
6. Confirmation of binding is performed by preparing the fragment binder with the protein and the STD acquired again.
7. Add known competitor and acquire the STD experiment again, to confirm the fragment binds in the site of interest. If there is good competition, the STD signal for the fragment should decrease with the addition of a tight inhibitor.
8. Monitor target integrity in the NMR sample by comparing the protein background signal with time.

3.2. X-Ray Screening

1. The fastest way to obtain co-structures with a protein and fragments is to soak the fragments into existing crystals. Since each protein is unique, trial and error will be necessary to deduce the conditions where your protein crystals are stable and the fragments are suitably soluble (referred to as the protein stabilization buffer) (*see* **Note 9**).
2. The organic solvent of choice is DMSO. Most fragments are soluble in DMSO and DMSO also has cryo-preservation qualities that assist in vitrifying crystals.
3. A suitable starting concentration for DMSO in a soaking experiment is 5%. Combine 9.5 μL of protein stabilization buffer with 0.5 μL fragment solution on a cover slip and mix thoroughly. Transfer protein crystal(s) to this solution and invert over a crystallization well (*see* **Note 10**).

4. If co-crystallization is indicated by properties of the protein or the fragment library, then prepare a solution of the protein with suitable concentration of fragment(s), (suggested 100 mM). Screen this solution around known crystallization conditions for the protein. If no crystals are observed a full screen using numerous conditions may be indicated.
5. Prepare a cryopreservation solution compatible with your crystals, starting with the protein stabilization buffer. Upon testing, the amount of cryo agent (DMSO, low molecular weight PEG, glycerol, ethylene glycol, etc.) used should produce a clear glass effect with no water rings when analyzed with X-rays.
6. Treat the crystal exposed to fragment(s) with the cryopreservation buffer and vitrify the sample with liquid nitrogen (*see* **Note 11**).
7. During data collection from crystals exposed to fragment(s), collect a data set that is complete in the low-resolution shells and has high redundancy. Also beneficial will be the highest resolution data possible, so examination of multiple crystals to select the one with suitable qualities is crucial (*see* **Note 12**).

4. Notes

4.1. NMR Screening

1. The NMR screening samples can be prepared in an automated fashion with a programmable platform such as Tecan (by Tecan) and samples can be automatically loaded into the spectrometer by using a Sample Rail (by Bruker Biospin). This allows for maximum spectrometer time and the sample is always freshly prepared prior to data acquisition.
2. The protein stock concentration should be in the NMR running buffer at concentrations slightly higher than what is used in the NMR samples. Alternatively, if the protein stability is better in a different buffer, the protein could be stored at high concentration (80 μM or higher) and a small aliquot diluted into the NMR running buffer for sample preparation.
3. Fragments in mixtures can sometimes precipitate due to the high total fragment concentration in solution, which could be up to 5 mM. In most cases where precipitation is observed, we have noted that the other fragments in the mixture are still soluble and give good NMR signals. Thus the mixture is still usable.

4. The higher the DMSO percentage used, the higher the fragment mixture solubility will be. The protein needs to be stable and active for at least a couple of hours under these conditions for data collection.
5. Competition experiments can be performed within the same NMR sample mixture used for screening if protein amounts are limited. The competitor is just added to the NMR solution in the tube and mixed well.
6. Fragments in the mixtures that bind to the protein target can easily be identified by comparing the NMR spectrum of the hit with the spectra of the individual fragments.

4.2. X-Ray Screening

7. The more concentrated the fragment sample the less a dilution effect is observed when added to the protein. For those proteins or crystals highly sensitive to DMSO concentration we have found that soaking is problematic and co-crystallization is indicated. Fragments are generally low-affinity compounds and in order for weakly binding compounds to be observed with X-ray crystallography they need to possess excellent solubility. A rule of thumb used is ten times the binding constant. Applying this to fragment screening, it is desirable to have the compound at 100 mM during the experiments. While this level of solubility is easily obtained in DMSO, the addition of an aqueous component will be an issue as precipitation of the small molecule often occurs. During soaking experiments it is not uncommon to have a successful experiment despite heavy precipitate or even crystallization of the small molecule upon addition of the fragment to the protein stabilization solution. When co-crystallizing protein with fragments, centrifugation prior to screening will be required in cases where precipitation is observed.
8. For automation in our lab, we use a Hamilton STAR for creating/dispensing crystallization solutions, a TTP LabTech mosquito liquid handling robot for setting up crystallization drops, a Formulatrix robotic storage/retrieval/imaging system for crystallization trays, and a Rigaku ACTOR robot for automatic crystal handling for testing of diffraction properties.
9. Prior to initiating the FBS soaking experiments a substantial amount of investigation needs to be completed on the methodology that will be used. The parameters that should be considered and optimized include
   - The length of time to soak the compound into the crystal

- The amount of DMSO (or other organic liquids) needed to maintain compound solubility as well as crystal integrity
- The protocol necessary to freeze the crystal for data collection
- The default values for the data collection software used
- The automation of structure solution

10. The Hampton Research VDXm Plate with sealant (Part HR3-306) is recommended. A 10 μL drop inverted on this tray will last for at least 7 days at 18 °C with no additional solution added to the well. For longer soaks or for security add 100 μL of the protein stabilization buffer to the well prior to inverting the cover slip containing the soaking experiment. The length of time to soak a crystal with compound is one of the, if not the most, critical steps. Too short and you may not find binding. Too long and you risk damaging the crystals. In addition, each protein and each crystal form for that protein are different. The successfully designed experiment will allow sufficient time for binding to occur as well as any remodeling of the protein required to accommodate the fragment(s).
11. To prevent backsoaking of the fragment(s) from the crystals swipe the crystal quickly through the cryopreservation buffer. If longer soaks in the cryopreservation buffer are required add fragment(s) to it so that equilibrium does not remove the weak binding fragments from the crystal.
12. It is not uncommon for data at low resolution to be inconclusive. Data collected from similarly treated crystals can often show inhibitor when the resolution is extended where it was not visible at low resolution.

References

1. Lipinski, C. A., Lombardo, F., Dominy, B. W., Feeney, P. J. (2001) Experimental and computational approaches to estimate solubility and permeability in drug discovery and development settings. *Adv Drug Deliv Rev* **46**, 3–26.
2. Keseru, G. M., Makara, G. M. (2009) The influence of lead discovery strategies on the properties of drug candidates. *Nat Rev Drug Discov* **8**, 203–212.
3. Jencks, W. P. (1981) On the attribution and additivity of binding energies. *Proc Natl Acad Sci U S A* **78**, 4046–4050.
4. Nakamura, C. E., Abeles, R. H. (1985) Mode of interaction of beta-hydroxy-beta-methylglutaryl coenzyme A reductase with strong binding inhibitors: compactin and related compounds. *Biochemistry* **24**, 1364–1376.
5. Bohacek, R. S., McMartin, C., Guida, W. C. (1996) The art and practice of structure-based drug design: a molecular modeling perspective. *Med Res Rev* **16**, 3–50.
6. Kuntz, I. D., Chen, K., Sharp, K. A., Kollman, P. A. (1999) The maximal affinity of ligands. *Proc Natl Acad Sci U S A* **96**, 9997–10002.
7. Hopkins, A. L., Groom, C. R. (2002) The druggable genome. *Nat Rev Drug Discov* **1**, 727–730.
8. Congreve, M., Carr, R., Murray, C., Jhoti, H. (2003) A 'rule of three' for fragment-based lead discovery? *Drug Discov Today* **8**, 876–877.

9. Hajduk, P. J. (2006) Fragment-based drug design: how big is too big? *J Med Chem* **49**, 6972–6976.
10. Bemis, G. W., Murcko, M. A. (1996) The properties of known drugs. 1. Molecular frameworks. *J Med Chem* **39**, 2887–2893.
11. Hann, M. M., Leach, A. R., Harper, G. (2001) Molecular complexity and its impact on the probability of finding leads for drug discovery. *J Chem Inf Comput Sci* **41**, 856–864.
12. Schuffenhauer, A., Ruedisser, S., Marzinzik, A. L., Jahnke, W., Blommers, M., Selzer, P., Jacoby, E. (2005) Library design for fragment based screening. *Curr Top Med Chem* **5**, 751–762.
13. Chessari, G., Woodhead, A. J. (2009) From fragment to clinical candidate–a historical perspective. *Drug Discov Today* **14**, 668–675.
14. Moriz, M., Bernd, M. (1999) Characterization of ligand binding by saturation transfer difference NMR spectroscopy. *Angew Chem Intl Ed* **38**, 1784–1788.
15. Dalvit, C., Pevarello, P., Tato, M., Veronesi, M., Vulpetti, A., Sundstrom, M. (2000) Identification of compounds with binding affinity to proteins via magnetization transfer from bulk water. *J Biomol NMR* **18**, 65–68.
16. Jhoti, H., Cleasby, A., Verdonk, M., Williams, G. (2007) Fragment-based screening using X-ray crystallography and NMR spectroscopy. *Curr Opin Chem Biol* **11**, 485–493.
17. Annis, D. A., Nickbarg, E., Yang, X., Ziebell, M. R., Whitehurst, C. E. (2007) Affinity selection-mass spectrometry screening techniques for small molecule drug discovery. *Curr Opin Chem Biol* **11**, 518–526.
18. Erlanson, D. A., Braisted, A. C., Raphael, D. R., Randal, M., Stroud, R. M., Gordon, E. M., Wells, J. A. (2000) Site-directed ligand discovery. *Proc Natl Acad Sci U S A* **97**, 9367–9372.
19. Neumann, T., Junker, H. D., Schmidt, K., Sekul, R. (2007) SPR-based fragment screening: advantages and applications. *Curr Top Med Chem* **7**, 1630–1642.
20. Hesterkamp, T., Barker, J., Davenport, A., Whittaker, M. (2007) Fragment based drug discovery using fluorescence correlation: spectroscopy techniques: challenges and solutions. *Curr Top Med Chem* **7**, 1582–1591.
21. Danziger, D. J., Dean, P. M. (1989) Automated site-directed drug design: the prediction and observation of ligand point positions at hydrogen-bonding regions on protein surfaces. *Proc R Soc Lond B Biol Sci* **236**, 115–124.
22. Gillet, V., Myatt, G., Zsoldos, Z., Johnson, A. (1995) SPROUT, HIPPO and CAESA: Tools for de novo structure generation and estimation of synthetic accessibility. *Perspect Drug Discov Design* **3**, 34–50.
23. Evensen, E., Joseph-McCarthy, D., Weiss, G. A., Schreiber, S. L., Karplus, M. (2007) Ligand design by a combinatorial approach based on modeling and experiment: application to HLA-DR4. *J Comput Aided Mol Des* **21**, 395–418.
24. Miranker, A., Karplus, M. (1991) Functionality maps of binding sites: a multiple copy simultaneous search method. *Proteins* **11**, 29–34.
25. Clark, M., Meshkat, S., Talbot, G. T., Carnevali, P., Wiseman, J. S. (2009) Fragment-based computation of binding free energies by systematic sampling. *J Chem Inf Model* **49**, 1901–1913.
26. Gillet, V., Johnson, A. P., Mata, P., Sike, S., Williams, P. (1993) SPROUT: a program for structure generation. *J Comput Aided Mol Des* **7**, 127–153.
27. Gillet, V. J., Newell, W., Mata, P., Myatt, G., Sike, S., Zsoldos, Z., Johnson, A. P. (1994) SPROUT: recent developments in the de novo design of molecules. *J Chem Inf Comput Sci* **34**, 207–217.
28. Nishibata, Y., Itai, A. (1991) Automatic creation of drug candidate structures based on receptor structure. Starting point for artificial lead generation. *Tetrahedron* **47**, 8985–8990.
29. Bohm, H. J. (1993) A novel computational tool for automated structure-based drug design. *J Mol Recognit* **6**, 131–137.
30. Bohacek, R. S., McMartin, C. (1994) Multiple highly diverse structures complementary to enzyme binding sites: results of extensive application of a de novo design method incorporating combinatorial growth. *J Am Chem Soc* **116**, 5560–5571.
31. Todorov, N. P., Dean, P. M. (1997) Evaluation of a method for controlling molecular scaffold diversity in de novo ligand design. *J Comput Aided Mol Des* **11**, 175–192.
32. Todorov, N. P., Dean, P. M. (1998) A branch-and-bound method for optimal atom-type assignment in de novo ligand design. *J Comput Aided Mol Des* **12**, 335–349.
33. Ishchenko, A. V., Shakhnovich, E. I. (2002) SMall Molecule Growth 2001 (SMoG2001): an improved knowledge-based scoring function for protein-ligand interactions. *J Med Chem* **45**, 2770–2780.

34. Wang, R., Gao, Y., Lai, L. (2000) LigBuilder: a multi-purpose program for structure-based drug design. *J Mol Model* **6**, 498–516.
35. Murray, C. W., Verdonk, M. L. (2002) The consequences of translational and rotational entropy lost by small molecules on binding to proteins. *J Comput Aided Mol Des* **16**, 741–753.
36. Thompson, D. C., Denny, R. A., Nilakantan, R., Humblet, C., Joseph-McCarthy, D., Feyfant, E. (2008) CONFIRM: connecting fragments found in receptor molecules. *J Comput Aided Mol Des* **22**, 761–772.
37. Eisen, M. B., Wiley, D. C., Karplus, M., Hubbard, R. E. (1994) HOOK: a program for finding novel molecular architectures that satisfy the chemical and steric requirements of a macromolecule binding site. *Proteins* **19**, 199–221.
38. Clark, D. E., Frenkel, D., Levy, S. A., Li, J., Murray, C. W., Robson, B., Waszkowycz, B., Westhead, D. R. (1995) PRO-LIGAND: an approach to de novo molecular design. 1. Application to the design of organic molecules. *J Comput Aided Mol Des* **9**, 13–32.
39. Lauri, G., Bartlett, P. A. (1994) CAVEAT: a program to facilitate the design of organic molecules. *J Comput Aided Mol Des* **8**, 51–66.
40. Yang, Y., Nesterenko, D. V., Trump, R. P., Yamaguchi, K., Bartlett, P. A., Drueckhammer, D. G. (2005) Virtual hydrocarbon and combinatorial databases for use with CAVEAT. *J Chem Inf Model* **45**, 1820–1823.
41. Maass, P., Schulz-Gasch, T., Stahl, M., Rarey, M. (2007) ReCore: a fast and versatile method for scaffold hopping based on small molecule crystal structure conformations. *J Chem Inf Model* **47**(2), 390–9.
42. Mori, S., Abeygunawardana, C., Johnson, M. O., van Zijl P. C. (1995) Improved sensitivity of HSQC spectra of exchanging protons at short interscan delays using a new fast HSQC (FHSQC) detection scheme that avoids water saturation. *J Magn Reson B* **108**(1), 94–8.

Chapter 13

LEAP into the Pfizer Global Virtual Library (PGVL) Space: Creation of Readily Synthesizable Design Ideas Automatically

Qiyue Hu, Zhengwei Peng, Jaroslav Kostrowicki, and Atsuo Kuki

Abstract

Pfizer Global Virtual Library (PGVL) of 10^{13} readily synthesizable molecules offers a tremendous opportunity for lead optimization and scaffold hopping in drug discovery projects. However, mining into a chemical space of this size presents a challenge for the concomitant design informatics due to the fact that standard molecular similarity searches against a collection of explicit molecules cannot be utilized, since no chemical information system could create and manage more than 10^8 explicit molecules. Nevertheless, by accepting a tolerable level of false negatives in search results, we were able to bypass the need for full 10^{13} enumeration and enabled the efficient similarity search and retrieval into this huge chemical space for practical usage by medicinal chemists. In this report, two search methods (LEAP1 and LEAP2) are presented. The first method uses PGVL reaction knowledge to disassemble the incoming search query molecule into a set of reactants and then uses reactant-level similarities into actual available starting materials to focus on a much smaller sub-region of the full virtual library compound space. This sub-region is then explicitly enumerated and searched via a standard similarity method using the original query molecule. The second method uses a fuzzy mapping onto candidate reactions and does not require exact disassembly of the incoming query molecule. Instead Basis Products (or capped reactants) are mapped into the query molecule and the resultant asymmetric similarity scores are used to prioritize the corresponding reactions and reactant sets. All sets of Basis Products are inherently indexed to specific reactions and specific starting materials. This again allows focusing on a much smaller sub-region for explicit enumeration and subsequent standard product-level similarity search. A set of validation studies were conducted. The results have shown that the level of false negatives for the disassembly-based method is acceptable when the query molecule can be recognized for exact disassembly, and the fuzzy reaction mapping method based on Basis Products has an even better performance in terms of lower false-negative rate because it is not limited by the requirement that the query molecule needs to be recognized by any disassembly algorithm. Both search methods have been implemented and accessed through a powerful desktop molecular design tool (*see* ref. (33) for details). The chapter will end with a comparison of published search methods against large virtual chemical space.

Key words: LEAP, PGVL, combinatorial chemistry, library design, similarity search, disassembly, Basis Product, symmetric similarity score, asymmetric similarity score, lead hopping.

J.Z. Zhou (ed.), *Chemical Library Design*, Methods in Molecular Biology 685,
DOI 10.1007/978-1-60761-931-4_13,

1. Introduction

The high attrition rate across multiple stages of the modern drug discovery process has significantly hampered the productivity of the pharmaceutical industry as a whole (1). One of the countermeasures implemented by pharmaceutical companies against this challenge is to build a large and diverse library of combinatorially enabled molecules to boost productivity in hit identification and lead optimization (2). Through a multi-year multi-million dollar investment in collaborations with ChemBridge, Tripos, ChemRx, and Arqule, Pfizer has developed validated reactions for parallel synthesis, and implemented those protocols, to expand its corporate compound collection for biological screening to ~3 million (3–5).

As an integral part of the parallel synthesis of these arrays of compounds, these collaborations and internal effort produced and validated ~2500 parallel synthetic protocols spanning across ~757 diverse chemical reactions. These combinatorial reactions, their synthetic procedures, and their reactant scope and limitations are well defined and have been captured electronically for future library production (6). Those experimentally validated synthetic protocols and their corresponding reactant sets compatible with their reaction conditions implicitly lead to a huge chemical compound space (PGVL, or Pfizer Global Virtual Library) with more than 10^{13} virtual yet synthetically feasible compounds. All starting materials are known, specified, and available. Only a very small fraction of PGVL has ever been synthesized (10^6 out of 10^{13}). Conceptually a medicinal chemist can use a query (or seed) molecule as input to search for similar molecules inside PGVL and thereby retrieve new analogs for lead optimization and scaffold hopping. Previous work has demonstrated that there are many lead- and drug-like molecules in this type of large virtual compound space spanned by combinatorial reactions selected by medicinal chemists and existing reactant sets (7). Yet there are significant challenges inherent in making the desired similarity search practical against such a huge chemical space. The standard similarity search methods require the construction of a file or database containing explicit molecules. However, as of today, no chemical information technology is known to enumerate and store more than 10^8 molecules, for example, both CAS (Chemical Abstract Service) (8) and Pubchem (9) have collections of substances in the 10^7 scale.

In the publications from the Tripos group, the authors had demonstrated that they could bypass the need for full enumeration of a huge virtual space and enable similarity search by extensively leveraging the reactant-level information (10). Even though the same authors went on to demonstrate that many drug-like molecules were found in their validation studies, their method did not gain wide usage within the community of medicinal chemists who are engaging in drug discovery. One could speculate that the long turnaround time (days instead of hours or even minutes) of a typical search session is the leading factor that has prevented this method from being widely adopted.

As combinatorial chemistry has been fully integrated into the modern drug discovery process, more computational search methodologies against large virtual combinatorial compound spaces have been steadily developed in recent years (11–16). A detailed summary and comparison of those published methods are reported in the Section **5** and in **Table 13.6**. A good review on this subject could also be found in the publication by Boehm and coworkers (16).

In this report, two methods (LEAP1 and LEAP2, *LE*ad-based *A*nalog ho*P*ping) for performing similarity search into PGVL are presented. The results of validation studies under controlled conditions are given to characterize their search performance profiles in terms of false-negative rate and search speed. Finally results from recent applications to advance two drug discovery projects are also included in this report to highlight the fact that LEAP1 and LEAP2 are fully integrated into chemists' molecular design workflow and have been in use for more than 5 years.

2. Methods

The standard similarity search problem is commonly defined as the following: *Given an input query molecule, find the molecules within a collection of compounds that are most similar (either top N or within a predefined similarity threshold) to the query molecule.* Of course a molecular similarity measure has to be given between a pair of molecules. The Tanimoto distance calculated on the basis of molecular fingerprints is the most commonly used similarity measure (17). Medicinal chemists routinely perform this type of similarity searches against molecular databases containing $\sim 10^6$–10^8 explicit molecules. The set of molecules returned by a similarity search is expected to be well defined by the search parameters such as the query molecule, the search domain, and the similarity measure in combination with the underlying molecular fingerprints. In this report, we refer to this set of molecules returned from such a search into a standard explicit database as the reference set to be used in comparison with search results obtained by new similarity search methodologies.

As stated before, PGVL is too large to be fully enumerated practically. Therefore our strategy is to find a way to focus in a just-in-time manner on much smaller sub-regions ($\sim 10^4$) of PGVL for subsequent on-the-fly enumeration followed by standard similarity search against the same query molecule. It is intuitively evident that a virtual compound space built from parallel synthesis reaction protocols has inherent array structures in the form of implicit arrays of related just-in-time enumerated compounds, even if those compounds do not have their molecular structures yet enumerated at the time this inherent array structure is exploited.

Hypothetically, if we could compare the set of molecules returned by such a high-speed approach with the reference set derived from the hypothetical search into fully enumerated PGVL, we would expect both false positives and false negatives. It is easy to understand the source of false negatives since only a much smaller sub-region of PGVL is searched and some true positives outside this sub-region of PGVL would be missed. If top *N* hits are returned, there would be false positives among them since some true top *N* positives are missed by the limited search and replaced by lower ranked false positives. On the other hand, there would not be any false positives if we ask for enumerated molecules with a similarity threshold with respect to the query molecule. The molecules, post-enumeration, are either similar or not, by that threshold. By accepting a tolerable level of false positives and false negatives, a similarity search strategy can be implemented to return interesting search results for practical usage by medicinal chemists.

Even with the same general just-in-time enumeration and search strategy, different methods for identifying and retrieving the required smaller sub-regions of PGVL can be developed. Their performance can be characterized by the rate of false positives and false negatives as well as search speed and ease of use. We have implemented two search methods (LEAP1 and LEAP2) which will be discussed in subsequent paragraphs. Yet this is an active research area open for further innovations. A summary comparison of LEAP1/2 with other published search methods is given in the Appendix.

2.1. LEAP1

If a query molecule can be disassembled into combinations of virtual reactants by in silico disconnection using one or more reaction schemes within PGVL, then those virtual reactants can be used to identify the most similar genuine reactants out of all suitable genuine starting materials for those known reactions. *For a given precise reaction*, similar reactant combinations always lead to similar product molecules. This is the basic principle used by the LEAP1 method to focus on smaller sub-region(s) of PGVL for explicit just-in-time enumeration and similarity search.

More explicitly, the four key steps of LEAP1 are depicted in **Fig. 13.1**:

(1) *Automatic scan over all PGVL reactions for retro-synthetic feasibility of the incoming query molecule and disassemble the query molecule into combinations of virtual reactants.* This step is carried out automatically using the known reaction cores from each reaction scheme within the PGVL reaction knowledge system. In the case where no reaction can be found by the in silico disconnection engine to break up the query molecule, then LEAP1 will fail to return any search result back to the user. This is the major limitation of LEAP1. For other cases, multiple reactions are identified to disassemble the same query molecule in different ways. This suggests that more than one sub-region of PGVL should be explicitly enumerated and searched. This is not a problem and is in fact a benefit. The output of this step is a list, each entry containing an explicit parallel reaction scheme and a specified combination of virtual reactants. By definition, one could use each reaction scheme and the corresponding virtual reactants to form the very same original query molecule.

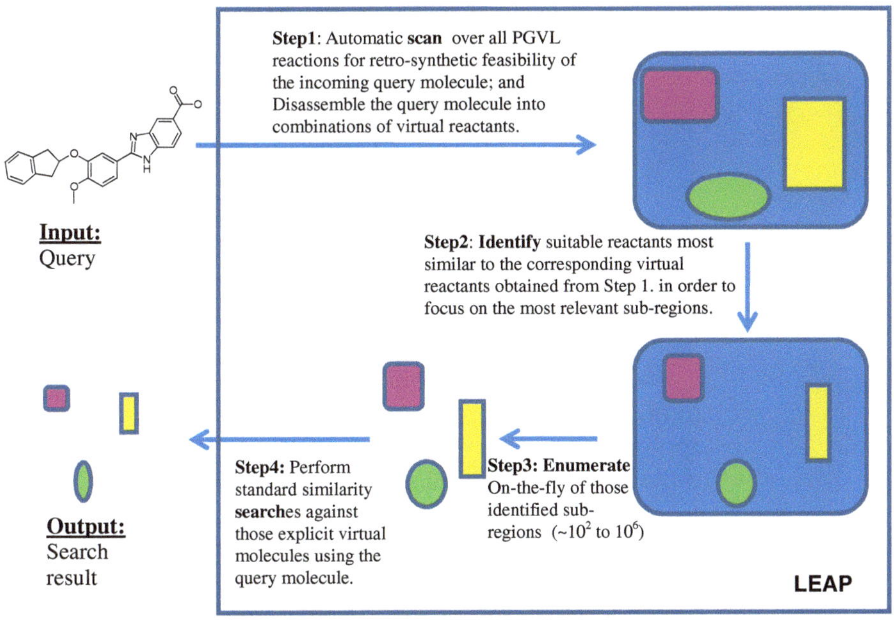

Fig. 13.1. Internal flowchart for the LEAP1 fully automated process. The diagram illustrates that there are three reactions identified whose chemical spaces are colored as *pink*, *green*, and *yellow*, which LEAP1 automatically identified as disconnection routes. LEAP1 then retrieves the most relevant sub-region within the chemistry space, followed by on-the-fly enumeration of those identified sub-regions. The final step can be any 2D/3D virtual screening algorithms. LEAP1 was implemented using Scitegic fingerprint technologies.

(2) *Identify suitable reactants most similar to the corresponding virtual reactants obtained from step 1 in order to focus on the most relevant sub-regions.* But the disconnection does not necessarily result in bona fide known and available starting materials, after just step 1. Consider as an example a two-component reaction which in the PGVL has *M* suitable bona fide reactants for the first reaction component and *N* suitable bona fide reactants for the second reaction component. Two similarity searches are used in the step to select *m* (out of *M*) and *n* (out of *N*) reactants based on two virtual reactants as seeds, which arose from the exact disconnection of the query molecule. In most cases, *M* and *N* are $\sim 10^3$, and *m* and *n* are $\sim 10^2$. Here extra search parameters need to be specified and/or optimized for each reaction component.

(3) *Enumerate on-the-fly the sub-region(s) using the optimized sets of bona fide reactants identified in step 2.* One can see that reduction in reactant set sizes makes explicit enumeration of product structures practical (for the same example used, $m \times n = 10^4$ vs. $M \times N = 10^6$ and both are << the total size of PGVL).

(4) *Perform standard similarity search using the original query molecule against the enumerated sub-region(s) obtained in step 3.* Looking at these four steps in the above discussion, one can see that steps 1, 3, and 4 are rather straightforward, whereas step 2 requires more tuning/optimization to get a balanced sampling of bona fide reactants for each reaction component to enable the precise and optimal enumeration sub-space to achieve the best overall search results. As mentioned above, either top-*N* or certain similarity threshold can be used to sample the reactant space. To balance the performance in terms of the adequate sampling and within reasonable runtime, top 20 reactants are used as the default value for each component list; users do have the flexibility to tune this number.

Since LEAP1 was built based on Pipeline Pilot technology, multiple molecular fingerprints and similarity methods can be applied at disposal, which currently include MDL Public Keys and different levels of FCFPs and ECFPs (18).

2.2. LEAP2

LEAP2 was developed to overcome the major limitation of LEAP1, the need to successfully disconnect the query molecule into combinations of virtual reactants using reaction schemes inside PGVL. Even though PGVL contains ~757 combinatorial reaction schemes, still experience has shown that there are many interesting hits and lead molecules whose structures could not be precisely disassembled. A fuzzy reaction mapping and reaction retrieval step is instead required. So in LEAP2, the identification of suitable candidate reactions and the subsequent focusing to their optimal corresponding bona fide starting materials is done with the help of Basis Products (BP) (19) as well as an asymmetric similarity measure of BPs using the query molecule. For any given query molecule, LEAP2 always returns search results to user. Before proceeding further, a short discussion on Basis Products and the asymmetric similarity measure between two molecules is given in the following paragraphs.

2.2.1. Basis Products

For a given combinatorial reaction and its associated fully enumerated product space spanned by all suitable reactants, Basis Products (BP) form a much smaller subset within the full product space and at the same time provide a systematic and efficient sampling of all reactants suitable for that reaction. **Figure 13.2** depicts an example for a two-component reaction. A Basis Product contains information about the R-groups as well as the reaction core, which can be expressed in the following statement (*see* **Fig. 13.2**):

BP = **R-groups** of one reaction component

+ Reaction **Core** + **CAP**(s)from other components(s)

where *CAP is the R-group of the smallest reactant from each reactant list.*

In **Fig. 13.2**, the first row and the first column of products are defined as the Basis Products for that reaction. Basis Products have an one-to-one relationship with their corresponding reactants. It can be seen also that in a two-component reaction, there are two sets of Basis Products; in a three-component reaction, there are three sets of Basis Products; always one set per reaction component. M plus N reactants lead to M plus N Basis Products, while the fully combinatorial product space is $M \times N$ in size. Currently there are $\sim 10^6$ Basis Products in PGVL, far smaller than the full PGVL space about 10^{13}–10^{14} in size.

Importantly, all Basis Products are real products, members within the product space, and, like the simple truncated R-groups, retain no transient reactant-only functional groups (reactive halides, aldehydes, etc.); in R-group methods these disappear by clipping, whereas in Basis Products these are transmuted in the reaction transformation preparing the Basis Products. Yet unlike truncated R-groups, Basis Products also incorporate the full reaction core (all of the newly formed bonds) as part of the BP structure. Furthermore the collection of available starting materials, e.g., aliphatic amines, aldehydes, acyl chlorides, benzyl halides, collapses to a fewer number of unique R-groups when clipped, whereas the same set of starting materials expands

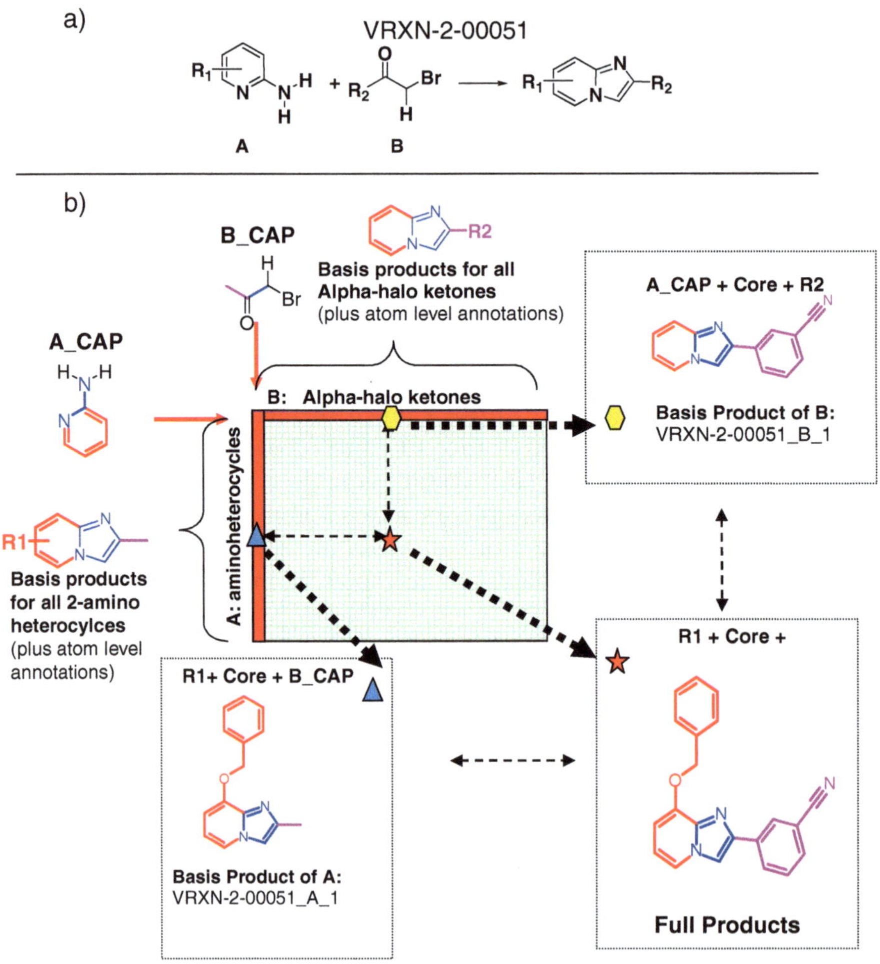

Fig. 13.2. Illustration of the basic concept of Basis Products. (**a**) The PGVL reaction scheme of VRXN-2-00051 (formation of the *H*-imidazo[1,2-*a*]pyridine ring system using aminoheterocycles and alpha-halo ketones) is used for the illustration; (**b**) The Basis Products of A are formed by all A reactants with one constant B reactant (B_CAP, 1-bromopropan-2-one). The Basis products of B are formed by all B reactants with a constant A reactant (A_CAP, 2-amino pyridine). The *blue* triangle and *yellow* hexagon represent two such basis products. The *red* star represents a product molecule which is related to those two corresponding basis products.

to many more unique BPs since each of these starting materials can typically participate in many different reactions yielding different reaction product cores, hence multiple BPs arise from the same starting material. Simply put, the structure of each Basis Product encodes within it, and through associated database fields, the precise combination of one reaction and the one starting material.

All Basis Products in the PGVL have been explicitly enumerated to support numerous molecular design, fragment-based design, and 2D and 3D methods; they also provide here a rigorous basis for the fuzzy reaction retrieval in the LEAP2 method.

In our previous publication, we have shown that knowledge of a useful set of physicochemical molecular properties of $(M+N)$

Basis Products can be used to provide a remarkably accurate and efficient estimation of the same molecular properties for any product molecule within a fully combinatorial product space *without enumeration* (19). Of course, with the just-in-time enumeration provided by LEAP2, such important ADMET molecular properties can also be explicitly and efficiently calculated now on product analogs rapidly mined by LEAP2. Additionally, Basis Products have been used to anchor structure-based library design methods when the 3D structure of a binding site of a target protein is known (20). There is a deep connection between Basis Products, fragment-based structure-based design (21), and parallel synthesis chemistry. In this report, we show that Basis Products are again instrumental for the implementation of LEAP2.

2.2.2. Asymmetric Similarity Between Two Molecules

Asymmetric similarity measure has been first described by Tversky (22) to provide a general mathematical framework for the perception of similarity and later adapted to molecular similarity by Bradshaw (23). The mathematical formula for both similarity measurements against BPs are shown below:

Symmetric similarity (SS) favors maximum common features and penalizes non-common features:

$$SS = \frac{\text{Number of features in both Query and Basis Product}}{\text{Number of features in either Query or Basis Product}} \quad [1]$$

Asymmetric similarity (AS) favors retrieval of basis products with the most features embedded within the query.

$$AS = \frac{\text{Number of feature in both Query and Basis Product}}{\text{Number of features in Basis Product}} \quad [2]$$

The well-known symmetric similarity measure rewards common features shared by two molecules and penalizes unique features present in either molecule which are not found in the other. Its value reaches 1 only when both molecules are identical. The asymmetric similarity measure focuses on the degree to which a test molecule (BPs in our case) can map into the original query molecule. When a BP molecule, which is typically smaller, is mapped into the query molecule, the asymmetric similarity measure can still reach 1.0 when the BP can be fully mapped into the query molecule, in another words the BP is a substructure of the query molecule. **Figure 13.3** uses a query molecule within PGVL and its corresponding Basis Products to highlight the difference between symmetric and asymmetric similarity measures. From the differences of the AS and SS scores of the same BP, it is seen that indeed the standard symmetric similarity measure penalizes any differences between two molecules, while the asymmetric similarity measure used in LEAP2 focuses on mapping the Basis Product into the query molecule, while ignoring the unique features

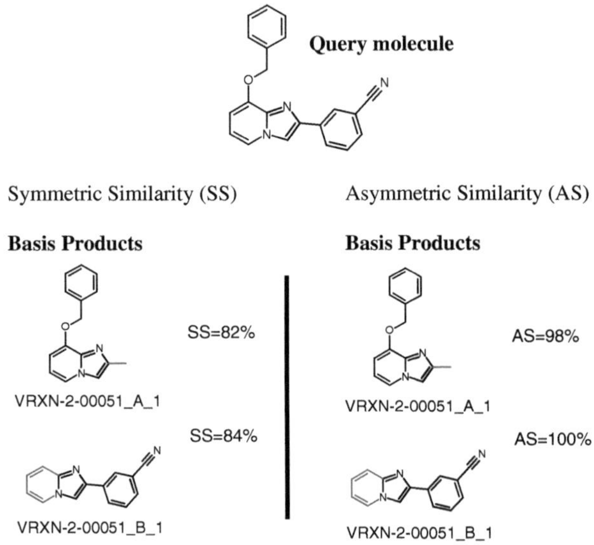

Fig. 13.3. Comparison of symmetric and asymmetric similarity scores. A virtual product from VRXN-2-00051 is used as a query molecule. The two corresponding Basis Products are VRXN-2-00051_A_1 and VRXN-2-00051_B_1. In reference to the query molecule, their corresponding similarity scores are listed under SS and AS (*see* equations [1] and [2] for details), respectively, depending on the similarity methods used.

in the query molecule which extend beyond the Basis Product structure – and those are analyzed using AS with the *other* BP sets from the other reaction components of the same reaction, which serve to scan these other R-group positions. Since AS is a similarity measure, high AS can be achieved without the need for precise substructure embeddability, hence this is still a fuzzy mapping.

We have hypothesized that when a Basis Product has a high asymmetric similarity value to a query molecule, there should be a higher probability that the candidate reaction and the specific available reactant encoded by the Basis Product will be associated with sub-regions of PGVL space where full-size product molecules most similar to the query molecule will be found. This is the principle based on which LEAP2 focuses the reaction search and retrieval. Sub-regions of PGVL spanned by reactions and reactants encoded by those Basis Products that map most favorably into the query molecule are detected by AS and these sub-regions advance to the next step.

2.2.3. Search Steps in the LEAP2 Algorithm

(1) *Search a database of Basis Products using Asymmetric Similarity measure.* Here this search is done using the query molecule against a database of 10^6 explicit enumerated Basis Products. The asymmetry similarity search in the BP database is implemented using MDL Keys finger print (24) with ISIS host technology (25).

The output is a set of Basis Products with high asymmetric similarity (AS) values (the default cutoff value is set to 90%) when they are mapped into the query molecule. The reaction schemes and reactants encoded by those Basis Products are then extracted, ranked, and used to form sub-regions of PGVL for subsequent just-in-time enumeration and symmetric similarity search against the query molecule.

Similarly to LEAP1, the top 20 similar molecules per reaction component list are used, as a default setting, to ensure balanced sampling of reactants for each reaction component and the reasonable performance. This is user adjustable.

(2) *Enumerate sub-region(s) using the smaller sets of reactants identified in step 1.* This on-the-fly enumeration step is identical to step 3 of LEAP1.

(3) *Perform standard similarity search using the original query molecule against the enumerated products from the sub-region(s) obtained in step 2.* This is identical to step 4 of LEAP1.

Since LEAP2 was also built based on Pipeline Pilot technology, multiple molecular fingerprints and similarity methods can be applied at disposal, which currently include MDL Public Keys and different levels of FCFPs and ECFPs (18).

3. Results and Discussion

3.1. Method Validation and Performance Profiling

As mentioned before, LEAP1 and LEAP2 are the results of conscious choices between accuracy and practical execution performance. Therefore it is important to conduct a set of controlled validation studies to assess the accuracy in terms of rates of false positives and false negatives in their search results and performance in terms of end-to-end search turnaround time.

To reach those objectives, we conducted validations to answer the following questions:

(1) Given a set of molecules known to be inside PGVL as query molecules, what is the success rate for returning the expected molecules identical to the query molecules (100% similarity threshold)? This is by definition a baseline test that a validated search strategy must pass.

(2) Given a sub-region of PGVL that can be enumerated explicitly and a query molecule, compare search results obtained by a LEAP search with the reference sets obtained by the exhaustive search against the fully enumerated

sub-region. What are the rates of false positives and false negatives?

The false negative is referred to those true positives outside the sub-regions of PGVL found by LEAP methods which would be missed as hits.

The false positives are due to the top *N* similar approach which resulted in some true top *N* positives missed by the limited search and replaced by lower ranked false positives.

(3) Given a set of drug-like molecules as query molecules, what is the success rate of a search method in returning interesting search results not only similar to the input queries but also pertinent to lead optimization and/or lead hopping?

Test One: Thirteen product molecules from 13 PGVL reactions (five 2-component VRXNs, three 3-component VRXNs, and five 4-component VRXNs) were randomly selected as query molecules for the first validation test. LEAP1 identified all 13 PGVL reactions and returned all 13 expected identical molecules. LEAP2 correctly located all 13 expected PGVL reactions and 12 expected molecules in its search results using the default setting (90% AS score and top 20 similar molecules per component list per VRXN). Due to the relative larger molecular size of CAP molecule vs. the reactant, the BP corresponding to the missing molecule shows only a modest 62% AS score, so it was not found at the BP level (LEAP2 step 1). If we consider lowering the AS score cutoff, then many more molecules per reaction component will have to be included in the interior of the processing (LEAP2 steps 2 and 3) which will result in impact on performance. This case also highlighted the nature of the balancing act between speed and accuracy.

Test Two: A much small sub-region of PGVL with only 224,700 product molecules was constructed explicitly. **Table 13.1** gives the details of the sub-region, which spans seven PGVL reactions (four of them are two-component reactions and three of them are three-component parallel synthesis reactions). At the time of the test, the full virtual space spanned by those seven reactions in combination with their suitable reactants was about 389 million in size. We randomly selected smaller sub-sets of those reactants to form this sub-region as a controlled environment for this validation study. The query molecule is also given in **Table 13.1**, which was chosen so that it can be disassembled by all seven VRXNs. The validation results are given in **Table 13.2** for LEAP1 with different similarity thresholds used. As the similarity threshold used in the searches dropped from 1.0 (for exact match) to 0.48, the number of returned molecules went from 1 to 1807 for the exhaustive search against the enumerated set of 224,700 explicit molecules. It is reassuring to see that 94% or more of expected molecules were correctly identified by LEAP1 while the

Table 13.1
Construction of a fully enumerated virtual library space (VL) for the second validation study

Seed Structure	Mapped VRXNs	Real VL size	Validation VL size
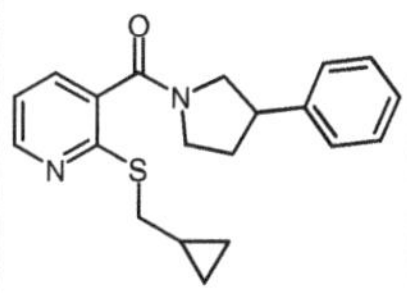	VRXN-2-00004	438 x 1171	60 X 60
	VRXN-2-00006	544 x 264	50 X 50
	VRXN-2-00010	3371 x 6635	60 X 60
	VRXN-2-00011	449 x 7044	50 X 50
	VRXN-3-00063	19 x 721 x 5697	18 X 50 X 50
	VRXN-3-00064	77 x 389 x 5175	50 X 50 X 50
	VRXN-3-00065	44 x 632 x 444	17 X 50 X 50
		Total: 388,798,585	Total: 224,700

Table 13.2
True-positive and false-negative rates of the LEAP1 method as a function of search threshold for molecular similarity

Similarity threshold	No of cpds retrieved by LEAP1	No of cpds retrieved by exhaustive search	No of true positives in LEAP1	% true positive in LEAP1	No of false negatives in LEAP1	% false negative in LEAP1
1	1	1	1	100	0	0
0.9	3	3	3	100	0	0
0.8	11	11	11	100	0	0
0.7	51	52	51	98	1	2
0.6	249	257	249	97	8	3
0.55	530	557	530	95	27	5
0.52	915	968	915	95	53	5
0.5	1331	1410	1331	94	79	6
0.48	1699	1807	1699	94	108	6

rate of false negatives remained less or equal to 6%. **Figure 13.4** graphically depicts the true-positive rate and false-negative rate of this validation test. Using a more common similarity threshold of 0.8, LEAP1 gave identical search results as those from the exhaustive search method. The false-positive rate is zero.

Table 13.3 shows the performance comparison of LEAP1 method vs. the exhaustive search. The speedup factor is the ratio between search times required by the exhaustive search and LEAP1. It is seen that in exchange for a ~6% false-negative rate we can get more than a 27,000-fold speedup. If we assume that

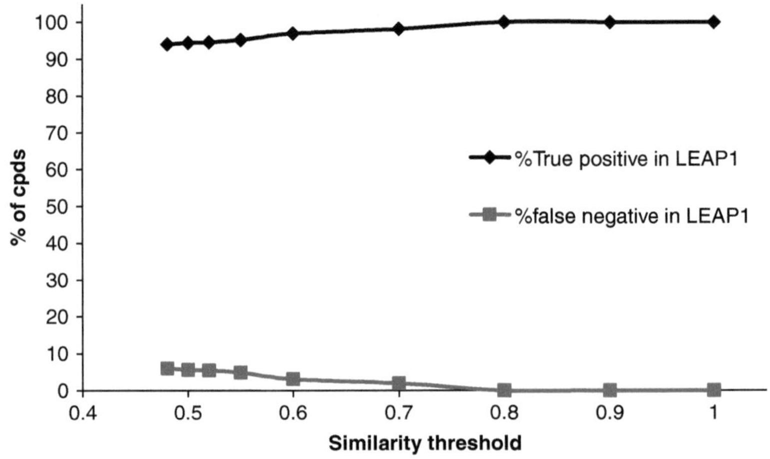

Fig. 13.4. Performance of LEAP1 when compared against the exhaustive search in the second validation study. *See* main text for details.

Table 13.3
Comparison of performance of LEAP1 vs. exhaustive search

Method	Validation VL (s)	Real VL (s)
LEAP1	446	602
Exhaustiv_search	9700	16,783,917[a]
Speedup factor	22	27880

[a]Estimated based on the reasonable assumption that standard search time is proportional to the size of the VL to be searched. The exhaustive search time against a smaller VL of 224,700 is 9700 s. Therefore we have estimated to the first approximation that an exhaustive search against the real VL of 388,798,585 would take 16,793,917 s (or about 194 days) to complete. *See* **Table 13.1** for VL sizes. Time is in the unit of second and based on single 3 GHz Pentium CPU.

search time required by the exhaustive search is proportional to the size of the VL to be searched, it is obvious that 194 days of exhaustive search is not practical; however, the 602 s LEAP1 search can be performed routinely.

Test Three: For the third validation test, we selected 24 known drugs on the market as query molecules (*see* **Fig. 13.5**). This is a very realistic and challenging set in terms of diversity in their molecular structures and complexities required for their synthesis. The top 10 most similar virtual compounds to each query molecule were identified and plotted as color dots in **Fig. 13.6** for both LEAP1 and LEAP2.

For every query molecule, LEAP2 is able to return top 10 molecules with best similarity scores ranging from ~0.4 for sertraline to 0.9 for celecoxib. Five of 24 searches return PGVL hits 80% or more similar to the query molecules for meaningful follow-up. If the threshold is relaxed to 0.7, then 11 of 24

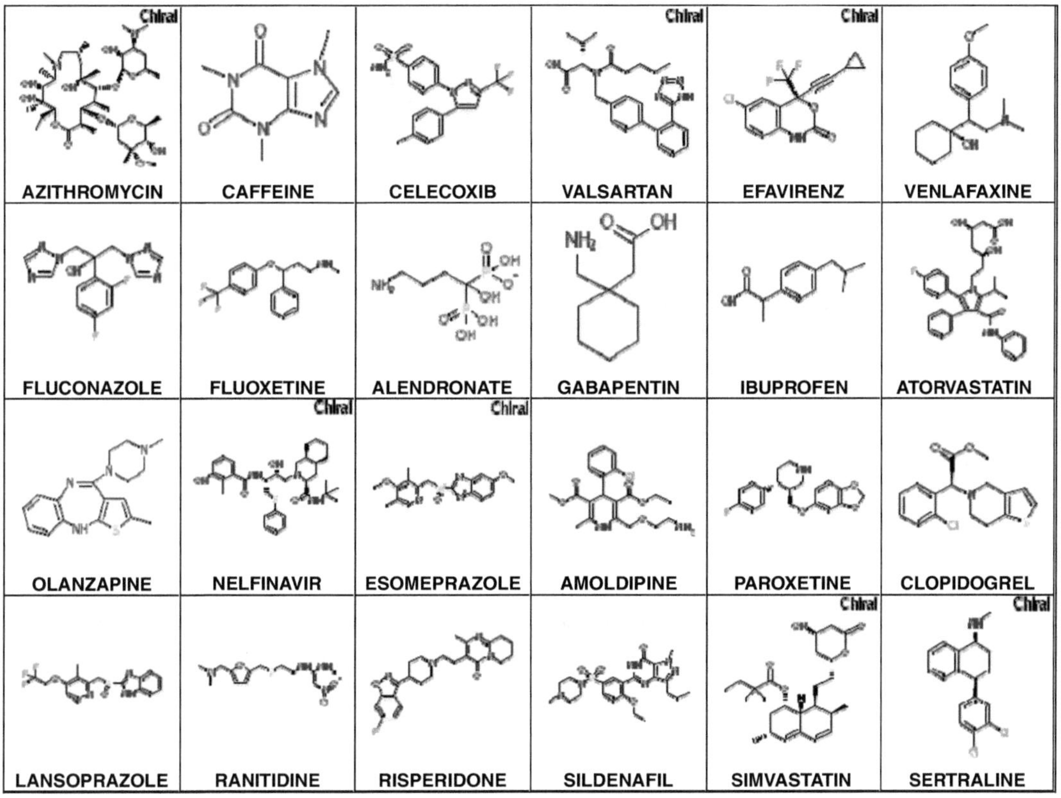

Fig. 13.5. A diverse set of 24 drug molecules on the market is compiled for the third validation study.

searches lead to PGVL hits for meaningful follow-up. The PGVL reactions identified at the same time can also be utilized by medicinal chemists to propose and evaluate new (not yet available) reactants for closer-in lead optimization and/or scaffold hopping, as indicated by the unique needs of the target active site. This point becomes even more significant given the observation that many (17/24) LEAP2 searches beneficially yielded top 10 hits originating from *more than two PGVL reactions*, while celecoxib, atorvastatin, amlodipine, lansoprazole, efavirenz, sildenafil, and sertraline are the exceptions to the trend.

As expected, due to the intrinsic requirement for precise disassembly of query molecules using PGVL reactions, LEAP1 failed to give any search results for fluconazole, alendronate, gabapentin, esomeprazole, and efavirenz. For the remaining 19 cases with hits, only 3 (3/19) LEAP1 searches lead to top 10 compounds originating from more than one PGVL reaction (fluoxetine/2, nelfinavir/2, and simvastatin/2). For the remaining 16 cases, precisely one PGVL reaction is identified. Three of 24 LEAP1 searches return PGVL hits 80% or more similar to the query molecules for meaningful follow-up. If the threshold is

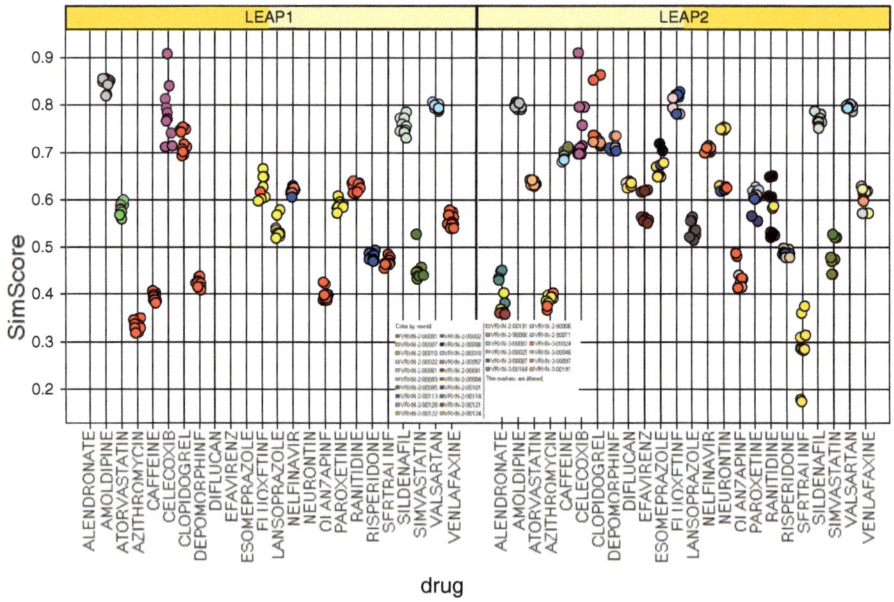

Fig. 13.6. Results from the third validation study. The *y*-axis represents the Tanimoto similarity score of returned hits with respect to their corresponding query molecule, calculated based on the FCFP4 molecular fingerprints (31). The *x*-axis are drug molecules in **Fig. 13.5**. Search hits are color coded by the PGVL reactions (VRXN) where they are originated from.

relaxed to 0.7, then 5 of 24 searches lead to PGVL hits for meaningful follow-up.

Based on the validation study it seems that the typical search time required for both LEAP1 (~15 min) and LEAP2 (~45 min) are short enough for routine and practical usage by medicinal chemists. On average LEAP2 is about three times slower than LEAP1, due to its larger VRXN coverage. LEAP2 in essence uses a "fuzzy" reaction retrieval strategy which returns more candidate PGVL regions of interest in the intermediate steps of the algorithm.

3.2. Application of LEAP1 and LEAP2

Two examples are included here to highlight how both LEAP1 and LEAP2 have been used, routinely, by chemists for idea generation and lead hopping.

3.2.1. LEAP1

In the first example, LEAP1 was used to help generate novel lead series against an anti-obesity target, MCH (melanin concentrating hormone) (26, 27). Fourteen query molecules consisting of both in-house and literature leads were used as search input, and the product similarity threshold was set as the top 50 final output molecules per VRXN per query, with default settings for remainder of the parameters. The LEAP1 search led to 7200 hits, all synthesizable based on PGVL chemistries and parallel synthesis protocols. An additional design step based on

Table 13.4
Two example virtual hits from among hundreds in the 14 LEAP1 searches

Structure	Name	monomerID	VRXN	simScore	Est_IC50 (nM)[a]
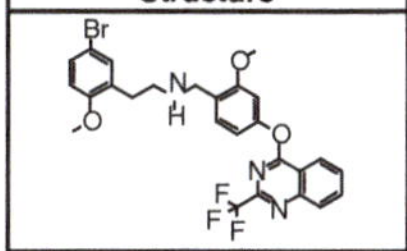	MCH-1	MFCD01443686: MFCD00238752	VRXN-2-00001	0.37	0.64
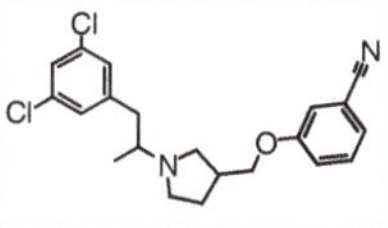	MCH-2	MN-011201:MN-017087	VRXN-2-00001	0.33	0.69

[a]IC50 estimated by 3D Pharmacophore model of MCH, their corresponding mappings are shown in Fig. **13.7**.

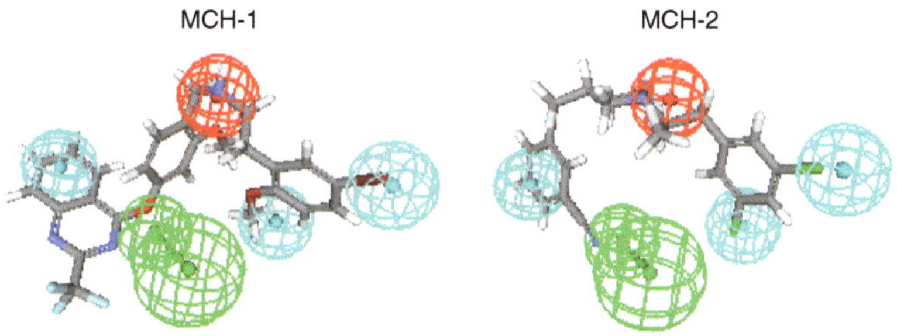

Fig. 13.7. Novel synthesizable compounds (*see* 2D structures in **Table 13.4**) produced by LEAP1 searches with high score judged by a project-specific 3D pharmacophore model (*red blob*: basic feature, *light blue blob*: hydrophobe, *green vector blob*: hydrogen bond acceptor).

a 3D pharmacophore model using Catalyst ™ (28) was used to further filter those virtual hits by MCH activity, thus leading to 21 final compounds. Two example virtual hits are shown in **Table 13.4** together with their similarity score to the respective query molecules and estimated IC_{50} based on their corresponding 3D pharmacophore mappings shown in **Fig. 13.7**. A targeted library using the corresponding chemistry, VRXN-2-00001, was launched, which resulted in 61 compounds prepared with an average 2D similarity around 30% to the original 14 query molecules. Thirteen hits from the pharmacophore-directed LEAP1 targeted library have shown greater than 60% inhibition at 1 μM in the MCH enzymatic assay. Due to the transfer of project and therapeutic area elimination, the lead development for those hits was stopped.

3.2.2. LEAP2

In the second example, LEAP2 was utilized to help generate novel leads against an anti-angiogenesis target, caspase-3 (29, 30). A literature compound (PAC-1 with reported 3.1 μM IC_{50}) and five of its analogs (31) were used as query molecules, and the product similarity threshold was set as top 50 final output molecules per VRXN per query molecule, with default settings for remainder of the parameters. The LEAP2 search resulted in 900 hits originating from 18 PGVL chemical reactions. Three targeted libraries were subsequently designed and synthesized based on those LEAP2 hits. The efforts resulted in 281 compounds synthesized, of which 13 yielded IC_{50} ranging from 1 to 20 μM (*see* **Table 13.5**). The result demonstrated that LEAP2 method is capable of generating multiple different design ideas which can be implemented quickly and fruitfully by the project team.

Table 13.5
Hits from the caspase-3 targeted libraries

Compound_Number	IC_{50} (μM)	VRXN_ID[a]
Cpd-1	1.01	VRXN-2-00086
Cpd-2	1.85	VRXN-2-00086
Cpd-3	3.36	VRXN-2-00086
Cpd-4	3.56	VRXN-2-00086
Cpd-5	5.82	VRXN-2-00086
Cpd-6	6.03	VRXN-2-00086
Cpd-7	7.46	VRXN-2-00086
Cpd-8	7.69	VRXN-2-00086
Cpd-9	12.5	VRXN-2-00010
Cpd-10	14.2	VRXN-2-00086
Cpd-11	16.7	VRXN-2-00086
Cpd-12	17.5	VRXN-2-00086
Cpd-13	19.5	VRXN-2-00010

[a]VRXN-2-00086 (hydrazone formation) and VRXN-2-00010 (amide formation)

4. Conclusion

It is very useful to emphasize systematic data capture within an organization as large as Pfizer. It has been beneficial to derive knowledge from those data in projects and sites different from the original settings which led to the original development of a given reaction protocol and most valuable if this knowledge can be reused in the essential operations on a regular basis.

The PGVL system is a large-scale knowledge system derived from rigorous multi-year systematic reaction knowledge capture, including the registration of large numbers of bona fide starting materials and validated parallel synthesis reaction protocols. LEAP chemistry space mining methodologies are ways to enable the efficient reuse of this knowledge in a practical manner, and this capacity is unleashed simply by entering the structure of the new lead at hand. In this sense, the usage of LEAP1 or LEAP2 is a lead-centric mining capability, as far as the user is concerned.

The validation studies show that the LEAP methods produce results reasonably comparable to exhaustive search and enable medicinal chemists with a practical method for the automated suggestion of synthesizable analogs for lead optimization and lead hopping.

In order to retrieve readily synthesizable virtual compounds from PGVL that are useful for virtual screening and for formulating the next synthesis plan, the LEAP-based methods can be used by itself or coupled together with other fundamental target-specific design methods, such as 3D pharmacophore modeling, docking, and SBDD, by simply replacing the final product similarity step with those well-known 3D design methods.

5. Notes on Comparison with Other Published Search Methods Against Large Virtual Chemical Space

Table 13.6 provides a comparison among several leading search methods in terms of their origin, search time, scope and nature of chemical space, format of input query ligand, and molecular similarity measure used.

1. *Origin*. Molecular similarity search into very large VLs is of great interest for drug discovery and is becoming a thriving research field. Methods 1–2 are from commercial software companies. The rest are developed by major pharmaceutical companies to facilitate their internal drug discovery. Four of the six are from Pfizer alone, all based on the PGVL chemical space (6).

2. *Turnaround time of a search*. The performances of most methods are within minutes for any single run with one query molecule, except for Method 1 and Method 2. The relatively long run time for Methods 1 and 2 is mainly due to the associated 3D searching technologies. For normal drug discovery process, run time within minutes or even hours are acceptable. For a computational technology to make a real impact, run time in month or even week scale is hard to justify the investment, at least in routine manner. With the current hardware and software advancement, one can imagine a coarse-grained parallelization for those methods with 3D searching need to significantly speed up those processes. In summary, minutes or even hours are acceptable but days

Table 13.6
Summary and comparison of representative methods to search into large virtual chemical space indexed by combinatorial libraries

No.	Method name	Ref#	Origin	Search turnaround time	# of chemical reactions	Source of chemical reactions	Size of virtual library space	Query input	Similarity measure
1	NA	7	Algodign	Month	400	Literature	1.00E+13	3D target	3D docking
2	AllChem	12	Tripos	Hour	100	Tripos Discovery Research (TDR)	1.00E+20	3D/2D ligand	3D Topomer
3	FTree-FS	13	Roche	Min	11	RECAP/TOPAS	1.00E+18	2D ligand	2D Feature-Tree
4	LEAP1	11, this report	Pfizer	Min	157[a]	PGVL (Pfizer Global Virtual Library)	1.00E+13	2D ligand	2D
5	LEAP2	11, this report	Pfizer	Min	534[a]	PGVL	1.00E+13	2D ligand	2D
6	MoBSS	14	Pfizer	Min	441[a]	PGVL	1.00E+13	2D ligand	2D Atom Pair (AP)
7	CoLibri/ FTrees-FS	16	Pfizer/BioSolvIT	Min	358[a]	PGVL	1.00E+13	2D ligand	2D Feature-Tree
8	CoLibri/ FTrees-FS	15	Boehringer Ingelheim (BI)/ BioSolvIT	Min	NA	BI CLAIM (Comprehensive Library of Accessible Innovative Molecules)	1.00E+11	2D ligand	2D Feature-Tree

[a]Those methods based on PGVL are implemented at different times, LEAP1 is the first among all four methods. The rest of the three are second-generation methods. Although there are 700+ VRXNs in the PGVL system, not all of them are registered to the full extent to enable the LEAP1 and LEAP2 search. For MoBSS and FTree-based methods, due to the assumptions made in the finger print additivity, some VRXNs, such as variable ring formation which depends on the reactant combinations used, were excluded from the implementation. For CoLibri/FTrees-FS method, the final enumeration step was implemented using CoLibri technology which is different from the PGVL foundation system, so certain VRXNs are excluded as well.

and beyond are not for practical application to impact drug discovery.

3. *Scope and nature of chemical space.* For Algodign, the entire chemical space is constructed based on chemistry from literature (7). For Tripos, Tripos Discovery Research (TDR), the former contract research division, provided most of the chemistry foundation for the virtual chemistry space (12). For Method 3, 11 simple reaction schemes implemented in RECAP (32) are used for both fragmentation and building block assembly. All four methods from Pfizer (LEAP1/2 in this report and two others in references (14) and (16)) are built based on PGVL (6) and enriched with library chemistry from File Enrichment (3, 4). Method from Boehringer Ingelheim is also built based on a collection of in-house library chemistries (15). The key differentiation factor here is synthetic feasibility of the result molecules. If the virtual space are constructed based entirely on a large pool of validated chemistry with step-by-step procedures for every library protocol and available starting materials, then it ensured that every hit found can be rapidly made and expanded synthetically. Size matters, but synthetic feasibility is even more significant. Methods with a large pool of validated chemistries, protocols, and starting materials, such as PGVL and BI CLAIM, have this advantage.
4. *Similarity measure.* Method 1 uses a structure-based de novo-like scoring function. The input query is not a ligand structure but a 3D structure of the active site for a target (7). For Method 2, it seems that search input can be either complete query structure or individual synthon; and the search result can be evaluated using any combination of filters such as (topomer) shape similarity (automatically generated topomer CoMFA), potency predictions, size, hydrophobicity, chemical reactivity, and synthetic accessibility (12). FeatureTree is used by Methods 3, 7, and 8 to compute molecular similarity. Two of them are implemented within the CoLibri library tool from BioSolvIT (15, 16). Method 6 employed atom pair (AP) descriptors derived from inter-atomic topological distances to compute molecular similarity (14). LEAP1 uses retro-synthetic analysis to break the input query molecule and applies similarity search at the fragment-level and product-level consecutively. LEAP2 uses the asymmetric similarity score between query molecule and Basis Products to focus on a subset of reactants. Then the standard symmetric similarity score between the query and the explicit product molecules is used to select final hits by LEAP2. Both LEAP methods are built based on Pipeline Pilot technology, so multiple molecular fingerprints

and similarity methods can be applied at disposal, which currently include MDL Public Keys and different levels of FCFPs and ECFPs (18). It is also expected that other similarity algorithms can be applied in the similar manner, as long as they can be integrated into the Pipeline Pilot (18) framework.

LEAP1/2 are easy to understand, implement, and have been in service since 2005 for idea generation and lead hopping inside a powerful and user-friendly molecular design tool called PGVL Hub (33). It also offers the general framework that encompasses all search methods against large combinatorial virtual spaces published so far in the literature (two main steps: 1. reactant/R-group focusing to reduce a large combinatorial virtual space into a much smaller and manageable one; 2. product-level similarity search within the reduced space). This framework suggests that the reactant/R-group focusing step is the major time saver while the extra saving harvested by estimation of product FPs based on additive nature of certain types of FPs (atom pair in MoBSS (14) and FTrees-FS in CoLibri (15, 16)) is only secondary with considerable cost introduced in terms of additional complexity to encode and ensure that combination rule is working properly for many combinatorial reactions and the restriction in choices of fingerprint and similarity measure. This framework also suggests that by working with explicitly enumerated products within that much reduced virtual space, one can apply any molecular finger print and any similarity measure without any approximation to the steps of reactant/R-group focusing as well as the final produce-level similarity search. This would allow users the flexibility to choose the more familiar similarity measures based on Tanimoto coefficient on top of FPs from MDL MACSS, Daylight, and SciTegic for close analogs or use atom pair FPs or FTree with higher abstraction for more aggressive and non-obvious lead hopping.

Finally we hope to see that more validation studies are conducted to compare any new search method with the reference exhaustive search (of course on a smaller validation virtual space of 10^4–10^6). Only through this type of rigorous validation studies, one can truly probe the rates of false positives and false negatives as well as the fold increase in search speed. This in turn allows end users to make informed decisions on which search method will be a best match for their specific tasks.

Acknowledgments

The authors would like to thank the following Pfizer colleagues for their generous help and support: Bo Yang, Thom Shulok,

Sarathy Mattaparti, Bo Chao, Tom Thacher, and Joe Zhou (for their work on the PGVL software and its reaction and starting material data foundation which enabled the development and deployment of LEAP1/2); Bob McDonough, Zi Yang, and Da Tse (for informatics support); and Gigi Paderes, Klaus Dress, Dilip Bhumralkar, and Michele Ramirez-Weinhouse (for being the early adopters of LEAP1/2 and applying them vigorously in their drug discovery projects); Ben Burke and Zhongxiang (Joe) Zhou for their valuable comments, suggestions, and proof reading the draft. We also appreciate the technical support from Derek Stonich and Anne Li-Zhong of SciTegic/Accelrys.

References

1. Kola, I., Landis J. (2004) Can the pharmaceutical industry reduce attrition rates? *Nat Rev Drug Discov* **3**, 711–715.
2. Milne, G. M. (2003) Pharmaceutical productivity: the imperative for new paradigms. *Annu Rep Med Chem* **38**, 383–396.
3. Estep, K. (2004) File Enrichment and Hit Follow Up: Evolution and Examples. Poster Presentations at the *ALA LabFusion*.
4. Smith, G. F. (2006) Enabling HTS Hit follow-up via Chemo informatics, File Enrichment, and Outsourcing. *High Throughput Medicinal Chemistry II*; MMS Conferencing & Events Ltd., Institute of Physics; London. This article is also available on-line via this web link (http://www.mmsconferencing.com/pdf/htmc/g.smith.pdf).
5. Borman, S. (2006) Improving efficiency. To eliminate R&D bottlenecks, drug companies are evaluating all phases of discovery and development and are using novel approaches to speed them up. *Chem Eng News* **84**, 56–78.
6. Peng, Z., Yang, B., Mattaparti, S., Shulok, T., Thacher, T., Kong, J., Kostrowicki, J., Hu, Q., Na, J., Zhou, J. Z., Klatte, K., Chao, B., Ito, S., Clark, J., Coner, C., Waller, C., Kuki, A. (2010) PGVL Hub: an integrated desktop tool for medicinal chemists to streamline design and synthesis of chemical libraries and singleton compounds. *Chemical Library Design*, in (Zhou, J. Z., ed.), Humana Press, New York, NY.
7. Nikitin, S., Zaitseva, N., Demina, O., Solovieva, V., Mazin, E., Mikhalev, S., Smolov, M., Rubinov, A., Vlasov, P., Lepikhin, D., Khachko, D., Fokin, V., Queen, C., Zosimov, V. (2005) A very large diversity space of synthetically accessible compounds for use with drug design programs. *J Comput Aided Mol Design* **19**, 47–63.
8. Chemical Abstract Service: http://www.cas.org/, under substances count
9. Pubchem: http://www.ncbi.nlm.nih.gov/sites/entrez?cmd=search&db=pccompound&term=all[filt].
10. Andrews, K. M., Cramer, R. D. (2000) Toward general methods of targeted library design: topomer shape similarity searching with diverse structures as queries. *J Med Chem* **43**, 1723–1740.
11. Hu, Q., Kostrowicki, J., Peng, Z., Kuki, A. (2008) LEAP into the Pfizer Global Virtual Library (PGVL) space – creation of the readily synthesizable design ideas automatically, *Scitegic Pipeline Pilot User Group Meeting*, San Diego, CA.
12. Cramer, R.D., Soltanshahi, F., Jilek, R., Campbell, B. (2007) AllChem: generating and searching 1020 synthetically accessible structures. *J Comput Aided Mol Des* **21**, 341–350.
13. Rarey, M., Stahl, M. (2001) Similarity searching in large combinatorial chemistry spaces. *J Comput Aided Mol Des* **15**, 497–520.
14. Yu, N., Bakken, G. A. (2009) Efficient exploration of large combinatorial chemistry spaces by monomer-based similarity searching. *J Chem Inf Model* **49**, 745–755.
15. Lessel, U., Wellenzohn, B., Lilienthal, M., Claussen, H. (2009) Searching fragment spaces with feature trees. *J Chem Inf Model* **49**, 270–279.
16. Boehm, M. Wu, T., Claussen, H., Lemmen, C. (2008) Similarity searching and scaffold hopping in synthetically accessible combinatorial chemistry spaces. *J Med Chem* **51**, 2468–2480.
17. Chen, X., Reynolds, C. H. (2002) Performance of similarity measures in 2D fragment-based similarity searching: comparison of

structural descriptors and similarity coefficients. *J Chem Inf Comput Sci* **42**, 1407–1414.
18. Pipeline Pilot from SciTegic: http://www.scitegic.com/
19. Shi, S., Peng, Z., Kostrowicki, J., Paderes, G., Kuki A. (2000) "Efficient combinatorial filtering for desired molecular properties of reaction products". *J Mol Graph Model* **18**, 478–496.
20. Zhou, Z., Shi, S., Na, J., Peng, Z., Thacher, T. (2009) Combinatorial library-based design with basis products. *J Comput Aided Mol Des* **23**, 725–736.
21. Lau, W., Hepworth, D., Magee, T., Du, J., Bakken, G., Miller, M., Hendsch, Z., Thanabal, V., Kolodziej, S., Xing, L., Hu, Q., Narasimhan, L., Love, R., Charlton, M., Hughes, S., Van Hoorn, W., Mills, J., Withka, J. (2010) Design of a multi-purpose fragment screening library using molecular complexity and orthogonal diversity metrics. *J Comput-Aided Mol Des.*
22. Tversky, A. (1977) Features of similarity. *Psycholog Rev* **84**, 327–352.
23. Bradshaw, J. (1997) Introduction to the Tversky Similarity Measure. Presented at *Daylight MUG Meeting, Laguna Beach, CA,* URL http://www.daylight.com/meetings/mug97/agenda97/Bradshaw/MUG97/tv¥tversky.html.
24. Durant, J. L., Leland, B. A., Henry, D. R., Nourse, J. G. (2002) Reoptimization of MDL keys for use in drug discovery. *J Chem Inf Comput Sci* **42**, 1273–1280.
25. ISIS host from Symyx: http://www.symyx.com/products/software/cheminformatics/isis-host/index.jsp
26. Qu, D., Ludwig, D.S., Gammeltoft, S. et al. (1996) A role for melanin-concentrating hormone in the central regulation of feeding behavior. *Nature* **380**, 243–247.
27. Saito, Y., Nothacker, H., Wang, Z., et al. (1999) Molecular characterization of the melanin-concentrating hormone receptor. *Nature* **400**, 265–269.
28. Li, H., Sutter, J., Hoffmann, R. (2000) HypoGen: an automated system for generating predictive 3D Pharmacophore Models. *Pharmacophore Perception, Development, and use in Drug Design*,in (Güner, O. F., ed.), International University Line, La Jolla, CA.
29. Nachmias, B., Ashhab, Y., Ben-Yehuda, D. (2004) The inhibitor of apoptosis protein family (IAPs): an emerging therapeutic target in cancer. *Semin Cancer Biol* **14**, 231–243.
30. Schimmer, A. D., Dalili, S., Riedl, S. J. (2006) Targeting XIAP for the treatment of malignancy. *Cell Death Different* **13**, 179–188.
31. Putt, K. S., Chen, G. W., Pearson, J. M., Sandhorst, J. S., Hoagland, M. S., Kwon, J. T., Hwang, S. K., Jin, H., Churchwell, M. I., Cho, M. H., Doerge, D. R., Helferich, W. G., Hergenrother, P. J. (2006) Small molecule activation of procaspase-3 to Caspase-3 as a personalized anti-cancer strategy. *Nat Chem Biol* **2**, 543–550.
32. Lewell, X. Q., Judd, D. B., Watson, S. P., Hann, M. M. (1998) RECAP—retrosynthetic combinatorial analysis procedure: a powerful new technique for identifying privileged molecular fragments with useful applications in combinatorial chemistry. *J Chem Inf Comput Sci* **38**, 511–522.
33. Peng, Z., Yang, B., Mattaparti, S., Shulok, T., Thacher, T., Kong, J., Kostrowicki, J., Hu, Q., Na, J., Zhou, J. Z., Klatte, K., Chao, B., Ito, S., Clark, J., Coner, C., Waller, C., Kuki, A. (2011) PGVL Hub: an integrated desktop tool for medicinal chemists to streamline design and synthesis of chemical libraries and singleton compounds, in (Zhou, J. Z. ed.) *Chemical Library Design*. Humana Press, New York, Chapter 15.

Section IV

Library Design for Kinase Family

Chapter 14

The Design, Annotation, and Application of a Kinase-Targeted Library

Hualin Xi and Elizabeth A. Lunney

Abstract

We present here a workflow for designing a kinase-targeted library (KTL) with the goal of capturing known kinase inhibitor chemical space. We validated our design retrospectively using recent, high-throughput screening data and found significant enrichment of kinase inhibitor hits while retaining majority of the active kinase inhibitor series. To further assist kinase projects in triaging KTL screen hits, we also developed a methodology to systematically annotate known kinase inhibitors in the KTL with regard to their binding modes.

Key words: Protein kinase, kinase-targeted library, library design, kinase chemical cores, substructure search, SMARTS Query, subsetting, binding mode annotation.

1. Introduction

The protein kinase family is one of the largest gene families encompassing almost 2% of the human genome. The enzymes phosphorylate proteins through the catalytic transfer of phospho groups from ATP to the protein substrates. Protein kinases play key roles in numerous cellular pathways that impact multiple cellular events such as growth, division, differentiation, and apoptosis. From a pharmaceutical perspective, kinases have been targeted in drug design across multiple therapeutic areas. The most prominent of these is oncology, for which eight small molecule kinase inhibitors have currently been approved in the USA.

This family of proteins exhibits a common fold that results in a two-lobe structure: a smaller N-terminal region connected by

J.Z. Zhou (ed.), *Chemical Library Design*, Methods in Molecular Biology 685,
DOI 10.1007/978-1-60761-931-4_14, © Springer Science+Business Media, LLC 2011

a hinge segment to the larger C-terminal portion. ATP binds in a well-defined pocket located between the two lobes and forms hydrogen bond interactions with the hinge. In addition, most protein kinases exist in both an active and an unactivated state, the latter of which can be very flexible. Targeting the ATP binding site in the active form plus the various unactivated conformations in drug design has led to discovery of numerous inhibitor cores or templates that can bind to members of the kinase family. Compiling a kinase-targeted library (KTL) of compounds representing this chemical landscape can greatly assist in jump-starting an early-stage project by affording a very efficient means of discovering lead matter and tool compounds. The collection can be readily screened and the analysis of assay results provides insights into structure–activity relationships and selectivity profiles. Annotation of inhibitors in the KTL in terms of chemical cores and potential binding modes further assists scientists in identifying compound series for hit-to-lead optimization.

2. Materials

Kinase protein/ligand crystal complexes were retrieved from the Pfizer Crystal Structure Database, an in-house X-ray structure repository that contains internally solved structures and selected ones imported from the Protein Data Bank (1). Kinase assay data were obtained by querying against Pfizer screening database for screens associated with any kinase target and tagged with IC_{50} or K_i as the endpoint type.

3. Methods

3.1. Kinase Domain and ATP Binding Site

The majority of protein kinases have a common fold (2) that includes a small N-terminal lobe, which is mainly beta sheet but contains a conserved α-helix C, connected by a hinge region to a larger more α-helical C-terminal lobe (**Fig. 14.1**). In the active conformation of the catalytic domain the large activation loop, A-loop, orients away from the ATP binding site to allow access to this pocket as well as the substrate docking region. This conformation is often stabilized through phosphorylation of one or more residues in the A-loop. Key residues are conserved in the binding region to align the ATP for catalytic transfer. These include a Glu in α-helix C that forms an ionic bond with a con-

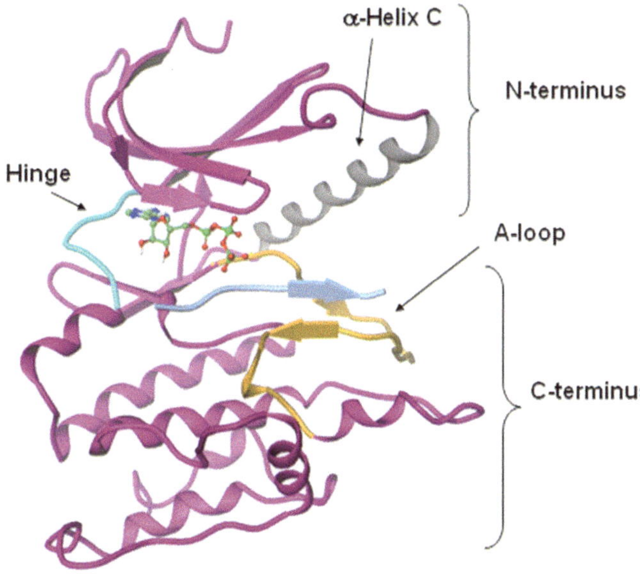

Fig. 14.1. Ribbon structure (*magenta*) of the phosphorylase kinase crystal structure 2PHK (20) bound with ATP (*green carbons*, colored by atom type) and substrate peptide (*light blue ribbon*). The N- and C-terminal lobes are highlighted; the hinge region is shown in *cyan*, the α-C helix in *gray*, and the A-loop in *orange*.

served Lys in β-strand 3, which in turn coordinates with the ATP α- and β-phosphate groups.

The ATP subunits (adenine, ribose, and tri-phosphate) bind in a cleft formed between the N- and C-lobes. Interactions are formed with the hinge backbone through hydrogen bonds: adenine N1 is an acceptor with a backbone NH and the 6-amino group is a donor to a backbone CO (**Fig. 14.2**, D1, A). Analogous polar interactions can be targeted in inhibitor design. Furthermore, a proximal hinge carbonyl group that does not interact with ATP (**Fig. 14.2**, D2) is positioned for hydrogen bond formation with a ligand. While residues in the ATP binding region that participate in the catalytic process are well conserved across the protein kinases, others vary and can be targeted to gain specificity. In addition, pockets and potential interaction points that exist beyond those utilized by ATP can be targeted in inhibitor design to enhance potency as well as selectivity (**Fig. 14.2**). These regions have been designated as NE (Northeast or Selectivity Pocket) near the Gatekeeper residue, SE (Southeast) delimited from the Phosphate pocket by the Asp of the DFG (Asp -Phe-Gly) segment at the beginning of the A-loop, W (West) extending toward solvent, and N (North), above the hinge segment. Therefore, although protein kinases bind a common endogenous ligand, the topology and the electrostatics of the ATP sites and adjacent regions vary and provide the opportunity for specificity.

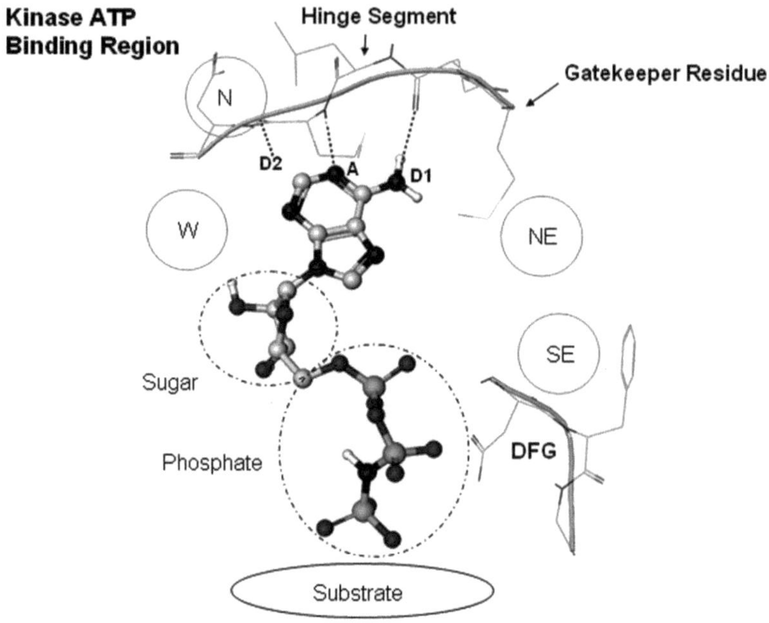

Fig. 14.2. Protein kinase active site of JNK3 bound with an ATP analogue extracted from an X-ray structure (1JNK) (21). Hinge region and DFG segment of the A-loop are shown; the sugar and phosphate binding areas are *circled*. Highlighted schematically: substrate binding region and the adjacent sites: NE, SE, W, and N. ATP hydrogen-binding interactions with the hinge region are labeled D1 (Donor1) and A (Acceptor). Another potential donor interaction (D2) with a hinge carbonyl is shown.

Unactivated states of the protein kinases have also been targeted in inhibitor design. In fact, five of the eight approved small molecule drugs have been reported to target the unactivated form of the enzyme (3). Two characterized, unactivated forms are the DFG-out (4, 5) and the αC-Glu-out conformations (6) (**Fig. 14.3**). In the DFG-out conformation, the DFG Phe, at the beginning of the A-loop, reorients from a buried pocket near the α-helix C and can extend to the ATP binding region. This in turn opens a pocket that can be accessed by inhibitor ligands. The multikinase inhibitors imatinib, which targets Abl in chronic myelogenous leukemia (CML) patients, and sorafenib, which targets angiogenesis in renal cell carcinoma, are approved drugs that probe this pocket (5, 7).

In the αC-Glu-out conformation the α-helix C moves away from the ATP site, such that the conserved Glu in the helix does not form an ionic interaction with the conserved Lys in the β-strand 3. Here again a pocket is formed for inhibitor binding (**Fig. 14.3**). The EGFR-targeted drug, lapatinib, takes advantage of this pocket in binding the tyrosine kinase (8), as do the MEK inhibitors identified by the Parke-Davis (PD) group (9). However, the latter compounds differ from lapatinib in that they do

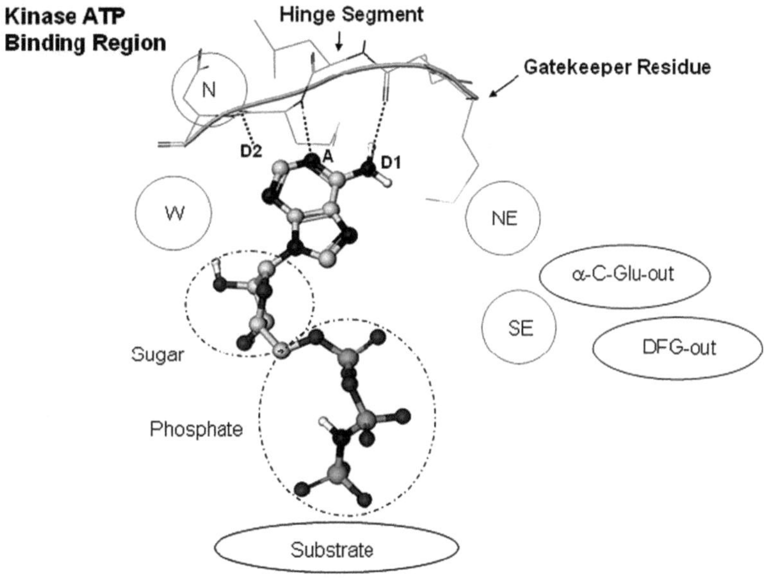

Fig. 14.3. An extension of Fig. 14.2 illustrating the DFG-out and αC-Glu-out binding regions. (Acknowledgment to J. F. Ohren for original mapping of the scheme).

not extend to the hinge region of the ATP binding site and are non-competitive with ATP.

Another category of kinase inhibitors include compounds targeting the substrate binding region (**Figs. 14.1** and **14.2**), which are also non-competitive with ATP (10). In addition, inhibitors that have been designed to target the related phosphoinositol kinases (PIK) can be characterized accordingly.

3.2. Compilation of the Kinase-Targeted Library

Our initial goal of the KTL is to compile a subset of compounds from Pfizer corporate compound collections that can comprehensively represent the existing kinase chemotypes. Several targeted library design methods including docking, use of privileged fragments and various ligand-based statistical models have been reported in literature in recent years (11–14). Here we applied a substructure query-based method to identify kinase inhibitor-like compounds in our corporate compound collection and implemented a series-based subsetting method to reduce the number down to a manageable collection of ~70K compounds. The substructure query-based approach has the advantage of being intuitive to medicinal chemists, therefore, enables computational scientists to effectively engage in design discussions with experimental scientists.

The overall design workflow for the KTL is schematically depicted in **Fig. 14.4**. In order to capture the majority of the known kinase chemotypes that represent the various binding modes, we first compiled a set of substructure queries based on

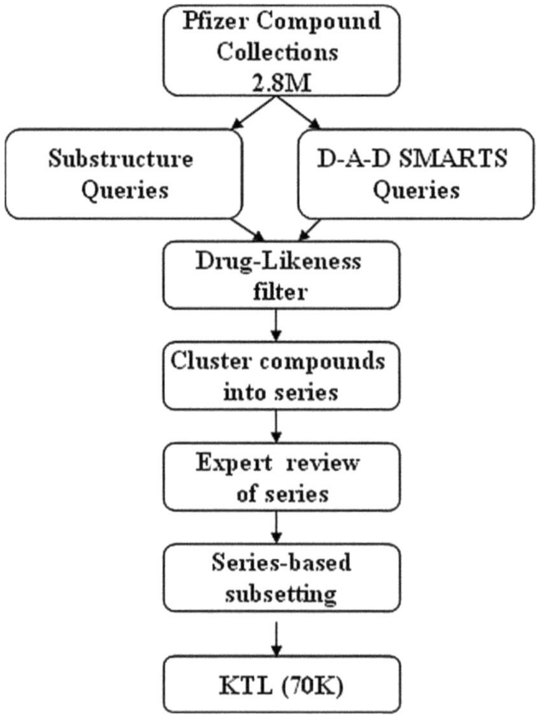

Fig. 14.4. Overall design workflow of the KTL.

the co-crystallized kinase ligands present in the corporate crystal structure database (CSDB). Our CSDB contains internally solved kinase structures and selected structures imported from the Protein Data Bank (1), Greater than 1000 kinase ligands from the CSDB representing a variety of binding modes were first clustered into ~150 major chemical series (as defined by compounds sharing common core structures) using Ward's clustering in combination with Daylight fingerprint and Tanimoto similarity metrics (15). Approximately 130 substructure queries (labeled as CSDB substructure queries) were then manually derived to capture the core structures of these chemical series. Some series such as staurosphorin-like structures were not included in the queries due to their known promiscuity or lack of interest from project teams. Then atoms in these substructure queries were replaced with query atoms while preserving the aromaticity of rings and hydrogen bonding potentials of heteroatoms. The use of query atoms allows the search to pick up additional series that resemble the known kinase chemotypes but are potentially novel. Some examples of the CSDB substructure queries are shown in **Table 14.1**. A test of these queries against the CSDB ligands correctly identified all known series of interests, while only 15% of compounds were found when the queries were run versus a set of randomly selected compounds. In order to capture additional kinase series

Table 14.1
Examples of substructure queries

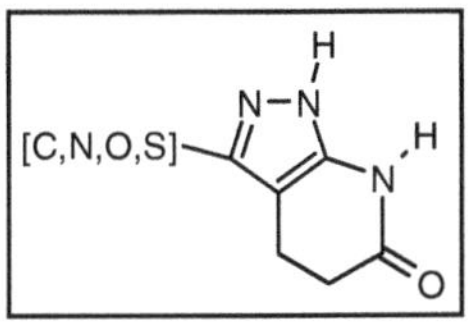
1

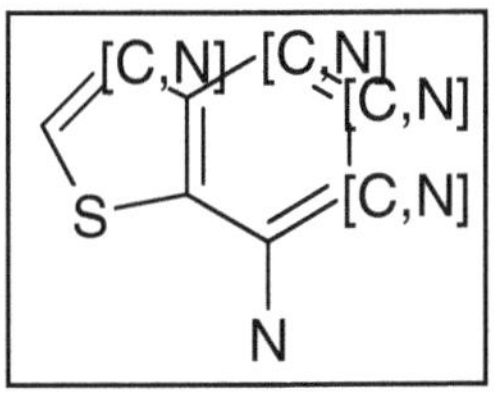
2

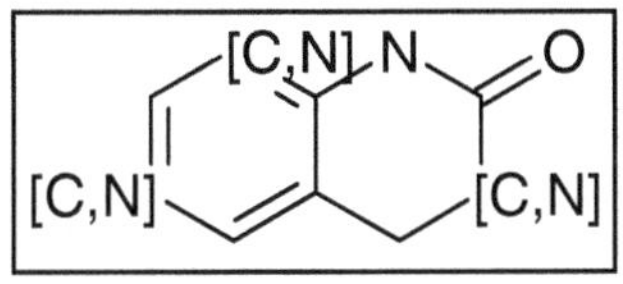
3

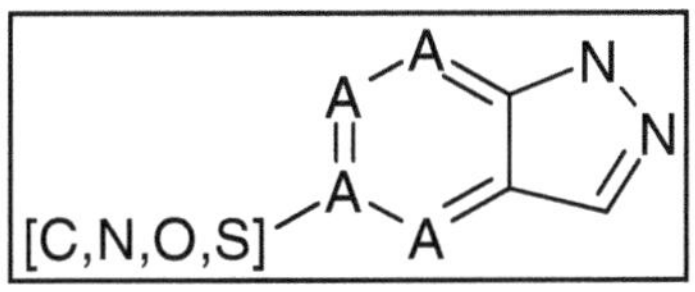
4

that did not yet have a solved structure in the CSDB, we mined our corporate screening database for any compounds with an IC_{50} less than 1 μM in either functional or enzymatic kinase assays. A total of 34K compounds from ~200 kinase assays were found. For these kinase active compounds, we filtered out compounds already represented by the CSDB substructure queries and then clustered the rest into structural series. From these, an additional ~100 substructure queries were derived from the maximal common substructure of each series.

In addition to these substructure queries, a small number of SMARTS (16) queries were derived to capture the more general hydrogen bond Donor-Acceptor-Donor (D-A-D) motif that is frequently observed in core structures interacting with kinases at the hinge regions. To make the search more specific, these SMARTS queries capture the presence of at least two out of three hydrogen bond features at the D-A-D motif (**Table 14.2**). Although single acceptor cores (for example, the pyridine moiety of Gleevec binding to Abl (17)) are missed by these SMARTS queries, the ones existing in the CSDB would be captured by the CSDB substructure queries. We tested the sensitivity and specificity of these SMARTS queries on CSDB ligands and the randomly selected compound set. While the SMARTS queries matched 75% of kinase ligands in CSDB, only 24% of the compounds in the random set were matched. Only ~40% of the hits from the random set are also found in the hits from CSDB substructure queries indicating that the SMARTS query could potentially identify additional kinase inhibitor-like compounds.

With these sets of substructures and SMARTS queries, we searched our corporate compound collections and identified

Table 14.2
Example of SMARTS queries

Motifs	Examples	SMARTS
A~D	Pyrazole	[N, n;!H0]~[nX2;H0; R1]
A~~D	Amide in a ring	[N, n;!H0;R1]~*~[$(O=[C, S])]
A~~D	Azaindole, adenosine, amino-pyrimidine, etc	[$([N, n;!H0]~*~[$([nX2; H0; R1])]);!$(n1*n**1);!$(NC=N)]
A~~~D	Pyrrolepyrmidine	[N, n;!H0]~*!:*~[nX2; H0]
A~~~D	amine, carbonyl attached to aromatic ring	[N; !H0]~a~[*; R1]~[$(O=[C, S])]
A~~~D		[n; !H0]~*~*~[$(O=[C, S])]
Other cases for amides	5-member-aryl-amide	[N; H2]C(=O)-a 1aaaa1
Other cases for amides	6-member-aryl-amide	[N; H2]C(=O)-a 1aaaaa1
Other cases for amides	Biaryl urea	a 1aaaaa1-[N;!H0]C(=O)[N;!H0]a

840K hits from a total 2.8 M compounds. Then a set of drug-likeness filters were applied to these compounds to reduce down the total number of compounds to 720K. We then split these 720K compounds into two collections – the library set (270K compounds) amenable to combinatorial synthesis with library synthesis protocols available and the "medchem" set (450K compounds) mostly made through traditional medicinal chemistry synthesis. To further prioritize these hits, compounds in the library sets were grouped by library protocol id and compounds in the "medchem" set were clustered into structural series using Ward's clustering with Daylight fingerprints. Then four representative structures from each library or series were selected. A panel of experienced kinase chemists were then asked to review and prioritize the represented compound library or series based on the physical properties, synthetic doability, as well as structural novelty. The review process focused on the chemical series as opposed to individual compounds. Although each kinase expert might unintentionally be biased toward a subset of chemical series that he or she worked on in the past, by having multiple experts in the review process, we were aiming to have a more unbiased representation of the kinase chemical space collectively. In the end, 310K compounds were retained after pooling the chosen series together. We validated this selection retrospectively using data from two recent kinase HTS projects (HGK and JNK1) that screened the full compound collection in Pfizer and data from Pfizer kinase selectivity panel screens (**Table 14.3**). For HGK, 82% of the Rule of 5 (Ro5) (18) compliant confirmed hits were recovered in the 310K collection representing an 8-fold hit rate

Table 14.3
Enrichment of kinase inhibitors in the initial compilation of the KTL (310K compound collection) using substructures and SMARTS queries prior to applying subsetting

	HGK	JNK1
# Compounds Screened in HTS	1,600,000	1,600,000
# confirmed actives	945	1455
# of actives passed filter (MW<=550, clogP<=7, RotB<12)	833	1376
# of actives passed filter and present in 310k collection	685	920
% recovered	82	67

enrichments (defined as hit rate in the 301K collection divided by the overall hit rate in the HTS). Similarly for JNK1, 67% of confirmed hits were recovered representing a 6.7-fold hit rate enrichment. For the kinase selectivity panel hits (compounds with 50% inhibition against any kinase on the panel at 10 μM concentration), 80% of Ro5 compliant compounds were recovered. Overall these validations indicate reasonable combination of sensitivity and specificity for this 310K compound collection.

To further reduce the total number of selected compounds to a manageable subset for screening, a final subsetting step was applied. We use a series-based subsetting method where compounds in each series were randomly sampled. The percentage of compounds selected from each series depends on the size of the series, ranging from 16% (1/6) for the largest clusters (clusters with >1000 compounds) to 100% (i.e., selecting all compounds) for the series with one or two compounds. This subsetting approach enabled us to remove overrepresentation of some large chemical series without significantly affecting representation of small series. This final step reduced the total number of compounds to 73K, an overall ~4-fold reduction of the collection. To evaluate the effect of the subsetting step, we analyzed the coverage of active series from the two actual kinase screens, HGK and JNK1. As shown in **Table 14.4**, 688 active compounds (representing 40 series) from the HGK screen and 730 active compounds (representing 44 series) from the JNK1 screen were found in the initial set of 310k compounds before the subsetting. After the subsetting, as expected only 25–30% of unique, active compounds were retained. In contrast, 85–90% of the series are retained indicating the subsetting step has minimal impact at the series level.

Table 14.4
Validation of the subsetting algorithm retrospectively using data from two HTS screening projects. Majority of the active series were retained after the subsetting

	HGK	JNK1
# compounds (series) before subsetting	688 (40)	730 (44)
# compounds (series) found in after subsetting	263 (37)	208 (37)

3.3. Kinase-Targeted Library Annotations

Individual inhibitors in the KTL with known kinase activity and related structural information can be annotated based on the types of binding interactions made with the protein. This process can greatly assist the project team in triaging the screening results and in identifying chemical series to prosecute. At the highest level, inhibitors can be defined by the site of binding or whether they are a known phosphoinositol kinase series: ATP site, DFG-out, PD-MEK-type inhibitor site, substrate site, and PIK inhibitors. The compounds can be further categorized by the key binding template or series core.

The majority of inhibitors in the KTL bind in the ATP site and the cores are defined by the rings or groups that form hydrogen bonds with the hinge segment. For example, the P38 inhibitor, SB203580 (19), shown in **Fig. 14.5** interacts with the hinge region through the pyridine ring nitrogen as the acceptor. This template or core would be designated "Pyri(mi)dine-5-MemberHeterocycle_4" with one interaction with the hinge region: "A". A compound with an amino group at the 2-position of the pyrimidine would have a second hydrogen bonding contact with the hinge and would be defined with a unique core, Pyri(mi)dine-5-MemberHeterocycle_3, with two hinge interactions: "D2, A". By analyzing bound cores in X-ray structures, the substitution sites can be annotated according to which pocket(s) they would probe and thus a specific compound could be so labeled. The inhibitor example in **Fig. 14.5** would be fully annotated as ATP site; A; Pyri(mi)dine-5-MemberHeterocycle_4; Phos, NE, indicating that the compound targets the ATP site, has one interaction with the hinge (A), which was made by the pyridine core Pyri(mi)dine-5-MemberHeterocycle_4 and extends to the phosphate and NE regions (Phos, NE). Examples of annotations for inhibitors that target the non-activated state of the protein are shown in **Fig. 14.6**: sorafenib (DFG-out binder) and the MEK inhibitor, PD318088 (9). In these cases, subsite binding would not be annotated.

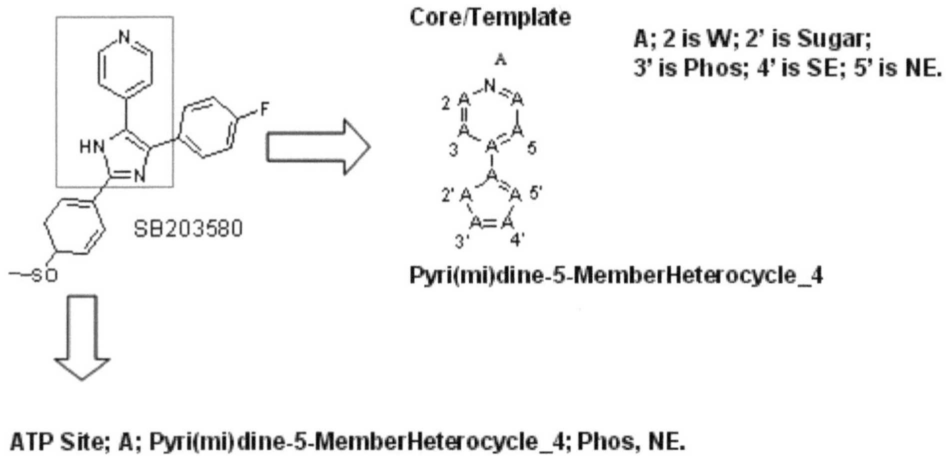

Fig. 14.5. Annotation for the ATP site inhibitor, SB203580. The core is Pyri(mi)dine-5-MemberHeterocycle_4, which is a hydrogen bond acceptor with the hinge region. The compound probes the Phosphate (Phos) and NE sites.

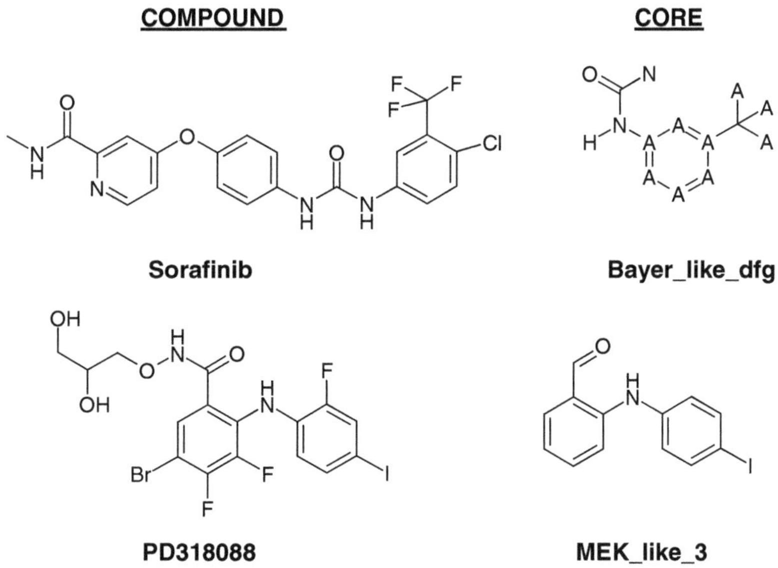

Fig. 14.6. Annotations for inhibitors that bind to unactivated kinase conformations. Sorafenib binds to the DGF-out conformation and its core is defined as Bayer_like_dfg. PD318088 binds to the αC-Glu-out conformation and its template is MEK_like_3.

3.4. Performance of KTL and Future Plans

Since the establishment of the KTL in Pfizer, many kinase projects have screened the KTL collection. Among the first seven screens completed, the hit rate (defined as retest confirmed hits divided by total number of compounds screened in the KTL) ranges from 0.5 to 3%, several fold higher than a typical full HTS campaign with confirmed hit rate in the range of 0.1–0.3%. For many of the projects, KTL screens have led to interesting chemical series for project teams to pursue hit-to-lead optimization.

While this version of the KTL successfully captured known kinase inhibitor series, the goal for our next-generation KTL would be to apply de novo design methods to incorporate novel chemotypes and to incorporate more nonclassical chemotypes that bind to a protein kinase beyond the typical ATP pocket.

4. Notes

4.1. Advantage of Using Substructure Query-Based Method for KTL Design

We presented here a workflow to design kinase-targeted library using a substructure query-based method. Compared to various de novo kinase-targeted library design methods, our approach has the advantage of ensuring a comprehensive coverage of known kinase inhibitor chemotypes.

In the past 10 years, there has been a large number of kinase projects conducted in Pfizer for a range of therapeutic areas. By deriving substructure queries from kinase ligands in our corporate crystal structure database as well as from active compounds identified in in-house kinase assays, we were able to capture the institutional knowledge on kinase inhibitors in the KTL design. It is a common observation that inhibitors against different kinases often share the same core structure. This is supported by the overall high similarity of active sites among kinases. The use of substructure searches in our design workflow guarantees coverage of all compounds containing any of these common kinase cores from our corporate compound collection . The subsequent series-based subsetting provides a sampling of each core series. Such diverse sampling is critical as the KTL will be screened against novel kinase targets.

4.2. Use of KTL Core Annotation for HTS Triage

The KTL core annotations have been integrated into several in-house desktop applications for compound design and HTS hit triage. The annotations provide key binding information for the inhibitors and can be used to cluster compounds or to search for inhibitors with a particular binding feature. Overall these insights can help accelerate the HTS triage process and allow project teams to advance chemical matter in a timely manner.

References

1. Berman, H. M., Westbrook, J., Feng, Z., Gilliland, G., Bhat, T. N., Weissig, H., Shindyalov, I. N., Bourne, P. E. (2000) The Protein Data Bank. *Nucleic Acids Res* **28**, 235–242.
2. Johnson, L. N., Lowe, E. D., Noble, M. E. M., Owen, D. J. (1998) The structural basis for substrate recognition and control by protein kinases. *FEBS Lett* **430**, 1–11.
3. Alton, G. R., Lunney, E. A. (2009) Targeting the unactivated conformations of protein kinases for small molecule drug discovery. *Expert Opin Drug Discov* **3**, 595–605.

4. Pargellis, C., Tong, L., Churchill, L., Cirillo, P. F., Gilmore, T., Graham, A. G., Grob, P. M., Hickey, E. R., Moss, N., Pav, S., Regan, J. (2002) Inhibition of p38 MAP kinase by utilizing a novel allosteric binding site. *Nat Struct Biol* **9**, 268–272.
5. Schindler, T., Bornmann, W., Pellicena, P., Miller, W. T., Clarckson, B., Kuriyan, J. (2000) Structural mechanism for STI-571 inhibition of Abelson tyrosine kinase. *Science* **289**, 1938–1942.
6. Levinson, N. M., Kuchment, O., Shen, K., Young, M. A., Koldobskiy, M., Karplus, M., Cole, P. A., Kuriyan, J. (2006) A Src-like inactive conformation in the Abl tyrosine kinase domain. *PLoS Biol* **4**, 753–767.
7. Reeves, D. J., Liu, C. Y. (2009) Treatment of metastatic renal cell carcinoma. *Cancer Chemother Pharmacol* **64**, 11–15.
8. Wood, E. R., Truesdale, A. T., McDonald, O. B., Yuan, D., Hassell, A., Dickerson, S. H., Ellis, B., Pennisi, C., Horne, E., Lackey, K., Alligood, K. J., Rusnak, D. W., Gilmer, T. M., Shewchuk, L. . (2004) A unique structure for epidermal growth factor receptor bound to GW572016 (lapatinib): relationships among protein conformation, inhibitor off-rate, and receptor activity in tumor cells. *Cancer Res.* **64**, 6652–6659.
9. Ohren, J. F., Chen, H., Pavlovsky, A., Whitehead, C., Zhang, E., Kuffa, P., Yan, C., McConnell, P., Spessard, C., Banotai, C., Mueller, W. T., Delaney, A., Omer, C., Sebolt-Leopold, J., Dudley, D. T., Leung, I.K., Flamme, C., Warmus, J., Kaufman, M., Barrett, S., Tecle, H., Hasemann, C.A. (2004) Structures of human MAP kinase kinase 1 (MEK1) and MEK2 describe novel noncompetitive kinase inhibition. *Nat Struct Mol Biol* **11**, 1192–1197.
10. Vanderpool, D., Johnson, T. O., Chen, P., Bergqvist, S., Alton, G., Phonephaly, S., Rui, E., Luo, C., Deng, Y. -L., Grant, S., Quenzer, T., Margosiak, S., Register, J., Brown, E., Ermolieff, J. (2009) Characterization of the CHK1 allosteric inhibitor binding site. *Biochemistry* **48**, 9823–9830.
11. Bradley, E. K., Miller, J. L., Saiah, E., Grootenhuis, P. D. (2003) Informative library design as an efficient strategy to identify and optimize leads: application to cyclin-dependent kinase 2 antagonists. *J Med Chem* **46**, 4360–4364.
12. Lowrie, J. F., Delisle, R. K., Hobbs, D. W., Diller, D. J. (2004) The different strategies for designing GPCR and kinase targeted libraries. *Comb Chem High Throughput Screen* **7**, 495–510.
13. Prien, O. (2005) Target-family-oriented focused libraries for kinases–conceptual design aspects and commercial availability. *Chembiochem* **6**, 500–505.
14. Stahura, F. L., Xue, L., Godden, J. W., Bajorath, J. (1999) Molecular scaffold-based design and comparison of combinatorial libraries focused on the ATP-binding site of protein kinases. *J Mol Graph Model* **17**, 1–9, 51–52.
15. Daylight. Daylight Cheminformatics Toolkits. http://www.daylight.com.
16. Daylight. Daylight SMARTS Theory. http://www.daylight.com/dayhtml/doc/theory/theory.smarts.html.
17. Nagar, B., Bornmann, W. G., Pellicena, P., Schindler, T., Veach, D. R., Miller, W. T., Clarkson, B., Kuriyan, J. (2002) Crystal structures of the kinase domain of c-Abl in complex with the small molecule inhibitors PD173955 and imatinib (STI-571). *Cancer Res* **62**, 4236–4243.
18. Lipinski, C. A., Lombardo, F., Dominy, B. W., Feeney, P. J. (2001) Experimental and computational approaches to estimate solubility and permeability in drug discovery and development settings. *Adv Drug Deliv Rev* **46**, 3–26.
19. Regan, J., Breitfelder, S., Cirillo, P., Gilmore, T., Graham, A.G., Hickey, E., Klaus, B., Madwed, J., Moriak, M., Moss, N., Pargellis, C., Pav, S., Proto, A., Swinamer, A., Tong, L., Torcellini, C. (2002) Pyrazole urea-based inhibitors of p38 MAP kinase: from lead compound to clinical candidate. *J Med Chem* **45**, 2994–3008.
20. Lowe, E. D., Noble, M. E., Skamnaki, V. T., Oikonomakos, N. G., Owen, D. J., Johnson, L. N. (1997) The crystal structure of a phosphorylase kinase peptide substrate complex: kinase substrate recognition. *EMBO J* **16**, 6646–6658.
21. Xie, X., Gu, Y., Fox, T., Coll, J.T., Fleming, M.A., Markland, W., Caron, P.R., Wilson, K.P., Su, M.S. (1998) Crystal structure of JNK3: a kinase implicated in neuronal apoptosis. *Structure* **6**, 983–991.

Section V

Library Design Tools

Chapter 15

PGVL Hub: An Integrated Desktop Tool for Medicinal Chemists to Streamline Design and Synthesis of Chemical Libraries and Singleton Compounds

Zhengwei Peng, Bo Yang, Sarathy Mattaparti, Thom Shulok, Thomas Thacher, James Kong, Jaroslav Kostrowicki, Qiyue Hu, James Na, Joe Zhongxiang Zhou, David Klatte, Bo Chao, Shogo Ito, John Clark, Nunzio Sciammetta, Bob Coner, Chris Waller, and Atsuo Kuki

Abstract

PGVL Hub is an integrated molecular design desktop tool that has been developed and globally deployed throughout Pfizer discovery research units to streamline the design and synthesis of combinatorial libraries and singleton compounds. This tool supports various workflows for design of singletons, combinatorial libraries, and Markush exemplification. It also leverages the proprietary PGVL virtual space (which contains 10^{14} molecules spanned by experimentally derived synthesis protocols and suitable reactants) for lead idea generation, lead hopping, and library design. There had been an intense focus on ease of use, good performance and robustness, and synergy with existing desktop tools such as ISIS/Draw and SpotFire. In this chapter we describe the three-tier enterprise software architecture, key data structures that enable a wide variety of design scenarios and workflows, major technical challenges encountered and solved, and lessons learned during its development and deployment throughout its production cycles. In addition, PGVL Hub represents an extendable and enabling platform to support future innovations in library and singleton compound design while being a proven channel to deliver those innovations to medicinal chemists on a global scale.

Key words: Drug discovery, chem-informatics, molecular design, combinatorial chemistry, combinatorial library, synthesis protocol, PGVL, reactant, product, enumeration, filtering integration, workflow, streamline, desktop tool, software deployment.

J.Z. Zhou (ed.), *Chemical Library Design*, Methods in Molecular Biology 685,
DOI 10.1007/978-1-60761-931-4_15,

1. Introduction

Like many new technologies introduced to drug discovery, combinatorial library design and synthesis have matured during the last 15 years (1–5). This is reflected by the fact that their practice has been shifted significantly from experts, such as computational and combinatorial chemists, to general medicinal chemists working in the pharmaceutical industry (6–27). In terms of library design technology, we have seen a similar shift of focus from methodology development to integration and deployment. With the maturing of this technology, three new needs have emerged. First, a significant amount of chemistry knowledge has accumulated in the form of detailed synthetic protocols. These protocols not only contain step-by-step synthesis instructions but also specify what is considered to be suitable chemical reactants compatible with the reaction conditions explored and validated experimentally (termed the "scope and limitation" of the given synthetic protocol). Systematic capturing, mining, and reuse of such knowledge would bring tremendous value and competitive advantage to a pharmaceutical company in hit generation, hit follow-up, and lead optimization. In a separate publication, we have described Pfizer's effort for the past 10 years in this type of systematic knowledge capture and reuse which led to the PGVL (Pfizer Global Virtual Library) chemistry knowledge base (28). Second, unlike the expert community which is more interested in the latest design algorithms or flexibility in customizable solutions, medicinal chemists working on drug discovery projects place more emphasis on ease-of-use, start-to-finish workflow management, and robust deployment as their top needs (22–27). For example, the ADEPT system was developed by scientists from Galaxo and Daylight and deployed to the Glaxo discovery chemistry community as an integrated suite of Web tools on its corporate intranet for reactant selection, library enumeration, and molecular property profiling and library design (25). The REAL-ISIS system from Johnson & Johnson essentially accomplished the same major goals of combinatorial library design, with more focus on medicinal chemists and utilization of more advanced software architecture and components (26). Finally, a modern medicinal chemist uses many software packages to increase productivity. To gain acceptance by the chemist, a new software package needs to work with existing software packages synergistically to reduce training effort and to enable richer and more powerful workflows. Over the years, both commercial vendor offers and in-house chem-informatics solutions have evolved to address these three aforementioned needs and have achieved varying degrees of success (6–27). Within Pfizer, we have witnessed a similar path starting from an expert-only tool running on the UNIX platform (23, 24), moving to a Web-based two-tier solution for a single department or a selective group of users, and eventually reaching a three-tier enterprise solution with a very capable and interactive Java client on the scientists' desktop computer.

In this chapter, we shall discuss the major requirements for PGVL Hub to meet the three needs listed above, the software architecture and key enabling data structures we have employed, and the major technical challenges encountered and solved. Some simple library design examples will be used throughout this report to showcase the main features of PGVL Hub and the enabled molecular design workflows. Finally, we will conclude by discussing the impact of PGVL Hub in terms of adoption and usage by medicinal chemists over the past several years.

2. Major Requirements

There are a few key design scenarios commonly requested by medicinal chemists. They are described in the following sections.

2.1. Singleton Design

In this workflow, chemists want to draw or import a set of molecules, profile their molecular properties, such as computed ADME&T (absorption, distribution, metabolism, excretion, and toxicity) properties, estimated activities against specific protein targets based on existing SAR models, and make selections based on the analysis of structural features and computed molecular properties of those singleton molecules.

2.2. Markush Exemplification

Similar to the singleton design scenario, but the input set of molecules is created automatically based on a user-supplied Markush drawing with R-groups attached to a molecular core structure along with sets of explicit examples of those R-groups. This design scenario is very popular with medicinal chemists in terms of expressing their chemistry ideas or in the analysis of compounds commonly represented in patent literature.

2.3. Library Design Using User Reaction and Reactant Sets

This is the standard library design scenario which is supported by most in-house and commercial vendor tools. The user-defined reaction is usually a Markush reaction drawing commonly in the format of MDL ISIS sketch or .rxn file (35). Chemists may also supply reactant sets for each reaction component either by loading pre-defined sets of molecules or by retrieving them via searches into chemical reactant databases. Molecular property calculations and analysis on the reactants are performed, and selections are made based on these results. This is commonly known as the "Reactant-Based Library Design," and the outcome is a fully combinatorial library. Conversely, chemists can also generate explicit products via a product enumeration tool and perform property calculations and analysis, then make decisions based on product properties. If the final product selections are done at the level of individual product molecules, then a cherry-picked (sparse matrix) library will be the outcome. On the other hand, if the final selections are still done at the reactant level, then the outcome remains a fully combinatorial library, even when explicit product properties are aggregated and used to guide and shape the reactant selection. Both cherry-picking and fully combinatorial approaches are used by medicinal chemists. The full combinatorial design has an advantage in terms of library production efficiency in maximizing the number of products synthesized with a given number of reactants to be handled during library production. However, the cherry-picked library design is more flexible and allows the designer to ensure that all products satisfy user-imposed design criteria. With an increased level of synthesis automation and popularity of small yet highly targeted libraries, the cherry-picked library design is becoming the dominant mode in pharmaceutical industry.

2.4. Library Design Using PGVL Reactions and Pre-mined Suitable Reactants

In this design scenario, users can take advantage of the captured chemistry knowledge inside the PGVL chemistry knowledge base by using various readily available components, ranging from high-quality product enumeration instructions to pre-mined reactants lists suitable to pre-validated synthetic protocols, in order to streamline library design and production (28). Since a significant portion of modern corporate screening compound collections originates from combinatorial chemistry, HTS hits resulting from this subset synthesized via combinatorial chemistry can be followed-up quickly and effectively with targeted libraries using the same pre-validated synthetic protocols and pre-mined compatible reactant sets. This is one of the unique objectives for PGVL Hub.

2.5. Initiating Library Design via Lead Centric Mining (LEAP)

Conceptually, the PGVL virtual product space is a collection of molecules that can be made via one or more of the registered synthetic protocols. While medicinal chemists routinely perform similarity searches into vendor and corporate molecular databases (approximately 10^7 in size) based on the given lead molecules to gather information on synthetic routes, formulate SAR relationships, or generate lead-hopping hypotheses, it is also desirable to perform similarity searches into the PGVL product space (approximately 10^{14} in size) (28). The only challenge is that the technology managing the current corporate compound collections cannot be directly used for this purpose due to the enormous size of the PGVL product space. One innovative solution has been developed via the lead centric mining tool, and its design concepts and application will be described in detail by a separate publication (29). The output of a LEAP search is a collection of PGVL virtual compounds, each linked to an available combinatorial synthesis protocol and a combination of explicit reactants that fully describes how this compound can be synthesized. Chemists can then use these results to further evaluate each hit and launch one or several targeted library designs to follow up on the LEAP-derived hits.

2.6. Additional Considerations

In addition to the above design scenarios which provide the core requirements for the design and implementation of PGVL Hub, there are several other requirements to be considered. As a desktop tool for medicinal chemists, PGVL Hub should be very graphical with good performance and robustness, easy to learn and use, and easy to deploy and update with minimum administrative effort. More specifically, it needs to manage a fairly rich set of hierarchical data structures (molecule, set of molecules, reaction, reactant, product, library, design, work session, etc.) and ensure their capture either as a saved file on a desktop computer or as a registration entry into downstream chemical information systems for library registration and synthesis. It should also have a set of design features and a pluggable framework to easily integrate new features to be developed in the future. Finally, PGVL Hub has to integrate closely and seamlessly with existing tools to realize synergies for a more streamlined and more powerful design experience. Among the software packages PGVL Hub interfaces with are ISIS/Draw (30) for structure and reaction drawing; Microsoft Excel (31) for list and table management; SpotFire (32) for data visualization, analysis, and decision making; and various other 2D and 3D molecular design tools.

3. PGVL Hub

3.1. Three-Tier Enterprise Architecture

We have chosen the J2EE (http://www.sun.com/java/) three-tier enterprise software architecture for PGVL Hub (**Fig. 15.1**). The client side is a J2SE GUI built as a Java Web Start (33) deployable application which enables easy and automated deployment and update of both prototype and production versions globally from a single central server machine. Chemists can easily install the client-side software component via a Web link available on an internal Web page. The Java Web Start technology also provides automatic version check and upgrade of the client-side component each time PGVL Hub is launched by the user. This mechanism was used to deploy PGVL Hub during all stages

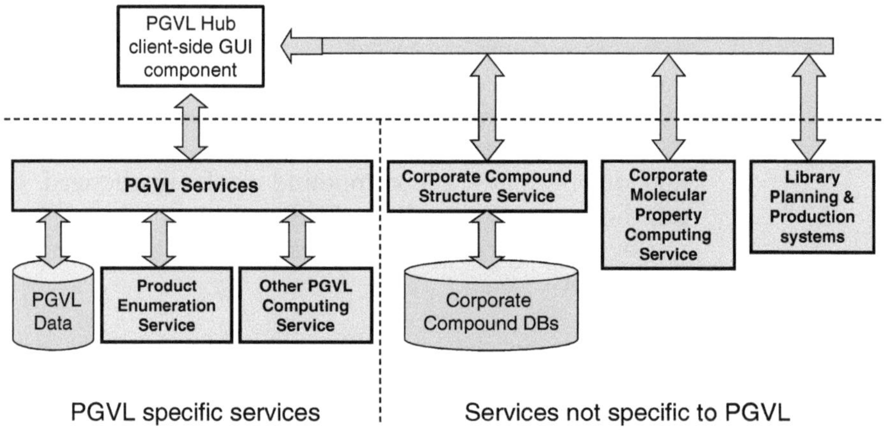

Fig. 15.1. Three-tier J2EE architecture used by PGVL Hub. The client is a Java Swing GUI deployed via the Java Web Start technology (33) to chemists' desktop. The middle-tier is WebLogic J2EE server (51). PGVL data are hosted by Oracle (52). And the product enumeration service and the PGVL computing service are hosted on SciTegic Pipeline Pilot server (34). The corporate compound structure service provides support to PGVL Hub client for compound ID to structure look-up, query searches into corporate databases, inventory checking, and compound duplicate checking. The corporate molecular property computing service (41) returns computed properties chosen by user for a set of submitted molecules.

of the software life cycle with maximum ease and flexibility while minimizing administrative cost.

The J2EE middle tier manages the interaction between client-side and server-side backend resources (e.g., various databases and molecular property computing services) and enables the client-side and server-side software components to be updated independently. The server-side backend has access to various corporate compound databases and the PGVL chemistry knowledge base which delivers captured combinatorial reactions, synthetic protocols, and pre-mined and indexed suitable reactants for these protocols (28). It also contains computational services that perform product enumeration and compute various molecular properties at the request of the client-side GUI component. Since several of the server-side services are not unique to PGVL Hub, we were able to leverage emerging general purpose services via the service-oriented architecture (SOA). Reciprocally, the PGVL server side is also designed and deployed as a service, so software packages other than the client-side component of PGVL Hub can tap into the PGVL services as a valuable resource. The underlying database engine for PGVL data is Oracle, and the underlying engine for product enumeration is SciTegic Pipeline Pilot (34). More discussion on the PGVL services and captured chemistry knowledge can be found in a separate publication (28).

3.2. Data Structures

A hierarchical data structure was designed to capture all data elements created during a design session, as shown in **Fig. 15.2**. At the very basic level of "Molecular Structure," a CTAB string

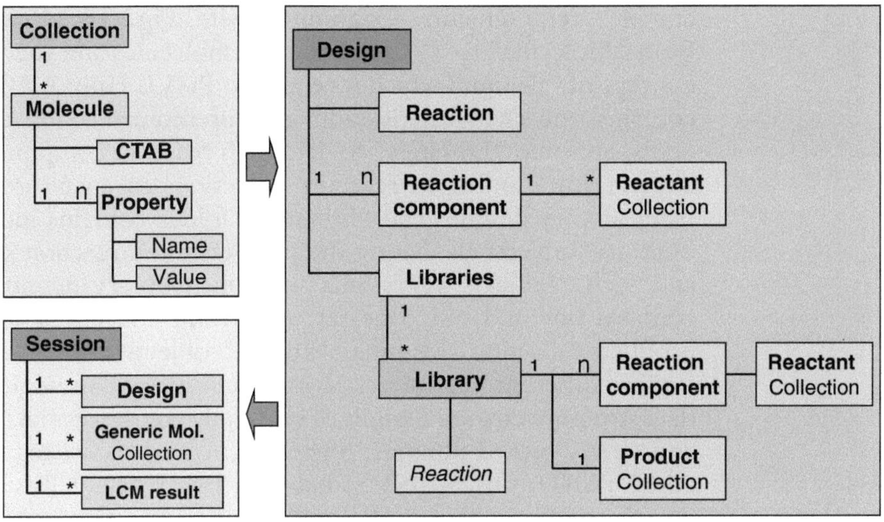

Fig. 15.2. Key data structures within PGVL hub for library design. The three data structures (*Collection, Design, and Session*) within PGVL Hub are hierarchical; each is built on top of previous ones sequentially. A "Collection" contains a set of molecules and their molecular properties. By default, one-to-one relationship between two data entities is assumed, unless marked as either "1*n*" which means for each parent entity, there are *n* (1, 2, 3, . . .) instances of child entities associated. "1 *" means that for one parent entity, there could be any number (0, 1, 2, 3, . . .) of child entities associated. MDL CTAB string (35) is used to represent the 2D structure of a molecule. "Design" encapsulates everything about a library design work using one chemical reaction scheme. The "Reaction" is either an Rxn ID pointing to a pre-registered PGVL reaction (28) or a MDL RDF string (35) for user-drawn reaction scheme for product enumeration, or even a Markush drawing with R-groups hanging off the core (also in MDL CTAB string format, *see* Ref. (35)) for the Markush exemplification workflow. The "Rxn Component" is created for each reaction component in reaction to be a place holder for various reactant sets (or R-group sets for the Markush exemplification design workflow) for that reaction component to enable reactant-based library design. The "Libraries" folder is a place to hold many explicit designed libraries. This folder also enables chemists to compare and combine individual libraries (designed based on different design objectives and protocols) within this folder into new ones. The "Library" concept contains all elements of a single explicit library designed, and is self-contained with the "*Reaction*" information used to form the library. The single set of reactant sets (one for each reaction component) now is fully synchronized with the product collection automatically during a library design to ensure self-consistency. Finally "Session" contains everything a user has worked on during a single working session after launching PGVL Hub and can be easily saved as an XML file for later use. Inside a "Session" folder, one can have multiple "Design" folders, each with a unique reaction. The "Generic Molecule Collection" is intended for simple singleton design workflow. It is also a convenience place to share sets of molecules between different "Design" folders via "Drag-and-Drop" or "Copy-and-Paste" operation. The "LEAP Result" folder is a specialized "Collection" which holds the Lead-Centric-Mining (LEAP) results as a collection of PGVL molecules with their combinatorial origin (reaction, protocol, and reactant combination to make each individual molecule).

(a molecular format published by MDL) with 2D atomic coordinates is used for molecular representation (35). PGVL Hub not only renders molecular structures graphically based on the 2D atomic coordinates inside the CTAB string but also allows chemists to update those 2D atomic coordinates via ISIS/Draw to improve the 2D layout of a molecule if desired. At the "Molecule" level, users can add any number of properties (textual or numerical) to a molecule, with the format being a combination of name and value. We have found this data structure to be adequate, although it could be extended to handle more complex requirements. Continuing with **Fig. 15.2**, we move from "Molecule" to a "Collection" of molecules and then to the concept of "Design" which is central to PGVL Hub. A "Design" contains one chemical reaction, one or more reaction components, and one "Libraries" folder. Each reaction component can contain any number of reactant collections to enable reactant-based library design. The "Libraries" folder contains individual "Library" objects all sharing the same chemical reaction scheme, and each "Library" represents an explicit library design which contains one and only one set of reactant collections and one product collection. Multiple "library" objects can be used to explore different hypotheses or molecular design strategies. This data structure enables designs of both fully combinatorial libraries and cherry-picked libraries. Moreover, it enables list-logic operations (AND, OR, MINUS) between two reactant lists or two libraries within a single "Design." Finally a "Session" object can contain any number of "Designs" plus other data elements of a design session. A single "Library," a "Design," and a full design "Session" can all be exported or imported back into PGVL Hub as data objects in three XML file formats for storage, sharing, and refinement.

We have taken several steps in working with the XML data format to ensure integrity of the data objects. To save a CTAB string as a data element into an XML file, it is encoded via the BASE64 scheme and compressed into an XML-safe string using gzip (36–38). A user-drawn or imported reaction scheme is represented as an MDL RDF (35) string, which is likewise encoded and compressed before saving into an XML file. The encoding and compression steps ensure robustness and compactness of the XLM files written by PGVL Hub. One special circumstance involves the use of special characters in molecular properties. These characters may have special meanings to the XML format and have the potential to corrupt the XML files. We have solved this problem by identifying all instances of these characters and encoding them into XML-safe strings before they are saved into XML files. All encoding and compression steps are reversed when importing an XML file back into PGVL Hub.

3.3. Performance

Library design, unlike singleton design, usually involves collections of many molecules. It is fairly common to encounter design cases starting with hundreds to thousands of reactants for each reaction component which could potentially lead to huge enumerated libraries. Such a large collection of molecules poses three performance challenges to PGVL Hub. First, how do we move them quickly between the client and server components of PGVL Hub? Second, how do we store and manage them once they arrive at a client machine? Lastly, how do we enhance the user experience when performing molecular property calculation on such large collections? Some molecular properties such as Rule-of-Five (39) can be computed quickly, but others like LogD (40) and docking scores can be computationally expensive and time consuming.

We initially encountered the network throughput bottleneck while moving a large number of molecules between the PGVL server and the client components. This problem was addressed by deploying multiple PGVL servers, one for each Pfizer research site. By having a PGVL server within close network proximity of the intended users, the throughput bottleneck was greatly reduced. Furthermore, having multiple PGVL servers provided redundancy which greatly improved the availability of the PGVL service. As network throughput improved, we eventually returned to the original deployment model with a single global server cluster running multiple instances of WebLogic Application Server (51). Nevertheless, the multiple-servers-at-multiple-sites approach has provided great value throughout the development, deployment, and user support of PGVL Hub.

At the early stage of the PGVL project, molecules were stored in the RAM of the client machine so that users could browse 20–100 molecular structures per viewing page quickly and efficiently. This speedy structure browsing capability of PGVL Hub was warmly received by medicinal chemists. However, when PGVL Hub encountered large molecule collections numbering in the tens of thousands or more (10 K molecules required approximately 1 GB of RAM for storage), its performance would drop significantly. This situation can be further exacerbated when other memory-intensive applications are running concurrently on the same client machine. To overcome this problem, we have implemented a hard disk-based cache system which operates with great efficiency and performance. All molecules either loaded into PGVL Hub or generated during a design session (e.g., an enumerated virtual library) are indexed and saved onto the disk file system which is much larger than the physical RAM space on a typical desktop computer. PGVL Hub would then use a fast lookup scheme to load only those molecules to be displayed on the screen. This solution enables chemists to work on extremely large library designs without experiencing any performance degradation. With a much reduced RAM footprint, PGVL Hub also becomes a good "desktop citizen" among the many software packages concurrently running on a typical machine.

The Pfizer molecular property computing service contains many properties related to ADME&T endpoints as well as project-specific SAR models (41). These computed molecular properties are often critical for medicinal chemists to ensure that the designed singleton and/or library compounds have good drug-like properties and to explore new SAR hypotheses against specific protein targets. While some molecular properties can be calculated quickly, others can be much more time consuming. Adding more computing power to speed up molecular property calculations was utilized as part of the solution to reduce turn-around time. More importantly, users should be allowed to continue working with PGVL Hub after submitting jobs for molecular property calculations. This requires an asynchronous computing model which is available from the Pfizer molecular property calculation service. To further improve performance, PGVL Hub also utilizes a "divide and conquer" approach which breaks up a large calculation job into smaller ones, each with ~500 molecules, and submit each smaller job to the molecular property calculation service. The service then sends a job ID for each job submission back to PGVL Hub, which it uses to check with the central computing service periodically for computed results. Whenever the job with a given job ID is finished, PGVL Hub then downloads the computed results and merges them with their corresponding molecules automatically. During this time, chemists can continue working with a design session without having to wait for the results to return. If a design session is ended before the computed results are returned, PGVL Hub would save the job IDs so that the computed results can be retrieved next time the design session is restored. This asynchronous job submission mode was proven to be absolutely essential for optimal user experience while working on designs with large molecule collections.

3.4. Workflow

One of the strategic goals of PGVL Hub was to streamline the most common workflows in combinatorial library design. In general, the design workflow begins with a chemical reaction that is either downloaded from the PGVL chemistry knowledge base, imported from a pre-drawn reaction file (in the MDL ISIS/Draw .rxn format), or the chemist can create a new reaction on the fly using the imbedded reaction drawing page (**Fig. 15.3**). Once the reaction is defined, PGVL Hub then creates a "Design" with folders for reactant lists of each reaction component. The reactant lists are defined by the chemists, either by importing pre-defined lists or by using the pre-registered reactant lists available within PGVL Hub. For list importation, PGVL supports the SDF file format as well as compound IDs from in-house corporate IDs or MFCD numbers. If the reaction selected comes from one of the several hundred pre-registered reactions in PGVL, users can choose the reactant lists associated with each synthetic protocol for that given reaction, with the advantage that each reactant list has been pre-filtered against the scope and limitations of the associated protocol to ensure the best synthetic success of the designed library. For virtual compounds, chemists also have the option to sketch them on-the-fly using MDL ISIS/Draw and use them in PGVL Hub.

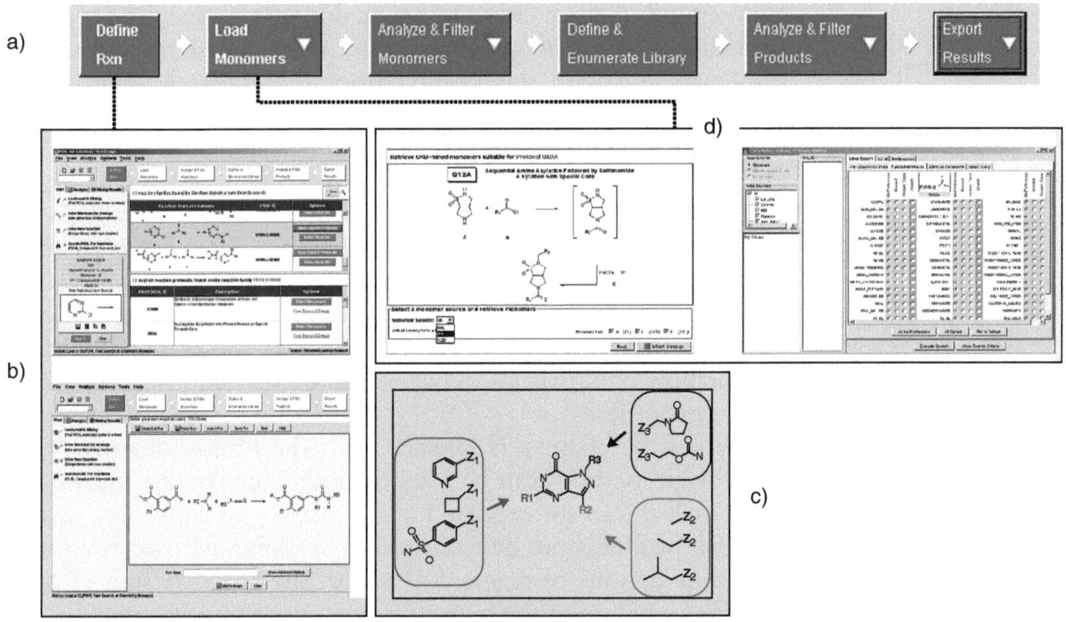

Fig. 15.3. Library design workflow: initiate design by defining reaction and gathering reactant sets: (**a**) Common workflows supported by PGVL Hub. Here the "Monomers" inside the figure means reactant sets. Clicking on a bottom that represents a step of the workflow leads to more detailed GUI panels for users to further specify what need to be done. (**b**) One can define a reaction either by search, browse, and select a pre-defined PGVL reaction or draw a new reaction on the fly using ISIS/Draw. (**c**) A Markush core capped by R-groups plus actual R-group fragment sets are used for the Markush exemplification workflow in place of reaction and reactant sets. (**d**) There are many ways to get reactant lists into PGVL Hub (*see* main text). Here one GUI component of ChemSelect (AQB) (50) was shown. Names of chemistry functional groups are listed in the menu for user to specify what functional group should and should not appear in the desired reactants. The query will be searched against in-house inventory systems and search results will be loaded into PGVL Hub as reactant lists.

Once the reactant lists are loaded, chemists can filter them through various molecular property calculations, substructure search/mapping, and similarity score against one or a set of lead structures. The substructure mapping and similarity scores against a collection of molecules are performed by a server-side SciTegic Pipeline Pilot component on the fly, eliminating the need for pre-registration. For library synthesis considerations, PGVL Hub also enables calculation for the amount of each reactant required for library synthesis and determines its availability from the corporate reagent inventory system. These calculations and filtering are in place to enable the designer to derive lists of desirable reactant for library synthesis.

Having the desired reactant lists, the chemist can now create a virtual library by enumerating the product structures in a fully combinatorial manner. Enumeration instructions are prevalidated for all PGVL registered reactions; for a user-specified reaction, PGVL Hub enables enumeration via the Markush representation of the reaction scheme. Once the products are enumerated, further filtering can be done via molecular property calculations, substructure mapping, similarity scoring, and structure matching with the Pfizer corporate databases for possible duplicates (product molecules already exist in the corporate collection). The enumerated products can be exported out of PGVL Hub and processed by additional 3D molecular design tools such as docking and scoring based on 3D pharmacophore models or target protein structures for further design refinement. Users can also prioritize and select desired product subsets for further consideration (cherry-picked library), perform list logic operations (AND, OR, MINUS) between two reactant sets or two libraries, which allow comparisons between and combinations of different design hypotheses.

When the chemist is satisfied with the library designs, various output methods can be used to export the results. Individual reactant lists can be exported as ID lists or as SDF files, while the final products can be exported as SDF files with corresponding combinations of reactant IDs. Or, the entire design or session file containing the library information can be exported in the XML format. PGVL Hub even provides the option to upload a design directly into a downstream chemistry informatics system to initiate library registration and synthesis. The XML format for library design has also been used as a preferred data exchange format between Pfizer and external chemistry outsource partners for library production. The Grid View in **Fig. 15.4** shows a screen shot of this basic workflow.

Using the same set of software capabilities, PGVL Hub also supports other design scenarios such as *Singleton Design* that is commonly practiced by medicinal chemists, *Markush Exemplification* where the reaction scheme becomes a Markush core structure, and reactant sets become R-group sets, and *Lead Centric Mining* (29).

3.5. Desktop Synergy

The desktop computer of a modern medicinal chemist usually contains software packages designed to increase productivity. Two such examples are Microsoft Excel (31) for data analysis and SpotFire (32) for data visualization, along with other 2D or 3D molecular design tools. In designing PGVL Hub, we recognized the importance of integration with some essential software packages to realize synergies as well as making PGVL Hub a more powerful design tool. Toward this goal, we have achieved seamless integration between PGVL Hub and MDL ISIS/Draw (30) and SpotFire (32). These two software packages are critical for structure drawing and SAR viewing, two of the most common practices by a medicinal chemist. From within PGVL Hub, chemists can launch ISIS/Draw and sketch a new molecule or reaction and then return the newly drawn molecule or reaction back to PGVL Hub via a single click (**Fig. 15.5**). This is the same synergistic

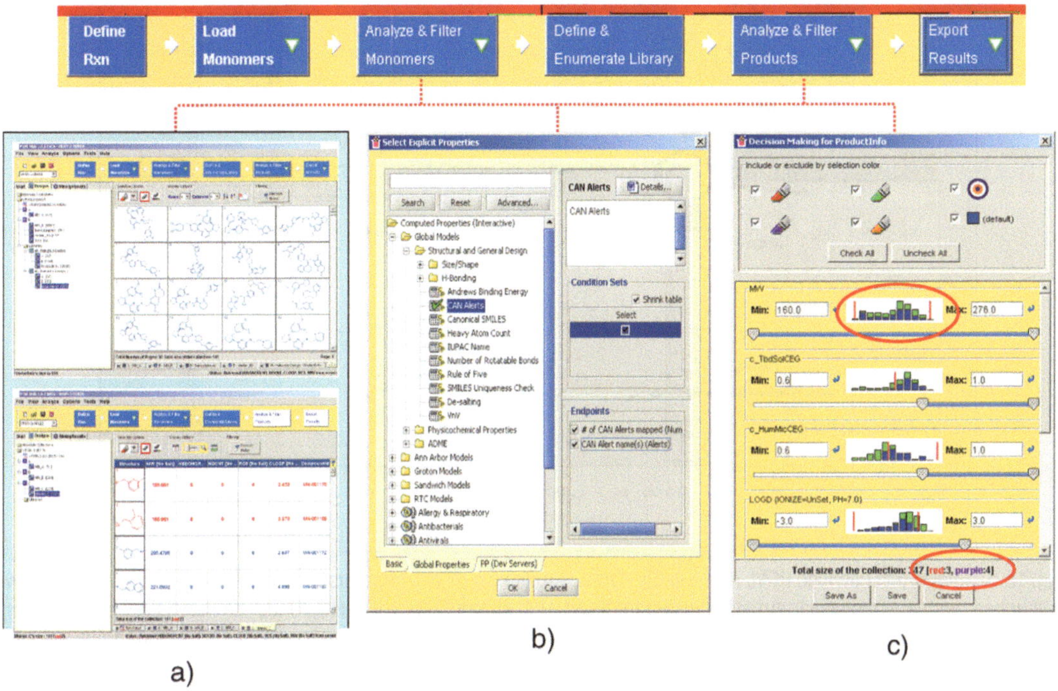

Fig. 15.4. Library design workflow: analysis and filtering of molecular collections: (**a**) A grid view and a table view panels are showed here displaying a collections of product and reactant molecules. Both views can be sorted by properties, and visibility and column location for each molecular property can be customized by user. Both views allow user to browse through large collections of molecules very efficiently and mark molecules of interest with various color pens. Also notice that the top workflow bar keeps track of design progress and highlights workflow steps already initiated. The first example in *Grid View* also offers a good view of all content within a single "Design." (**b**) User finds, selects, and submits computation jobs to the Pfizer molecular property calculation service (41). This service contains many in silico models for ADME&T and even project-specific SAR activity models. (**c**) User makes all filtering decisions within this "Decision Maker" panel. All molecular properties available can be used by "Decision Maker." The user hand-marking using color pens or selection returned from a SpotFire session are used as binary input. Numerical data are filtering using range slider bars. The color histograms display the property distributions of the starting set (in *green color*) and the current set (in *blue color*) before a filtering action is fully committed. This gives user an immediate and dynamic feedback on possible consequences of the current filtering setting. Textual properties can also be used for decision making (not shown here) via string searches such as "Exact Match," "contains," "Starts with," and "Ends with" in comparison with user-specific string. Also one can use SpotFire for data visualization and selection.

behavior between ISIS/Base and ISIS/Draw, already familiar to most medicinal chemists. The integration between PGVL Hub and SpotFire offers an even richer set of behaviors (**Fig. 15.6**). To use the SpotFire viewer within PGVL Hub, simply select a set of molecules (either reactants or products), then launch SpotFire directly within PGVL Hub. SpotFire would then display the molecular properties of the molecule collection, such as compound ID and various imported and/or computed ADME&T properties. Mousing over a data point displayed within the SpotFire window will display the associated molecular structure inside a structure viewing panel in PGVL Hub. Selections made within the SpotFire window will be automatically passed back into the

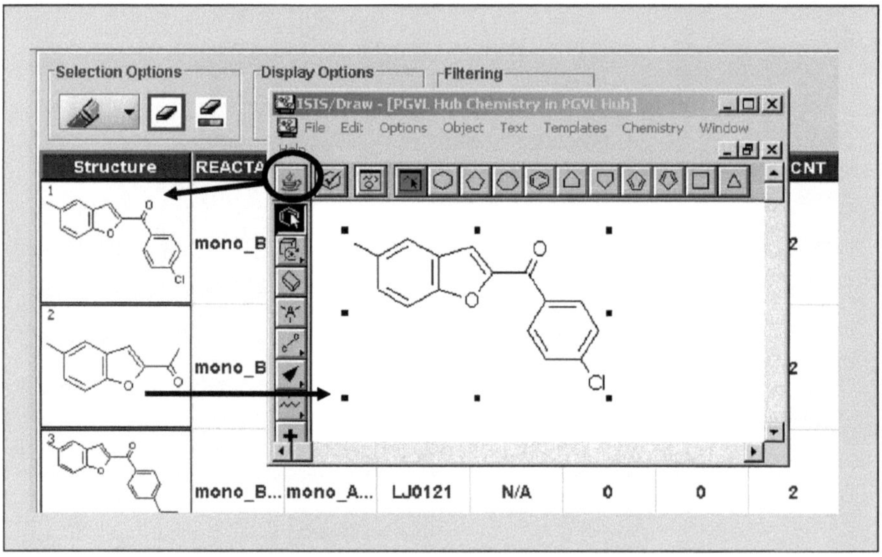

Fig. 15.5. Integration between PGVL Hub and ISIS/Draw. By clicking on any structural box in PGVL Hub window, ISIS/Draw will appear with any molecular structure inside the PGVL Hub structural box if exists. Once the user is done creating or polishing a structure drawing inside ISIS/Draw, a single click on ISIS/Draw will transfer the new structure drawing back to the PGVL Hub structural box, and the ISIS/Draw window will disappear automatically.

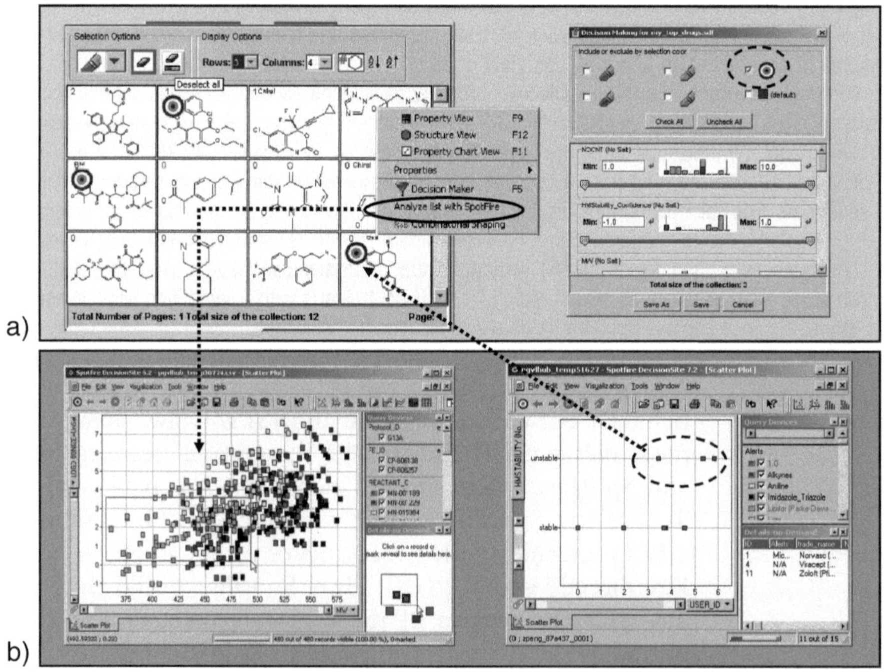

Fig. 15.6. Seamless integration between PGVL Hub and SpotFire. User can launch SpotFire within PGVL Hub to visualize the molecular properties associated with a molecule collection. Any selection done within SpotFire is dynamically passed back to PGVL Hub as marking on individual molecules. And user then can use the "Decision Maker" within PGVL Hub to make selections based on the SpotFire marking.

PGVL Hub window. Such integration allows seamless behavior between the two tools so that the user experience feels like a single piece of software. Additionally, PGVL Hub also has a one-way connection with Microsoft Excel and several in-house molecular design tools which can be launched within PGVL Hub while passing appropriate data to those applications (e.g., textual and numerical data for Excel, and SDF files for other 2D and 3D tools capable of reading such file format. Details on the MDL SDF file format can be found in reference (35)). By realizing these synergies, chemists can easily access and combine the best features from several applications to realize an even more powerful design experience. Furthermore, by not having to manually shuttle data between several applications, chemists can enjoy a more streamlined workflow. From a software development point of view, these synergies also allow PGVL Hub to leverage the best features from other software packages without re-inventing the wheel.

3.6. Design Features

Over a period of several years, we had worked closely with user communities and implemented many singleton and library design features. The most basic but heavily utilized is the *Browse and Mark* capability (*see* **Fig. 15.4**), where a chemist can display many molecular structures and their properties in grid or table view, page through them quickly and efficiently, and mark molecules of interest using different color makers for later decision making. Both the grid and the table views can be sorted based on molecular IDs or properties. We also implemented a SpotFire-like Decision Maker component (**Fig. 15.4**), where filtering can be performed on user-selected molecules as well as on numerical and textural properties. For numerical properties, a range filter is implemented where user can enter numerical values as well as using a sliding bar to perform the filtering process. For textual properties, we have implemented "Exact Match," "Starts with," "Ends with," and "Contains" to allow very flexible filtering operations. Color marking created by manual "Browser and Mark" steps as well as selection marking created from a SpotFire session can also be used by the decision maker component for molecule filtering. The histograms inside the decision maker provide chemists with immediate feedback on the consequence of the filtering action before committing to a filtering operation. For product collections, filtering on the products will result in a cherry-picked library (sparse matrix, not fully combinatorial). This design approach maximizes flexibility to ensure all product molecules are compliant with desired property ranges, although the consequence is lower production efficiency since not all products within the combinatorial matrix are made. An alternative approach is to use aggregated product properties to shape and optimize the reactant selection so that the library outcome is still fully combinatorial while maintaining a desired profile of product molecular properties. Previous publications (27) have provided several possible solutions to the design objective described above. We have incorporated within PGVL Hub a combinatorial shaping design feature that is simple, graphical, and intuitive for medicinal chemists. For each product molecule in a given library, we calculate a user-customizable "Pass/Fail" score. Then for each reactant molecule, the number of "Failed" products associated with this reactant is calculated and used as an aggregated property called "Failed Score," which can then be used to sort the reactant list and reorder the product matrix according to this score. Removing the reactants with the highest "Failed Scores" would make the most impact to improve the quality of the remaining library in terms of molecular property compliance. As shown in **Fig. 15.7**, users can simply use the slider bars within the combi-shaping panel to remove the reactants with the highest failed score, and

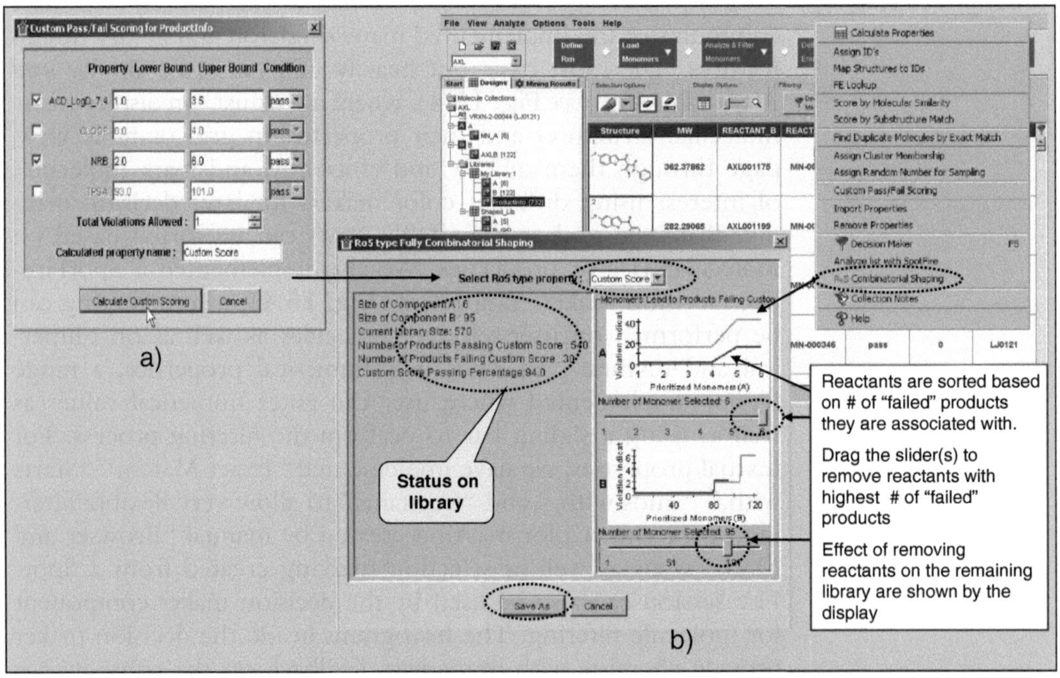

Fig. 15.7. Interactive combinatorial library shaping (**a**) A user-defined Pass/Fail score (such as Rule-of-Five) can be constructed and computed for product molecules based on existing molecular properties. This type of Pass/Fail scores can be used for combinatorial library shaping. (**b**) After user selects a library for combinatorial shaping, PGVL Hub allows user to pick the appropriate Pass/Fail scores as input, then sort reactants for each reaction component from low to high based on number of "Failed" product molecules a reactant molecule is associated with and plot the library status visually. User then uses the slider bars (one for each reaction component) to remove the worst reactant(s) and get an immediate feedback from the status report. The *green* curves are static based on the input reactant sets; the *blue* curves are updated dynamically to indicate the possible outcome. User can explore various strategies in reducing worst offenders in reactant sets to reduce the number of "Failed" products while still maintaining a fully combinatorial library of good size and production efficiency (number of products to be synthesized vs. number of reactants to be handled). User then commits the shaping by creating a new library.

PGVL Hub provides immediate feedback of the action in terms of updated property ranges and library size. This instant graphical feedback enables chemists to review the results before committing to the library shaping steps, and have the option to either updating the existing library or creating a new library with modified product properties and updated reactant sets.

One other useful feature of PGVL Hub is substructure searching and mapping enabled by SciTegic Pipeline Pilot (34). This feature allows chemists to determine what substructures are mapped to a target molecule and where as well as how many times each substructure is mapped. The substructure query can be entered by a user via a pop-up panel or run as a set of pre-built substructure queries as a part of a molecular property calculation service. An example is shown in **Fig. 15.8**, where substructure fragments are compiled at the corporate level to flag undesirable structural elements (41b) so they can be flagged and avoided in molecular designs. Such practice provides another example where PGVL Hub enables the reuse of valuable knowledge collectively captured within the Pfizer drug discovery community.

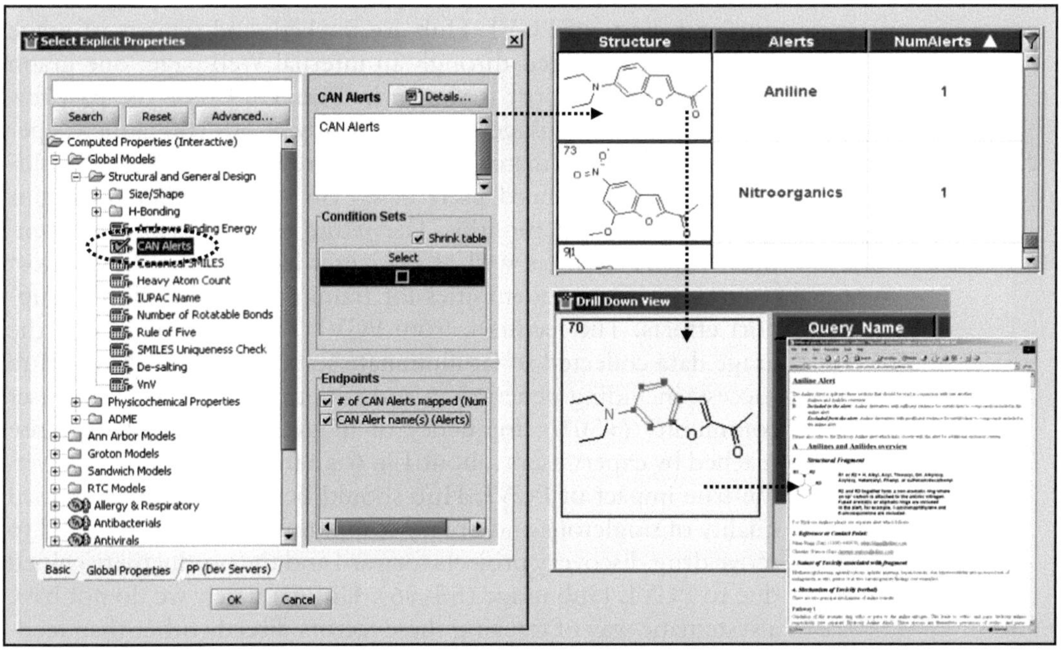

Fig. 15.8. Substructure mapping, highlighting, and drill-down. Based on on-the-fly substructure query and mapping capability within SciTegic Pipeline Pilot, PGVL Hub allows user to perform substructure queries into a set of target molecules. In the example shown, a set of substructure queries globally collected and validated as undesirable substructure features to be avoided are mapped into target molecules (41b).

4. Remarks

4.1. Deployment, Usage, and Impact

Since PGVL Hub client side is Java Web Start deployable and Pfizer-supported desktop computers all have the Java Web Start utility installed, the installation of PGVL Hub is easily done by users themselves directly via a Pfizer internal web site. This web site also contains other resources to facilitate the distribution, training, and support of PGVL Hub. Training materials include the user's guide, presentation slides, animated tutorials, and scientific literatures pertaining to singleton and library designs. Support is provided by a global support team, a network of local power users at various Pfizer research sites, and a global steering committee comprising PGVL champions.

The initial deployment began around 2003 targeting a small yet highly motivated community of approximately 100 beta testers comprising mainly medicinal chemists and computational chemists. Based on feedback from these users and results of two formal software usability studies conducted at various Pfizer sites involving medicinal chemists with some or no prior exposure to PGVL Hub, we were able to add further enhancements to make the software even better and more user-friendly. A full deployment was initiated around 2005 to the global discovery chemistry community of over 2000 potential users at that time. The adoption and usage of PGVL Hub are tracked, and reports of usage statistics are provided through an internal Web page; one graph within a typical report is shown in **Fig. 15.9**. For the past few years, PGVL Hub usage has been steady at 60–100 launches per day. Of the approximately 1000 registered users, 30% are considered to be experienced users based on their numbers of logins during the last 12 months. The usage tracking tool also identified expert users as well as novices which helped the support team recognize opportunities for training and allocation of support efforts. The feedback from PGVL local champions and the usage data collected so far illuminate some aspects of PGVL Hub success, including penetration and adoption by the intended user community (>50%), frequency of usage, and level of expertise reached by expert users (about 1 in 6 is a frequent user). However, the true impact of PGVL Hub should be measured by increased quality of singletons and library compounds chemists designed to move drug discovery projects forward and the productivity gained due to PGVL Hub usage (53–56). Unfortunately we do not have a systematic way of tracking these factors directly other than feedback from medicinal chemists and their research leadership. Ultimately, the success of PGVL Hub to enable smarter designs and higher productivity should be better assessed by successful drug candidates coming out of drug discovery projects.

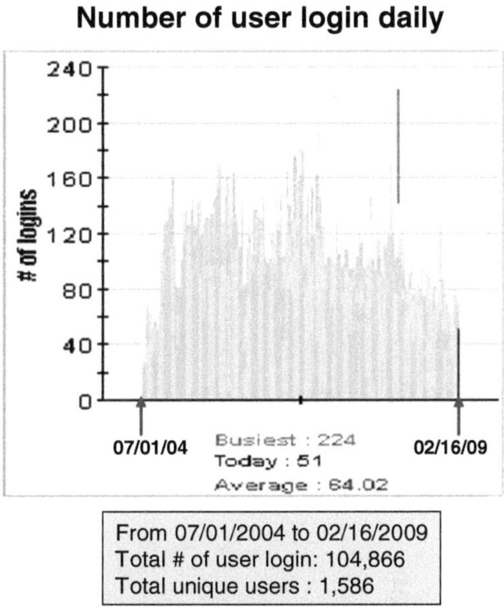

Fig. 15.9. Tracking usage of PGVL Hub. Each time a user logins into PGVL Hub, information about user ID and time stamp is recorded into a tracking database. A usage report can be generated via a Web reporting tool.

4.2. Comparison with Published Integrated Library Design Tools Used in Pharmaceutical Industry

It is likely that every pharmaceutical company would have an integrated library design tool in place as part of the strategy to incorporate the combinatory library approach into its drug discovery process. Since most of them were not published, we could only make comparisons among ADEPT, REALISIS, and PGVL Hub (*see* **Table 15.1**). Here we shall leave the details for readers to explore further while making a general statement that all three are designed to address essentially the same set of major questions in library design. They only differ in level of software engineering, GUI capability and intuitiveness, and scope of feature coverage.

4.3. A Proven Platform for Future Enhancement and Innovation

PGVL Hub has been well entrenched as one of the key desktop molecular design tools used by Pfizer medicinal chemists. Its solid three-tier enterprise architecture and powerful client-side component easily deployed by Java Web Start provide a very attractive platform with a proven track record for future enhancement and innovations in singleton and library design. There are many possibilities for further enhancement based on user requests as well as attractive methodologies and algorithms already published in the literature (6–27). Here we would like to list a few, with some already being prototyped.

4.3.1. Multiple Property Optimization

This is a well-known area of research with significant practical implications for molecule design (11, 12, 14, 18, 19, 42–46).

Table 15.1
Comparison of three integrated library design tools from the pharmaceutical industry

Feature or capability	ADEPT (25) (Glaxo & Daylight, 1999)	REALISIS (26) (J & J, 2004)	PGVL Hub (Pfizer, fully deployed to 1200 users in 2005)
Source for reactant	On-the-fly searching using SMARTS either pre-defined or user provided	On-the-fly searching using ISIS query	One-the-fly searching using ISIS queries either pre-defined or user provided; Load lists of compound IDs; Load pre-mined lists of reactants suitable for registered PGVL reaction reactions; Load SDF files; User drawn
Reaction encoding/ enumeration method	SMIRKS either pre-defined or user entered/chemical transformation based	ISIS reaction scheme either pre-defined or user entered/chemical transformation based	ISIS reaction scheme entered by user or pre-defined reaction object for ~500 PGVL registered combinatorial reactions/reactant clipping and assembly of a Markush core and R-groups
Product property prediction	Limited set mainly available in daylight	Limited set	Very large collection, including many vendor-supplied and internally developed *in silico* models for ADME&T end points and target SAR
Support for fully combinatorial shaping using product properties	Yes	No	Yes
Cherry-picking library design based on product properties	Yes	Yes	Yes
Support for Markush exemplification	No	No	Yes

Table 15.1 (continued)

Similarity search into pre-defined virtual library space for idea generation and lead hopping	N/A since there is no pre-defined virtual library space	N/A since there is no pre-defined virtual library space	Yes, virtual library space of 10^{14} molecules (PGVL) spanned by ~500 combinatorial reactions are searchable directly through PGVL Hub (*see* Ref. (29) for details)
Structure and property viewing, sorting, and exporting	Limited	Limited	Very general and powerful. Both table- and grid-viewers are fully configurable by user. User can also select and re-order property columns for both display and export
Integrated decision making	Numerical property filters and histograms	Numerical property filters and histograms	Powerful SpotFire-like filter for both numerical and textural properties; also integrated with SpotFire directly for data exploration and decision making
Session file for persistence and design sharing	No	Limited. SDF file of various reactants and products are created and stored in user directory during design	Yes, intermediate results of a design session can all be saved into a single XML-based session file for later use or share with collaborators
Software architecture	Web GUI with CGI scripts on the server side for integration. two-tier	Java GUI, two-tier, access to various DBs and SciTegic Pipeline Pilot via SOAP and ODBC	Java GUI deployed via Web Start for easy deployment and update, J2EE three-tier. Access to various DBs, SciTegic Pipeline Pilot, and other molecular property predicting services
Software performance enhancements		Multi-thread during reactant mining against multiple reactant databases	Multi-thread, batch data fetching or batch job submission, asynchronous mode for ADME/Tox property prediction. Client-side disk-based fast cache for molecular structures
Targeted users	Computational and medicinal chemists	Mainly for medicinal chemists (70%)	Mainly for the medicinal chemists, but it has also been used extensively by computational chemists

The key challenge is to make it intuitive so that it is easy to use and easy to interpret the results. Significant progress in this area has been made and reported in the literature (47).

Genetic-Algorithm (GA)-Driven Library Design and Lead Centric Mining: An abundance of literature on singleton and library design utilizing genetic algorithms already exists (6, 8, 16–19). The usage of GA is based on a "Goodness" score function, for example, a combination of similarity to a known set of lead compounds, various ADME&T molecular properties, or computed activity scores based on specific project SAR models against a protein target. The GA methodology will act as an agent, explore automatically the vast compound space either defined by the user or by PGVL, and return a set of molecules with enhanced or optimized "Goodness" scores. In essence it is a virtual screening methodology against a virtual chemical space not fully enumerated. In the context of similarity search, it is a generalized form of our current Lead-Centric-Mining methodology (29).

4.3.2. Structure-Based Library Design (SBLD)

For discovery projects where target protein structures are available, a structure-based library design strategy would be highly desirable. In the current version of PGVL Hub, 2D library molecules are exported out of PGVL Hub and imported to SBDD-enabled molecular design tools; it would be desirable to have a more integrated workflow. The challenge is how to deal with the one-to-many relationship between a 2D molecule inside PGVL Hub and its many potential 3D conformers in complex with the target protein binding sites. By utilizing an aggregation step to reduce the one-to-many relationship into a one-to-one relationship (e.g., keeping just the best docking score of the best 3D conformation), the SBDD aspect is reduced to the best docking score, and a very simple molecular property is returned to PGVL Hub for decision making. This approach is very simple and intuitive, but is best for smaller libraries due to the computation-intensive nature of docking and scoring. One way to extend the range of SBLD coverage to a much larger virtual space is through the usage of Basis Products (21). The detail of this SBLD effort using Basis Products has been described in a separate publication (48).

All of the more advanced library design capabilities described above can be integrated into the PGVL Hub platform in an intuitive way and utilized by medicinal chemists routinely to impact progression of drug discovery projects.

5. Conclusions

PGVL Hub is an integrated desktop tool which has been developed and globally deployed throughout Pfizer discovery research units for singleton and library design and synthesis. It has a highly intuitive and interactive GUI, an excellent performance profile, and is easy to install and update. For the past several years it has been routinely accessed by hundreds of medicinal chemists and other scientists for compound and library design work. This tool provides direct access to Pfizer's proprietary PGVL chemistry knowledge base to enable fast HTS hit follow-up and lead optimization. It offers a very rich and intuitive set of design capabilities and covers a wide range of workflows commonly used by medicinal chemists. PGVL Hub also has the advantage of being integrated with other desktop tools, such as ISIS/Draw, Microsoft Excel, SpotFire, and other 2D and/or 3D molecular design tools, and it leverages the best features of those tools to provide synergies and an integrated workflow. Its three-tier J2EE enterprise architecture and a powerful GUI provide a proven platform and delivery mechanism for future enhancements. Beyond its usage statistics, the true measure of PGVL Hub's positive impact in design quality and productivity should be an increase of attractive chemical leads emerging from drug discovery projects.

Acknowledgments

Over the years, the PGVL development team has received strong support and help from Pfizer research management, PGVL site champions and steering committee, user communities in medicinal chemistry and computational chemistry, research informatics, and sister software development projects. We would like to express our deepest gratitude and apologize for not being able to list all their names explicitly here.

References

1. Hogan, J. C. Jr. (1997) Combinatorial chemistry in drug discovery. *Nat Biotechnol* **15**, 328–330.
2. Hall, S. E. (1997) The future of combinatorial chemistry as drug discovery paradigm. *Pharm Res* **14**(9), 1104–1105.
3. Salemme, F. R., Spurlino, J., Bone, R. (1997) serendipity meets precision: the integration of structure-based drug design and combinatorial chemistry for efficient drug discovery. *Structure* **5**, 319–324.
4. Floyd, C. D., Leblanc, C., Whittaker, M. (1999) Combinatorial chemistry as a tool for drug discovery. *Prog Med Chem* **36**, 91–163.
5. Beeley, N., Berger, A. (2000) A revolution in drug discovery. Combinatorial chemistry still needs logic to drive science forward. *BMJ [Br Med J]* **321**(7261), 581–582.
6. Singh, J., Ator, M. A., Jaeger, E. P., Allen, M. P., Whipple, D. A., Soloweiij, J. E., Chowdhary, S., Treasurywala, A. M. (1996) Application of genetic algorithm to combinatorial synthesis: a computational approach to lead identification and lead optimization. *J Am Chem Soc* **118**, 1669–1676.
7. Blaney, J. M., Martin, E. J. (1997) Computational approaches for combinatorial library design and molecular diversity analysis. *Curr Opin Chem Biol* **1**, 54–59.

8. Brown, R. D., Martin, Y. C. (1997) Design combinatorial library mixtures using a genetic algorithm. *J Med Chem* **40**, 2304–2313.
9. Agrafiotis, D. K., Myslik, J. C., Salemme F. R. (1998) Advances in diversity profiling and combinatorial series design. *Mol Diversity* **4**, 1–22.
10. Bures, M. G., Martin, Y. C. (1998) computational methods in molecular diversity and combinatorial chemistry. *Curr Opin Chem Biol* **2**, 376–380.
11. Zheng, W., Cho, S. J., Waller, C. L, Tropsha, A. (1999) Rational combinatorial library design. 3. Simulated annealing guided evaluation (SAGE) of molecular diversity: a novel computational tool for universal library design and database mining. *J Chem Inf Comput Sci* **39**, 738–746.
12. Gillet, V. J., Willett, P., Bradshaw, J., Green, D. V. S. (1999) Selecting combinatorial libraries to optimize diversity and physical properties. *J Chem Inf Comput Sci* **39**, 169–177.
13. Spellmeyer, D. C., Grootenhuis, P. D. J. (1999) Computational approaches to combinatorial chemistry. *Annu Rep Med Chem* **34**, 287–296.
14. Brown, R. D., Hassan, M., Waldman, M. (2000) Combinatorial library design for diversity, cost efficiency, and drug-like characters. *J Mol Graph Model* **18**, 427–437.
15. Jamois, E. A., Hassan, M., Waldman, M. (2000) Evaluation of reagent-based and product-based strategies in the design of combinatorial library subsets. *J Chem Inf Comput Sci* **40**, 63–70.
16. Douguet, D., Thoreau, E., Grassy, G. (2000) A genetic algorithm for the automated generation of small organic molecules: drug design using an evolutionary algorithm. *J Comput Aided Mol Design* **14**, 449–466.
17. Sheridan, R. P., SanFeliciano, S. G., Kearsley, S. K. (2000) Designing targeted libraries with genetic algorithm. *J Mol Graph Model* **18**, 320–334.
18. Gillet, V. J., Khatib, W., Willett, P., Fleming, P. J., Green, D. V. S. (2002) Combinatorial library design using a multiobjective genetic algorithm. *J Chem Inf Comput Sci* **42**, 375–385.
19. Chen, G., Zheng, S., Luo, X., Shen, J., Zhu, W., Liu, H., Gui, C., Zhang, J., Zheng, M., Puah, C. M., Chen, K., and Jiang, H. (2005) Focused combinatorial library design based on structural diversity, drug likeness and binding affinity score. *J Comb Chem* 7 (3), 398–406.
20. Mason, J. S., Beno, B. R. (2000) Library design using BCUT chemistry-space descriptors and multiple four-point pharmacophore fingerprints: simultaneous optimization and structure-based diversity. *J Mol Graph Model* **18**, 438–451.
21. Shi, S., Peng, Z., Kostrowicki, J., Paderes, G., Kuki, A. (2000) Efficient combinatorial filtering for desired molecular properties of reaction products. *J Mol Graph Model* **18**, 478–496.
22. Gobbi, A., Poppinger, D., Rohde, B. (1997) Developing an in-house system to support combinatorial chemistry. *Perspect Drug Discov Des* **7/8** (Combinatorial Methods for the Analysis of Molecular Diversity), 131–158.
23. Polinsky, P., Feinstein, R. D., Shi, S., Kuki, A. (1996) LiBrain: software for automated design of exploratory and targeted combinatorial libraries, in (Chaiken, I. M., Handa, K. D., eds.) *Molecular Diversity and Combinatorial Chemistry*. American Chemical Society, Washington, DC, pp. 219–232.
24. Shi, S., Kuki, K., Zhou, Z., Na, J., Thacher, T., Yanovsky, A., Polinsky, P. LiBrain^{TM}, An Intelligent System for the High-Throughput Design of Combinatorial libraries in Drug Discovery. Poster Presentations at the *Fifth International Conference on Chemical Structures* (1999) and the *Second European Conference on Strategies and Technologies for Identification of NOVEL BIOACTIVE COMPOUNDS* (1998).
25. Leach, A. R., Bradshaw, J., Green, D. V. S., Hann, M. M., Delany, J. J., III (1999) Implementation of a system for reagent selection and library enumeration, profiling, and design. *J Chem Inf Comput Sci* **39**, 1161–1172; also see a review article from Leach, A. R., Hann, M. M. (2000) The *in silico* world of virtual libraries. *Drug Discovery Today* **5**, 326–336.
26. Yasri, A., Berthelot, D., Gijsen, H., Thielemans, T., Marichal, P., Engles, M., Hoflack, J. (2004) REALISIS: a medicinal chemistry-oriented reagent selection, library design, and profiling platform. *J Chem Inf Comput Sci* **44**, 2199–2206.
27. (a) Truchon, J., Bayly, C. I. (2006) GLARE: A new approach for filtering large reagent lists in combinatorial library design using product properties. *J Chem Inf Model* **46**, 1536–1548. (b) Stanton, R. V., Mount, J., Miller, J. L. (2000) Combinatorial library design: maximizing model-fitting compounds within matrix synthesis constraints. *J Chem Inf Comput Sci* **40**, 701–705.
28. Peng, Z., et al. PGVL: a vast virtual space of synthetic feasible compounds based on

captured knowledge of combinatorial chemistry synthesis protocols at the enterprise level. Manuscript in preparation.
29. Hu, Q., Peng, Z., Kostrowicki, Kuki, A. (2011) LEAP into the Pfizer Global Virtual Library (PGVL) space – creation of readily synthesizable design ideas automatically, in (Zhou, J. Z. ed.) , *Chemical Library Design*. Humana Press, New York, Chapter 13.
30. ISIS-Draw, MDL Information Systems, Inc. http://www.mdli.com.
31. Microsoft Excel: http://office.microsoft.com/en-us/excel/FX100487621033.aspx
32. SpotFire for data visualization and decision making: http://spotfire.tibco.com/
33. Java web Start from Sun Microsystems, Inc: http://java.sun.com/products/javawebstart/
34. Pipeline Pilot from SciTegic: http://www.scitegic.com/
35. Dalby, A., Nourse, J. G., Hounshell, W .D., Gushurst, A. K. I., Grier, D. L., Leland, B. A., Laufer, J. (1992) Description of several chemical structure file formats used by computer programs developed at molecular design limited. *J Chem Inf Comput Sci* **32**, 244–255.
36. A good resource for BASE64 encoding could be found in: http://en.wikipedia.org/wiki/Base64
37. Resource on data compression in Java could be found in: http://java.sun.com/developer/technicalArticles/Programming/compression/
38. Resource about XML (http://en.wikipedia.org/wiki/XML) and its special characters (http://www.devx.com/tips/Tip/14068)
39. Lipinski, C. A., Lombardo, F., Dominy, B. W., Feeney, P. J. (1997) Experimental and computational approaches to estimate solubility and permeability in drug discovery and development settings. *Adv Drug Deliv Rev* **23** (1–3), 3–25.
40. More info on the LogD prediction tool from of the ACD Lab could be found: http://www.acdlabs.com/products/phys_chem_lab/logd/
41. a) A Pfizer in-house service and framework for development, validation, and deployment of *in silico* models for ADME-Tox and target specific SAR prediction. New ADME-Tox models published into this service will be automatically visible to all client software packages (such as PGVL Hub) subscribed into this service. b) Kalgutkar, A. S., Gardner, I., Obach, R. S., Shaffer, C. L., Callegari, E., Henne, K. R., Mutlib, A. E., Dalvie, D. K., Lee, J. S., Nakai, Y., O'Donnell, J. P., Boer, J., Harriman, S. P. (2005) A comprehensive listing of bioactivation pathways of organic functional groups. *Curr Drug Metab* **6**, 161–225.
42. Steuer, R. E. (1986) *Multiple Criteria Optimization: Theory, Computations, and Application*, John Wiley & Sons, Inc., New York, ISBN 047188846X.
43. Sawaragi, Y., Nakayama, H., Tanino, T. (1985) *Theory of Multiobjective Optimization (vol. 176 of Mathematics in Science and Engineering)*. Academic Press Inc., Orlando, FL, ISBN 0126203709.
44. Messac, A., Ismail-Yahaya, A., Mattson, C. A. (2003) The normalized normal constraint method for generating the pareto frontier. *Struct Multidis Optim* **25**(2), 86–98.
45. Das, I., Dennis, J. E. (1998) Normal-boundary intersection: a new method for generating the pareto surface in nonlinear multicriteria optimization problems. *SIAM J Optim* **8**, 631–657.
46. Deb, K., Pratap, A., Agarwal, S., Meyarivan, T. (2002) A fast and elitist multiobjective genetic algorithm. *NSGA-II. IEEE Trans Evol Comput* **6**(2), 182–197.
47. Two examples: **Mobius** from Coalesix, Inc.: A GA driven, R-group based compound/library design tools (http://www.coalesix.com/FAQ.html); and **C2-LibX** from Accelrys, Inc. contains a GA based library design module (more from http://accelrys.com/).
48. Zhou, Z., Shi, S., Na, J., Peng, Z., Thacher, T. (2009) Combinatorial library-based design with basis products. *J Comput Aided Mol Des* **23**, 725–736.
49. SciTegic's re-implementation of Canonical SMILES string based on the original publication: Weininger, D., Weininger, A., Weininger, J. L. (1989) SMILES. 2. Algorithm for Generation of Unique SMILES Notation. *J Chem Inf Comput Sci* **29**(2), 97–101.
50. **ChemSelect AQB** (Advanced Query Builder): A Pfizer in-house developed reusable Java component that allows users to query various molecular structure databases within Pfizer based on molecular structural information and/or other properties, retrieve, manage, and export retrieved hits. It has a functional role similar to MDL ISIS/BASE, but with many enhanced capabilities. PGVL Hub has embedded this reusable Java component within itself so that users can search for suitable reactants from various corporate reactants databases and inventory houses and return the hits seamlessly back into PGVL Hub design session. From user's point of view, the **ChemSelect AQB** component is just part of PGVL Hub.

This is one of the successful examples that re-usable components are developed and shared among multiple software development projects within Pfizer.
51. **WebLogic** is a J2EE middle-tier software suit from the former BEA Systems, now part of Oracle (http://www.oracle.com/appserver/weblogic/weblogic-suite.html)
52. Oracle is a database application software from Oracle (http://www.oracle.com/index.html)
53. Smith, G. F. (2006) Enabling HTS Hit follow-up via Chemo informatics, File Enrichment, and Outsourcing. *High Throughput Medicinal Chemistry II*; MMS Conferencing & Events Ltd., Institute of Physics; London, 2006. This article is also available on-line via this web link (http://www.mmsconferencing.com/pdf/htmc/g.smith.pdf).
54. Clark, J. D., Hu, Q., Kuki, A., Peng, Z., Sciammetta, N., Smith, G. F., Ramirez-Weinhouse, M., Van Hoorn, W. (2006) Pfizer global virtual library: one-stop-shop for design on the desktop. A poster given by Nunzio Sciammetta during the *2006 Gordon Conference on Combinatorial Chemistry*, August 20–25, 2006 in The Queen's College Oxford, United Kingdom.
55. Teng, M., Zhu, J., Johnson, M. D., Chen, P., Kornmann, J., Chen, E., Blasina, A., Register, J., Anderes, K., Rogers, C., Deng, Y., Ninkovic, S., Grant, S., Hu, Q., Lundgren, K., Peng, Z., Kania, R. S. (2007) Structure-based design of (5-Arylamino-2H-pyrazol-3-yl)-biphenyl-2′,4′-diols as Novel and Potent Human CHK1 Inhibitors. *J Med Chem* **50** (22), 5253–5256.
56. Peng, Z., Hu Q. (2011) Design of targeted libraries against the human Chk1 kinase using PGVL Hub, in (Zhou, J. Z. ed.) *Chemical Library Design*. Humana Press, New York, Chapter 16.

Chapter 16

Design of Targeted Libraries Against the Human Chk1 Kinase Using PGVL Hub

Zhengwei Peng and Qiyue Hu

Abstract

PGVL Hub is a Pfizer internal desktop tool for chemical library and singleton design. In this chapter, we give a short introduction to PGVL Hub, the core workflow it supports, and the rich design capabilities it provides. By re-creating two legacy targeted libraries against the human checkpoint kinase 1 (Chk1) as a showcase, we illustrate how PGVL Hub could be used to help library designers carry out the steps in library design and realize design objectives such as SAR expansion and improvement in both kinase selectivity and compound aqueous solubility. Finally we share several tips about library design and usage of PGVL Hub.

Key words: PGVL Hub, combinatorial chemistry, library design, reaction, synthesis protocol, reactant, product, enumeration, filtering, Chk1, kinase, inhibitor, SAR, ADME&T (Adsorption, Distribution, Metabolism, Excretion, and Toxicity), selectivity, solubility, protein–ligand complex.

1. Introduction

1.1. PGVL Hub

PGVL Hub was developed for and deployed within Pfizer global chemistry communities (1). The main goal of PGVL Hub is to offer bench chemists a very capable desktop tool to (a) access Pfizer's proprietary chemistry knowledge database containing information about many experimentally validated combinatorial chemistry synthesis protocols; (b) support and streamline the full cycle of library design, synthesis, and registration; and (c) harness the power of synergy with many other desktop software packages (ISIS/Draw, MS-Excel, SpotFire (http://spotfire.tibco.com/), and additional 2D/3D molecular design tools). PGVL Hub has

J.Z. Zhou (ed.), *Chemical Library Design*, Methods in Molecular Biology 685,
DOI 10.1007/978-1-60761-931-4_16, © Springer Science+Business Media, LLC 2011

been used by more than 1000 users within Pfizer with more than 110,000 user logins accumulated since 2004. Its application to targeted library design is the focus of this chapter.

1.2. A Sample Case of Targeted Library Design Against the Chk1 Kinase Domain

In 2007, Teng and coworkers published a potent and selective lead series of human Chk1 inhibitors (2). Their work was initiated by two advanced lead matters **1808-1** and **1819-1** obtained through two rounds of targeted library design and synthesis based on a high-throughput screening (HTS) hit **Cpd-1** (*see* **Fig. 16.1** for the progress of this hit through the library-based lead optimization). This report distills the essence of those two legacy targeted libraries to showcase the design process using PGVL Hub. The PGVL Hub screen shots used in this report are recreations of our legacy design efforts in the past.

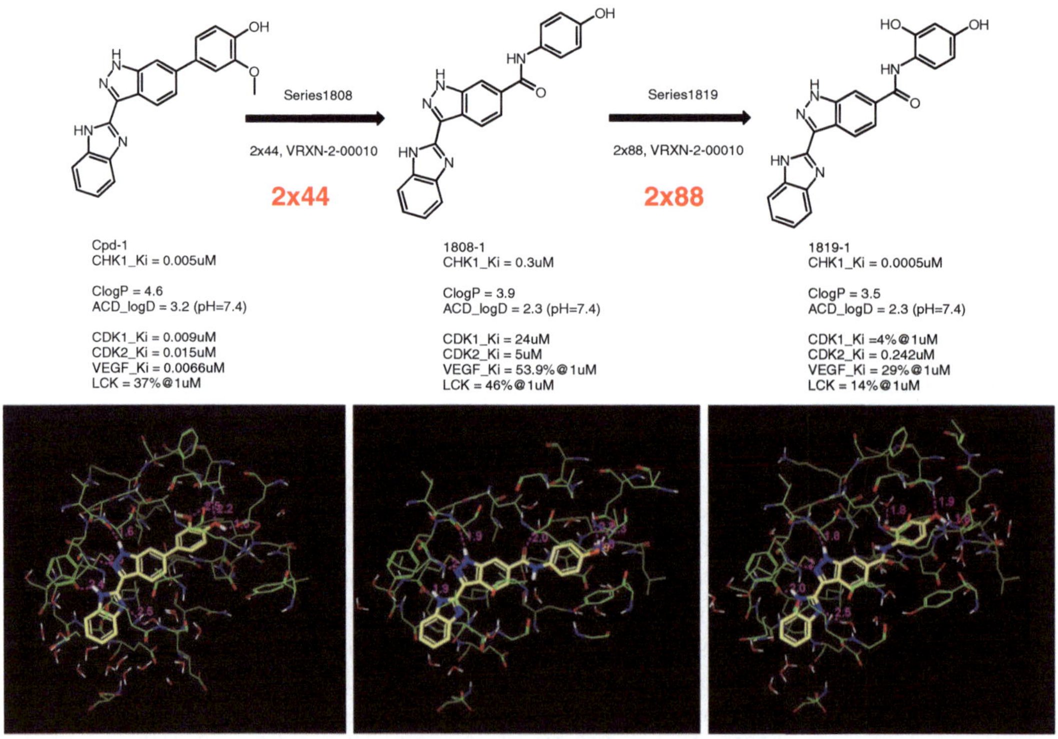

Fig. 16.1. Progress of two rounds of Chk1 targeted libraries. **Cpd-1** is the original HTS hit with a broad kinase inhibition profile and based on which the first round library was designed and synthesized. **1808-1** is the best hit from the first round targeted library with improved kinase selectivity profile, based on which the second round library was designed and synthesized. **1819-1** is the best lead with improved potency, kinase selectivity, and solubility. Co-crystal structures of Chk1 kinase domain and corresponding lead compounds were solved and extensively utilized in structure-based singleton and library designs. For details of the X-ray co-Crystal structures, please refer to the publications from Ming and et al (4a) and Foloppe and et al for details (4b).

2. Materials

More information about the biological roles of the Chk1 gene and its protein product can be found in Ref. (3). In short, DNA damages are sensed and passed to the Chk1 kinase to activate the checkpoint between the G2 (secondary gap) and the M (mitosis) phases of the cell cycle. The cell cycle rest at this checkpoint allows the cell to repair its DNA damages before proceeding to the M phase (*see* **Fig. 16.2**). Cells entering into the M phase with un-repaired DNA damages tend to suffer from mitotic catastrophe, which ultimately leads to cell death via the apoptosis pathway. Since the anti-cancer drugs used in standard chemotherapies are mostly DNA-damaging agents intended to induce cancer cell death in the M phase, it has been hypothesized that a Chk1 kinase inhibitor could synergistically enhance the anti-cancer effect of those DNA-damaging agents. What makes this approach even more attractive is the observation that normal cells tend to arrest at various checkpoints in both G1 and G2 phases after DNA damage, while many cancer cells tend to heavily rely on the G2/M checkpoint to repair DNA damages where activity of Chk1 is critical. This implies selective kill of cancer cells vs. normal cells (3). The first X-ray crystal structure for the Chk1 kinase domain was solved by Pfizer scientists at the Pfizer La Jolla site (4a), and the key structural features around its hinge region in the ATP binding site are also depicted in **Fig. 16.2**. Due to its above-mentioned

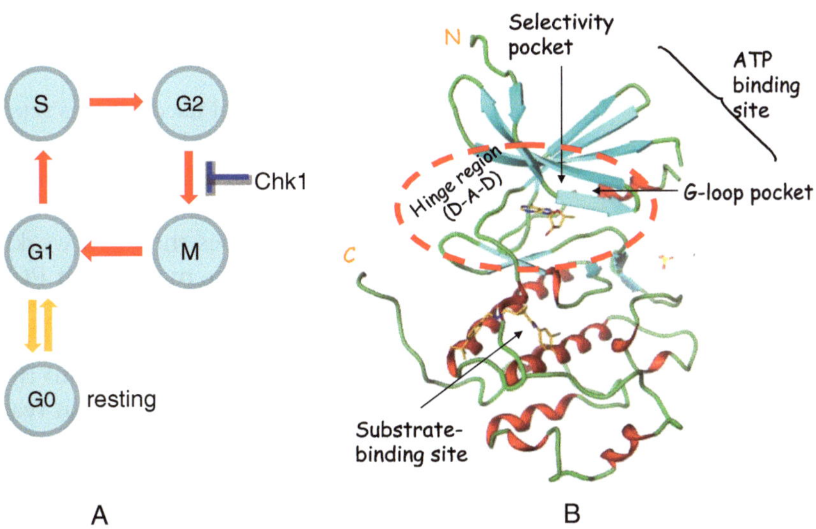

Fig. 16.2. Cell cycle, Chk1 at the G2/M checkpoint, and key structural features of the Chk1 kinase domain (4a). The *highlighted* location is the hinge region of the Chk1 kinase domain, which also corresponds to the same regions of protein–ligand structures shown in **Fig. 16.1**.

connections with cancers, Chk1 has been identified as an attractive oncology target (5) for inhibition by small organic molecules (6). For details on the biological assays and solutions of protein–ligand X-ray crystal structures cited in this report, please refer to Refs.(2) and (4a) directly.

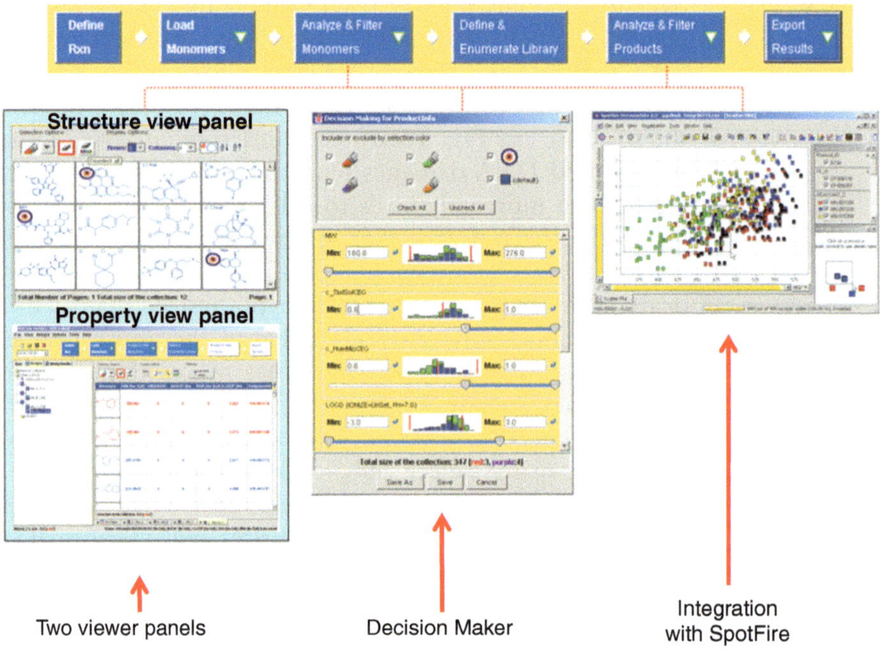

Fig. 16.3. Screen shots of PGVL Hub. It has two ways to display molecules and their properties (Structural Viewer panel and Table viewer panel). It has been integrated with SpotFire for data visualization. It also has a decision maker capable of handling numerical and textual data as well as user selections by hand.

The library design tool used for this report is PGVL Hub (1). **Figure 16.3** contains several screen shots of PGVL Hub, highlighting its capabilities in viewing molecules and their properties, decision making, and integration with SpotFire for data visualization. **Figure 16.4** describes the main workflow of library design seen through the workflow manager of PGVL Hub and the key questions a library designer would ask and address with the help of a library design tool such as PGVL Hub. Those key questions in library design can be summarized into the following list: What chemical reaction should be used? What reactants are available to start the library design? Which reactants (also called monomers) or products should be chosen for testing design hypotheses as well as satisfying various constraints such as ADME&T compliance? How can the library design be communicated to collaborators as well as a downstream synthesis process? **Figure 16.5** describes more about the design strategies and PGVL Hub's capabilities that allow a library designer to analyze the possible

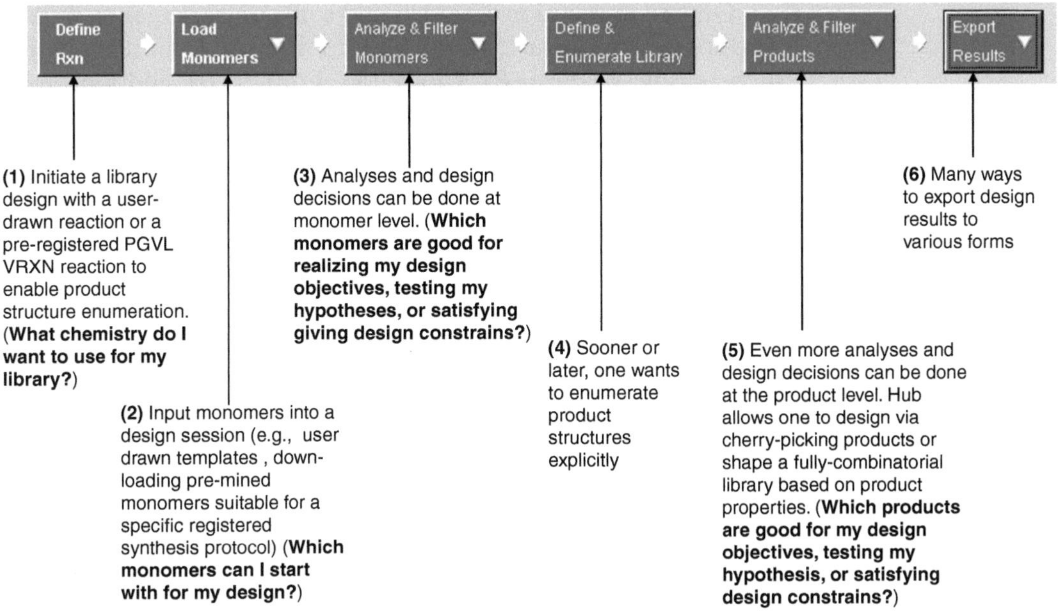

Fig. 16.4. The basic library design workflow enabled in PGVL Hub.

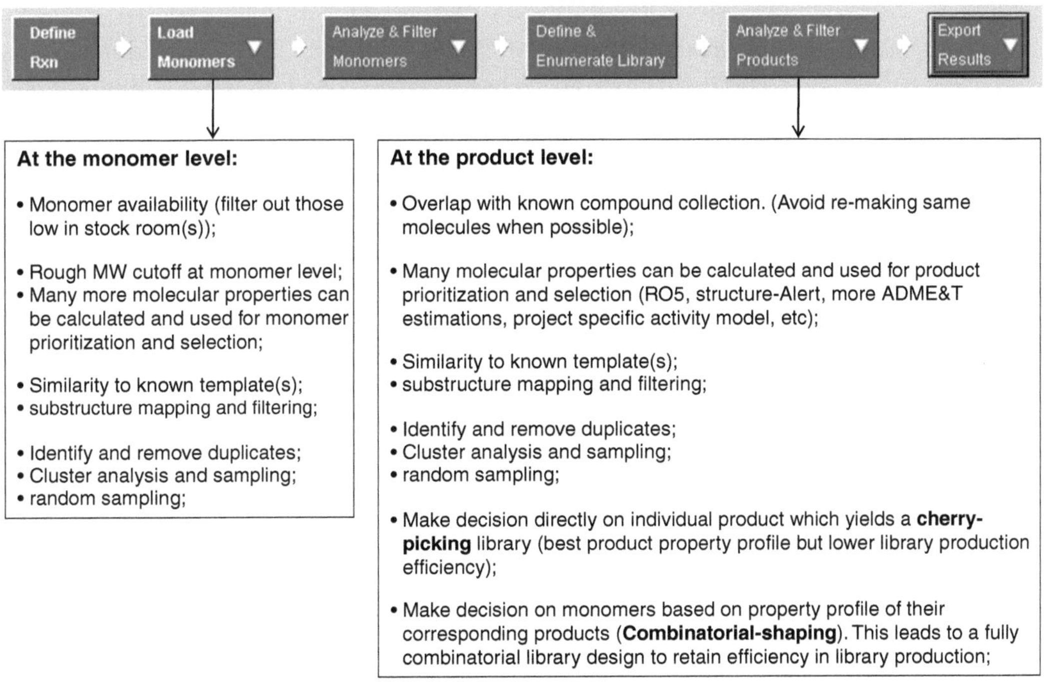

Fig. 16.5. Library design strategies and features enabled in PGVL Hub.

reactants/products one can use/synthesize and select a subset from which to realize the intended objectives of a library. For more detailed information, please refer to our report dedicated to PGVL Hub (1).

Other computational models and software packages were also used for design of the targeted libraries showcased in this report, such as Rule-Of-Five (8), ClogP (9), LogD (10), and a protein structure-based docking and scoring tool called AGDOCK (12) and its associated protein–ligand score function called HT-Score (13). For the library design, we have also used the Tanimoto coefficient (11) computed based on the molecular fingerprints from SciTegic Pipeline Pilot (14) as the measure of molecular similarity.

3. Methods

3.1. Information Known Before the Design of the First Targeted Library

The initial HTS hit **Cpd-1** was originally synthesized as an inhibitor against another human kinase called VEGF (7) with a measured K_i value of 6.6 nM. Nevertheless, it is not very selective and possesses a broad inhibition profile against many kinases such as Chk1 (5 nM), CDK1 (9 nM), CDK2 (15 nM), and LCK (37% at 1 μM) (*see* **Fig. 16.1**). **Cpd-1** is also known to have low aqueous solubility, as indicated by the high values of calculated ClogP (4.6) and LogD (3.2 at pH 7.4) values (9, 10). The X-ray crystal structure of **Cpd-1** with the Chk1 kinase domain was solved through an in-house effort (*see* **Fig. 16.1**). Binding site comparison of this protein–ligand complex with others containing different kinase domains and different ligands in Pfizer's crystal structure knowledge database (those Pfizer proprietary complex structures are not shown in this report) was conducted. This analysis suggested that the Chk1 region accessed by the right-hand side of **Cpd-1** might provide opportunities to gain specificity toward Chk1. Therefore the objectives for the first targeted library focused on improving selectivity and aqueous solubility while expanding the body of SAR information around the "selectivity pocket" with $2 \times 44 = 88$ product molecules (*see* **Fig. 16.6**).

3.2. Design Steps of the Targeted Library

3.2.1. Selection of Reactions

As shown in **Fig. 16.6**, we planned to use a pre-registered combinatorial chemistry protocol (LJ0194) to synthesize the targeted library. PGVL Hub allowed us to easily search and load this pre-registered reaction scheme into a design session without the need to draw a reaction scheme required for product enumeration (*see* **Fig. 16.7**). Even for this simple reaction, a simple reaction scheme drawn by users may not be sufficient to ensure proper formation of product structures in the case where bonds associated with chiral centers are near the reactive sites on the reactants.

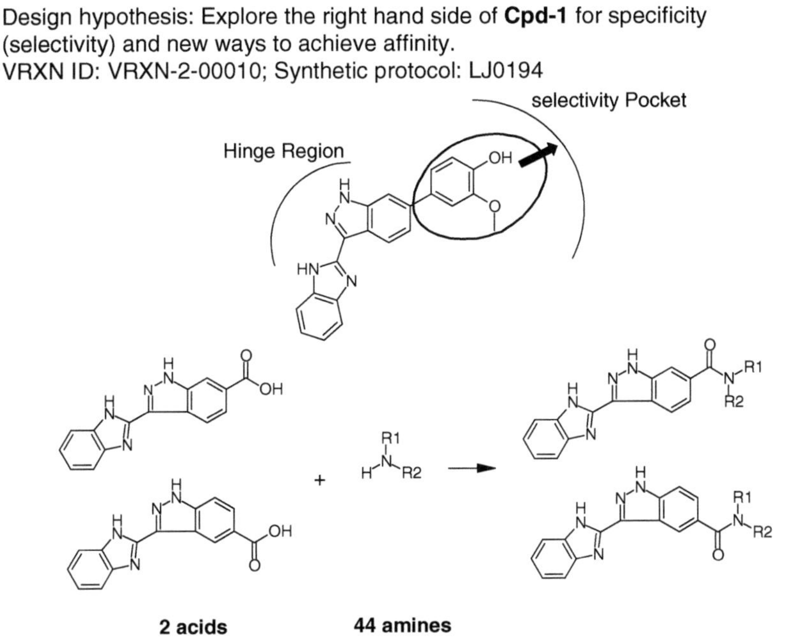

Fig. 16.6. Objectives of the first round library. The main goal is to explore the protein pocket probed by the right-hand side of **Cpd-1** to improve kinase selectivity and further build SAR knowledge. The single aryl–aryl bond was replaced by three bonds containing an amide group with more flexibility. A registered combinatorial synthesis protocol (LJ0194) was used for this library and a 2×44 plate format was planned before the design of the library.

3.2.2. Selection of Reactants

In addition to the pre-registered reaction scheme, the pre-mined reactants (acids and amines) suitable for the reaction conditions used in the library synthesis protocol LJ0194 were also available for download directly through PGVL Hub (also *see* **Fig. 16.7**). Since the two special acids of the library are already chosen, the task of library design now is to select 44 amines from the 8449 amines that are compatible with the synthesis protocol (*see* **Fig. 16.8**).

3.2.3. Reactant Analysis and Selection

Due to the small size of the binding pocket accessed by the right-hand side of **Cpd-1**, we first hypothesized that smaller amines would have a higher chance of fitting to that binding pocket. By applying the MW and ClogP calculations and filtering, we significantly reduced the possible choices of amines from the original set of 8449. Using molecular similarity score (11, 14) with respect to 4-amino-2-methoxyphenol, we ensured that the neighborhood of the right-hand side of **Cpd-1** was well sampled (*see* **Fig. 16.1**). Finally we looked up the reactant amount available in our chemical inventory for each reactant and only used those with sufficient amount for library production so that the designed library could be synthesized without further delays. With those two considerations, we were able to focus the remaining choices further

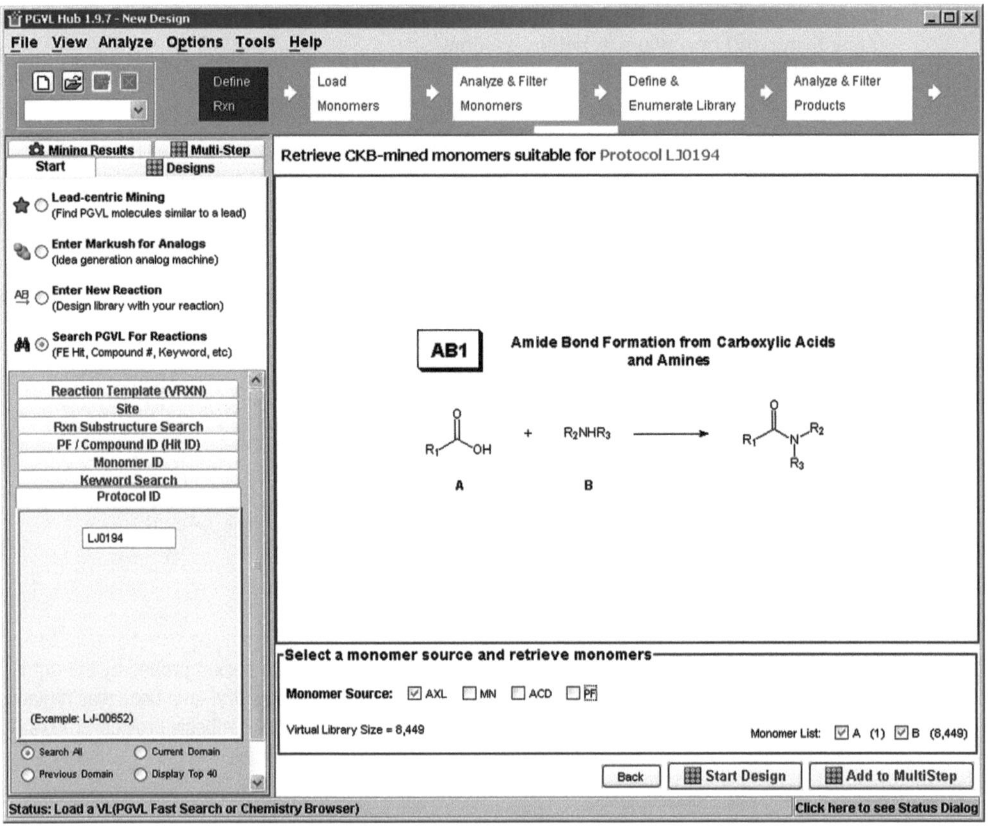

Fig. 16.7. Accessing the pre-registered reactions and pre-mined suitable reactants for synthesis protocol LJ0194 to initiate library design. PGVL Hub makes it simple to load the pre-registered reaction scheme and suitable reactants for the reaction conditions specified in synthesis protocol LJ0194. The product structure enumeration and the synthetic feasibility of those product molecules are taken care of by the PGVL Hub through extensive knowledge capturing and reusing, so that the library designers can focus their efforts on design issues such as target binding, selectivity, and ADME&T.

to a subset of a few hundred amines. Then we used the Structural Viewer panel of PGVL Hub (*see* **Fig. 16.3**) to display many amines in a single page and browsed through them visually. Each molecule was examined in terms of possible hypothesis it could help to form and validate. Molecular diversity of the final library is also a consideration. Desirable ones were marked with color markers provided by PGVL Hub and included in the final set of 44 (plus a few backups). Even though the actual legacy design also contained input from the project medicinal chemists and library production chemists, the first target library was designed mainly based on the reactant-level considerations described so far. The first targeted library yielded several hits with weaker potency than **Cpd-1** (*see* **Fig. 16.9**); however, assay data on kinase selectivity suggested that the top hit **1808-1** had a much improved kinase selectivity profile and some improvement in solubility (*see* **Fig. 16.1**). This prompted the project team to solve the co-crystal

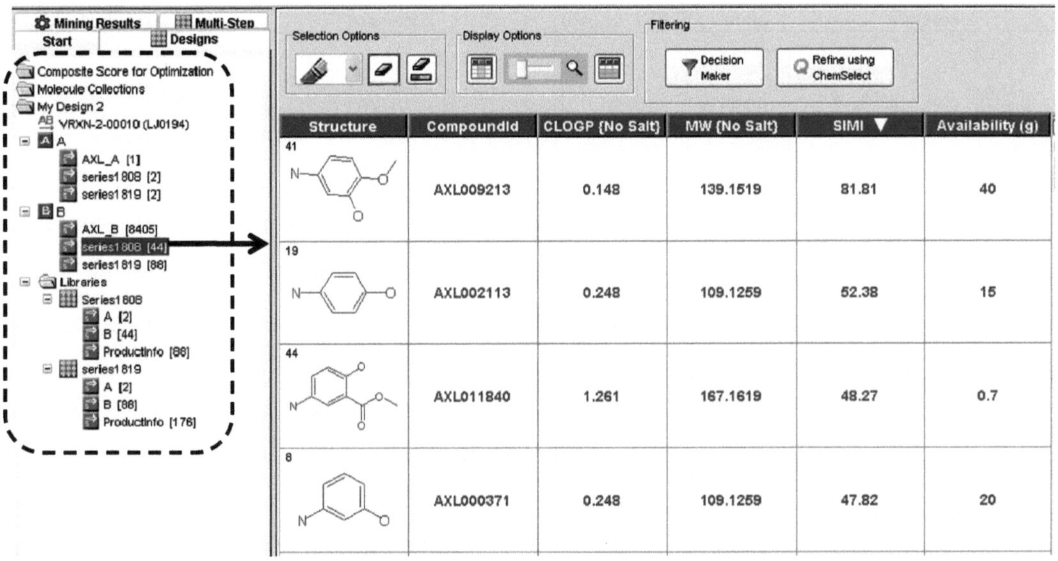

Fig. 16.8. Reactant-level (pre-enumeration) design steps. This is a screen shot of PGVL Hub during the design of the two libraries. The reactant sets for this two-component reaction and the generated explicit libraries and products are all captured during the design session (*see* the left-hand side). The A-component is for acids and the B-component for amines. The molecular structures of the two special acids are shown in **Fig. 16.6**. Many annotations can be added to reactants to aid their analysis and selection. Here ClogP, molecular weight (MW), similarity (SIMI) with respect to a user-defined reactant, and reactant amount available in the inventory house are just a few examples.

structure of **1808-1** bound to the kinase domain of Chk1 (*see* **Fig. 16.1**) and plan another round of targeted library using a 2×88=176 format (*see* **Fig. 16.10)** to further refine **1808-1**.

3.2.4. Product Analysis and Selection

For the second targeted library, we focused our attention on the enumerated products directly. After certain initial selection steps for reactants similar to what were used for the first targeted library, a few thousand product molecules were enumerated. Then we subjected them to various property calculations for additional analysis and filtering. MW, ClogP (9), and LogD (10) values were computed to shape molecular size and solubility. Molecular similarity score (SIMI, *see* (11) and (14)) with respect to **1808-1** was computed to ensure that the chemical neighborhood of **1808-1** was well sampled. A protein binding pocket was created based on the experimental X-ray structure of **1808-1** bound to the Chk1 kinase domain, and all product molecules were exported out of PGVL Hub and docked and scored using the in-house protein–ligand docking software (12, 13). The numeric docking scores (HT_Score, (13)) were then imported back into PGVL Hub and displayed along with the product structures (*see* **Fig. 16.11**). We also leveraged activity models built by project teams working on other kinase targets to provide some early read about kinase selectivity (e.g., Predicted_CDK2 activity, *see* **Fig. 16.11**), even though no such model existed at the time when we designed

Structure	name	CHK1_Ki (uM)	Structure	name	CHK1_Ki (uM)
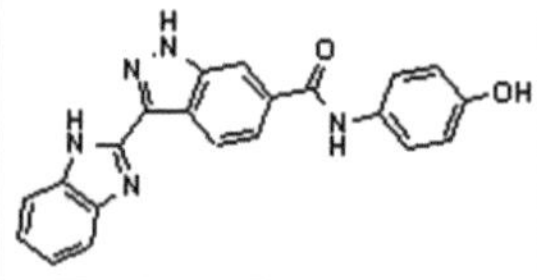	1808-1	0.301	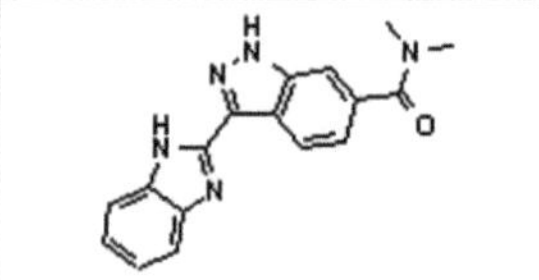	1808-5	5.2
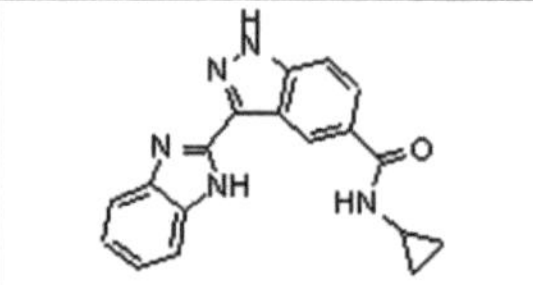	1808-2	0.645	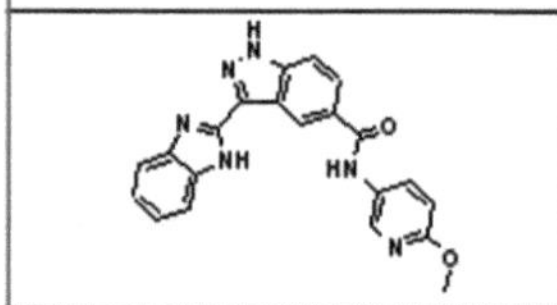	1808-6	14
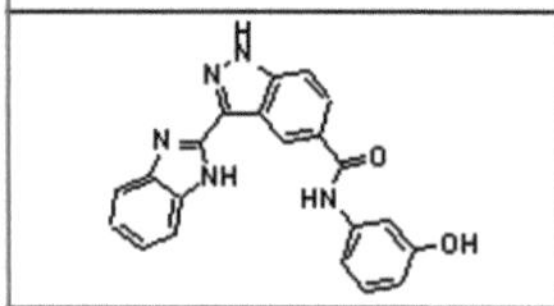	1808-3	1	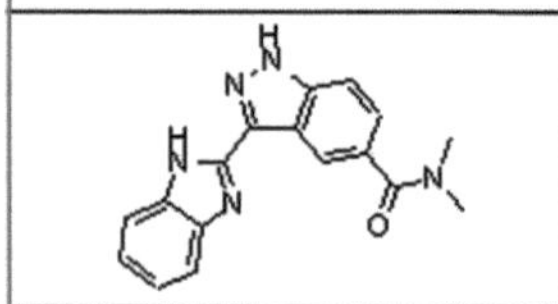	1808-7	15
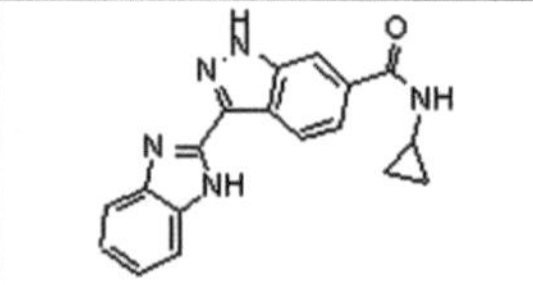	1808-4	1.4	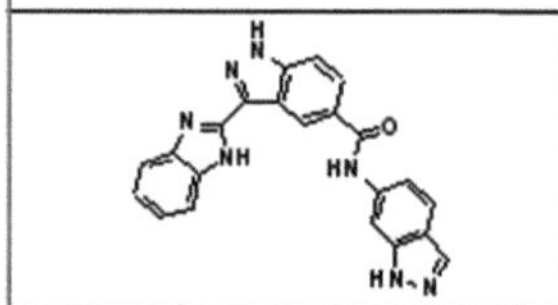	1808-8	16

Fig. 16.9. Top hits from the first targeted library. One can see that a fairly diverse set of small amines are all tolerated by the binding pocket but with a significant reduction in potency when compared with the initial HTS hit **Cpd-1** (5 nM). On the other hand, the top hit **1808-1** shows significant improvement in kinase selectivity and improved solubility (*see* data given in **Fig. 16.1**).

Design hypothesis: Explore the right hand side of **1808-1** for specificity (selectivity) and further improve potency.
VRXN ID: VRXN-2-00010; Synthetic protocol: LJ0194

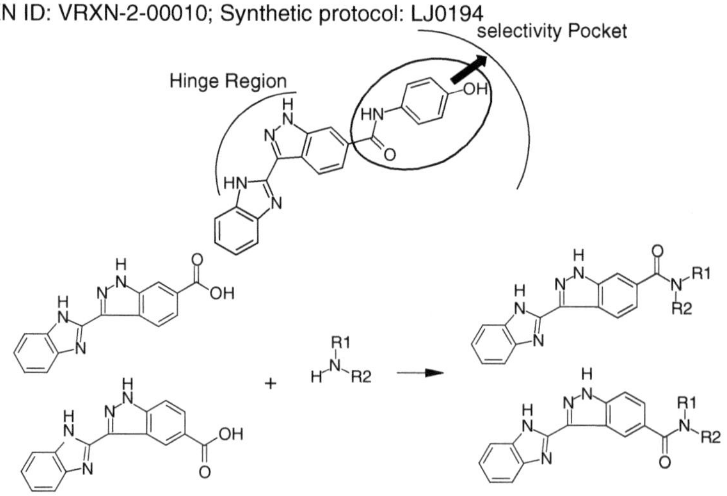

2 acids **88 amines**

Fig. 16.10. Objectives of the second round targeted library. The main goal is to further expand the SAR knowledge around the right-hand side of **1808-1** with the aim to improve potency while retaining kinase selectivity. A same combinatorial synthesis protocol (LJ0194) was used and a 2×88 plate format was planned before the design of the library.

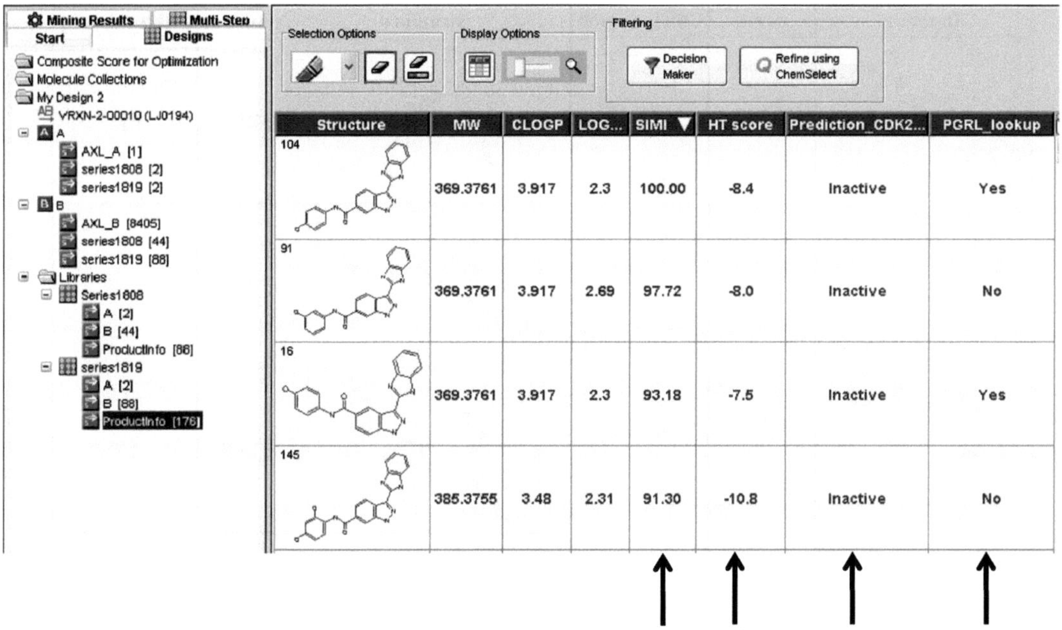

Fig. 16.11. Product-level design steps. In this screen shot, product structures and their calculated properties are listed in the table. Those annotations are key to implementing various design considerations such as ADME&T profile (e.g., Rule of Five (8, 9), LogD (10)), molecular similarity with respect to lead compound **1808-1** (SIMI), protein–ligand complementation (docking score again the binding pocket in the Chk1 kinase domain initially occupied by **1808-1**, such as HT score (13)), kinase selective (predicted CDK2 activity based on an in silico model), and finally checking for duplicates against the corporate compound database (PRGL_Lookup).

our legacy libraries in the past. Finally we wanted to know if any of our designed molecules had already been registered into Pfizer's corporate compound database (PGRL_lookup, *see* **Fig. 16.11**). If duplicates were found, we could either remove them from the design or strategically include a few of them in our design as internal controls for the library production and biological assays. With all those desired molecular properties calculated and combined, we conducted further focusing and filtering to reduce our choices to about a few hundreds, then finalized our 88 amines (plus some backups) through visual inspection of product structures using PGVL Hub's Structural Viewer panel. Again, input from project medicinal chemists and library production chemists was included in the actual legacy library design. The result of the second round library is given in **Fig. 16.12**. One compound **1819-1** has a much higher potency than **1808-1**. Additional data also suggested a very sharp SAR among **1819-1**, **1808-1**, **1819-2**, **1819-5**, and **1819-6**. The X-ray structure of **1819-1** bound to the kinase domain of Chk1 was solved subsequently to provide significant molecular insights for the observed sharp SAR. It showed that the OH group at the *ortho* position was able to replace two tightly bound waters and induce a significant rearrangement in that region of the protein binding pocket (*see* **Fig. 16.1**). This

Structure	name	CHK1_Ki (uM)
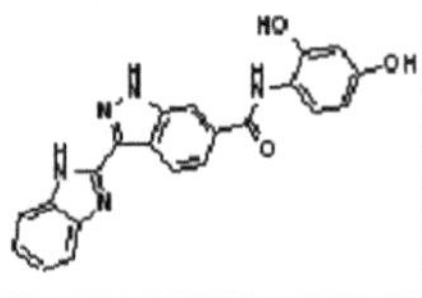	1819-1	0.0005
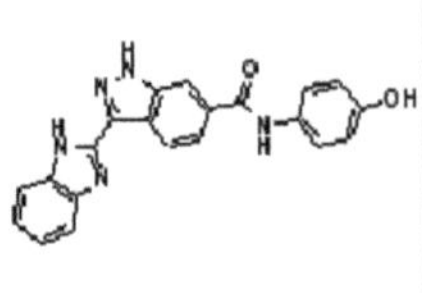	1808-1	0.301
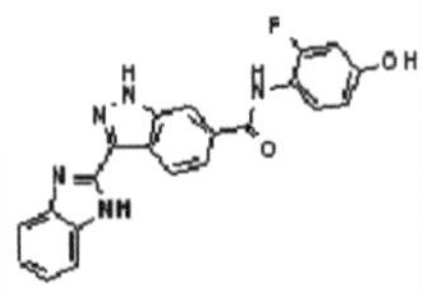	1819-2	0.492
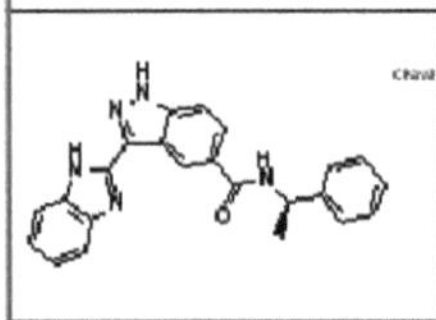	1819-3	0.755

Structure	name	CHK1_Ki (uM)
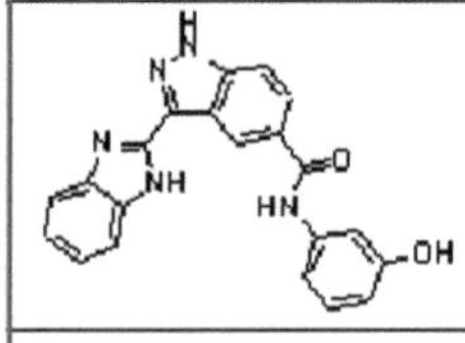	1819-4	1
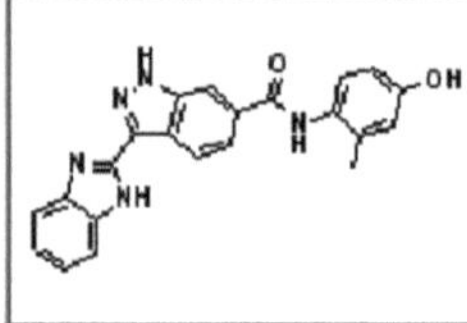	1819-5	1.1
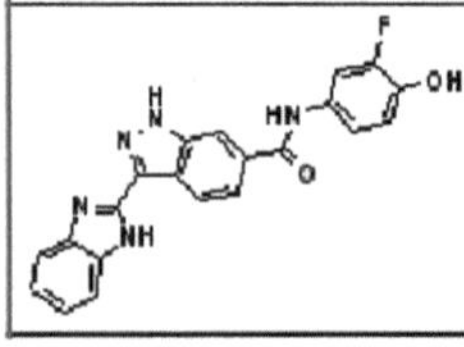	1819-6	1.5

Fig. 16.12. Top hits from the second round library. One compound **1819-1** (0.5 nM) shows significant improvement over the lead **1808-1** (300 nM) in terms of Chk1 inhibition results. More data show (*see* **Fig. 16.1**) that it is even more potent than the initial hit **Cpd-1** (5 nM) while having a much improved kinase selectivity profile and better solubility.

in-depth structural knowledge generated by **1819-1** was further used in the additional lead optimization effort through one-on-one synthesis (2).

3.2.5. Integration of Final Design (List Operation at the Reactant/Product Level)

The library workflow depicted in **Fig. 16.2** seems to imply that library design is a straightforward step-by-step linear process. In reality, it involves multiple iterations and revisions through collaborations with other stakeholders. It is quite common for a designer to simultaneously pursue several design hypotheses and strategies that yield multiple intermediate library designs. PGVL Hub offers a list of logic operations (AND, OR, and MINUS) at the reactant as well as the library/product levels so that the designer can easily compare and/or combine two reactant lists or two libraries. The final design submitted for library production is usually a combination of several individual designs intended to test multiple hypotheses.

3.3. Result Summary and Project Impact

Two rounds of targeted libraries (2×44 and 2×88) were designed, synthesized, and assayed within 6 months. This effort led to a new lead matter **1819-1** with improved potency (~10 folds), kinase selectivity (>100 folds), and better solubility (~1 log unit or ~10 folds) when compared with the original HTS hit **Cpd-1**. The extensive SAR information spanned by those 264 library compounds, supplemented by the two co-crystal structures associated with **1801-1** and **1819-1** showing significant protein flexibility around the ligand binding site, provided a solid foundation for additional lead optimization effort on this project (2).

There are many design requirements, constraints, and hypotheses for a given library design case. In our example, we have touched upon several reactant as well as product-level design considerations. These considerations include but not limited to ADME&T properties, similarity with respect to one or more lead molecules, docking and scoring against a given protein binding site, activity models for selectivity profiling, and even practical issues such as reactant availability in chemical inventory systems and duplication check against corporate compound collections. Historical usage tracking strongly suggests that PGVL Hub is a proven, streamlined, and highly effective design environment to fulfill many of those diverse library design objectives in the hands of Pfizer medicinal and computational chemists.

4. Notes

1. *Shorten the cycle time*: The design–synthesis–test cycle of lead optimization is the dominant workflow in drug discovery projects. The library design–synthesis–test cycle should be short enough to be compatible with the progression of the project so that relevant project questions can be proposed and answered by targeted libraries in a timely manner. Effective communication and coordination among the library designer, his/her project collaborators, and the library production team is essential to reduce the cycle time. PGVL Hub allows one library designer to save a full design session into a file and share it with another collaborator to enable effective communication and coordination. Selecting only reactants that are readily available from chemical inventory systems is another way to bypass the wait required to restock missing reactants, and the inventory check feature of PGVL Hub makes this check straightforward.
2. *Complementary to singleton synthesis*: In terms of lead optimization, library synthesis works as a shot-gun approach, with multiple shots on the goal while its resolution is limited by availability of types of combinatorial chemistries and suitable reactants. On the other hand, the one-on-one singleton synthesis practiced by standard medicinal chemistry offers the highest resolution, yet at a lower throughput. So for a well-explored SAR region, the singleton approach is the best way to further refine project leads effectively, while

the library approach is best for exploring a new SAR region, as in the example showcased in this report. In addition, the singleton approach could be used to synthesize a few special template reactants (like those two special acids shown in **Fig. 16.6**), which are then subsequently amplified extensively by targeted libraries.

3. *Control the size of the enumerated library*: As the name implies, combinatorial libraries can explode in size very quickly. Therefore one must perform reactant-level selections before product enumeration in most design cases. As shown in the example library, molecular weight (MW) is an effective filter to cut down number of reactants, so is reactant availability inside the reactant inventory system. As a matter of principle, more expensive computational approaches (e.g., protein–ligand docking and scoring) should be applied only to smaller subsets of reactants or products.
4. *The importance of visual inspection:* Visual inspection of reactants and product molecules offers a library designer tremendous value in terms of what product molecules can or cannot be synthesized to help formulate and address SAR hypotheses. Project medicinal chemists have fondly called this popular approach "cerebral processing." PGVL Hub has provided a capable environment to enable this approach (e.g., Structural Viewer panel with many molecules per page for fast browsing, sorting of displayed molecules by molecular properties, and multiple color markers to label molecules for further processing).
5. *Leverage externally computed molecular properties:* Many molecular properties pertinent to ADME&T and project-specific activity and/or selectivity models are available within PGVL Hub for use by library designers. If a computational model is not available within PGVL Hub, one could export pertinent molecules out from PGVL Hub, compute the desired molecule properties using the external software package, and then import the computed results back into PGVL Hub for further decision making in an integrated manner. The docking and scoring calculation used in this report (HT_Score, *see* **Fig. 16.11**) exemplifies this type of use case.

Acknowledgments

We would like to express our gratitude toward other members of the Chk1 project team for their design input and their efforts in library production (Haresh Vazir and Dr. Ming Teng), bio-assays

(James Register), X-ray structures of protein–ligand complex (Dr. Ping Chen, Dr. Chun Luo, and Yali Deng), and in-depth medicinal chemistry follow-ups (Dr. Ming Teng and her team). We also appreciate the strong Chk1 project leadership provided by Dr. Karen Lundgren. Finally we would like to thank Drs. Joe Zhongxiang Zhou and Ben Burke for their valuable comments and suggestions and Dr. David Simon for proof reading the draft.

References

1. Peng, Z., Yang, B., Mattaparti, S., Shulok, T., Thacher, T., Kong, J., Kostrowicki, J., Hu, Q., Na, J., Zhou, J. Z., Klatte, K., Chao, B., Ito, S., Clark, J., Coner, C., Waller, C., Kuki, A. (2011) PGVL Hub: an integrated desktop tool for medicinal chemists to streamline design and synthesis of chemical libraries and singleton compounds, in (Zhou, J. Z., ed.) *Chemical Library Design*. Humana Press, New York, Chapter 15.
2. Teng, M., Zhu, J., Johnson, M. D., Chen, P., Kornmann, J., Chen, E., Blasina, A., Register, J., Anderes, K., Rogers, C., Deng, Y., Ninkovic, S., Grant, S., Hu, Q., Lundgren, K., Peng, Z., Kania, R. S. (2007) Structure-Based Design of (5-Arylamino-2H-pyrazol-3-yl)-biphenyl-2',4'-diols as Novel and Potent Human Chk1 Inhibitors. *J Med Chem* **50**(22), 5253–5256.
3. (a) Zhou, B., Elledge, S. J. (2000) The DNA Damage Response: Putting Checkpoints in Perspective. *Nature* **408**, 433–439.
3. (b) Melo, J., Toczyski, D. (2002) A unified view of the DNA-damage checkpoint. *Curr Opin Cell Biol* **14**, 237–245.
4. (a) Chen, P., Luo, C., Deng, Y., Ryan, K., Register, J., Margosiak, S., Tempczyk-Russell, A., Nguyen, B., Myers, P., Lundgren, K., Kan, C. C., O'Connor, P. M. (2000) The 1.7 Å Crystal structure of human cell cycle checkpoint kinase Chk1: implications for Chk1 regulation. *Cell* **100**, 681–692.
4. (b) Foloppe, N., Fisher, L. M., Francis, G., Howes, R., Kierstan, P., Potter, A. (2006) Identification of a buried pocket for potent and selective inhibition of Chk1: Prediction and verification. *Bioorganic & Medicinal Chemistry* **14**, 1792–1804.
5. Li, Q., Zhu, G.D. (2002) Targeting serine/threonine protein kinase B/Akt and cell-cycle checkpoint kinases for treating cancer. *Curr Top Med Chem.***2**, 939–971.
6. (a) Jackson, J. R., Gilmartin, A., Imburgia, C., Winkler, J. D., Marshall, L. A., Roshak, A. (2002) An indolocarbazole inhibitor of human checkpoint kinase (Chk1) abrogates cell cycle arrest caused by DNA damage. *Cancer Res* **60**, 566–572.
6. (b) Prudhomme, M. (2006) Novel checkpoint 1 inhibitors. *Rec Pat Anti-Cancer Drug Discov* **1**, 55–68.
6. (c) Tao, Z. F., Lin, N. H. (2006) Chk1 inhibitors for novel cancer treatment. *Anti-Cancer Agents Med Chem* **6**, 377–388.
6. (d) Foloppe, N., Fisher, L. M., Francis, G., Howes, R., Kierstan, P., Potter, A. (2006) Identification of a buried pocket for potent and selective inhibition of Chk1: prediction and verification. *Bioorg Med Chem* **14**, 1792–1804, and references cited therein.
6. (e) Tao, Z. F., et al. (2007) Structure-based design, synthesis, and biological evaluation of potent and selective macrocyclic checkpoint kinase 1 inhibitors. *J Med Chem* **50** (7), 1514–1527.
6. (f) Tong, Y., et al. (2007) Discovery of 1,4-dihydroindeno[1,2-*c*]pyrazoles as a novel class of potent and selective checkpoint kinase 1 inhibitors. *Bioorg Med Chem* **15,** 2759–2767.
7. Kania, R. S., Bender, S. L., Borchardt, A., Braganza, J. F., Cripps, S. J., Hua, Y., Johnson, M. D., Johnson, T.O.J., Luu, H.T., Palmer, C. L., Reich, S. H., Tempczyk-Russell, A. M., Teng, M., Thomas, C., Varney, M. D. (2001) Wallace, M. B. Patent WO 0102369.
8. Lipinski, C.A., Lombardo, F., Dominy, B. W., Feeney, P. J. (1997) Experimental and computational approaches to estimate solubility and permeability in drug discovery and development settings. *Adv Drug Deliv Rev* **23** (1–3), 3–25.
9. CLOGP v 4.0 for Unix, BioByte Corp., 201 W. Fourth Street, Suite 204, Claremont, CA 91711; see Leo, A. J. (1993) Calculating LogPoct from structures. *Chem Rev* **93,** 1281–1306 and references cited within for details.
10. More info on the LogD prediction tool from of the ACD Lab could be found: http://www.acdlabs.com/products/phys_chem_lab/logd/

11. (a) Tanimoto, T. T. (1957) IBM Internal Report 17th Nov. 1957; and (b) Wikipedia entry on the Tanimoto coefficient: http://en.wikipedia.org/wiki/Tanimoto_coefficient#Tanimoto_Coefficient_.28Extended_Jaccard_Coefficient.29
12. (a) Gehlhaar, D. K., Verkhivker, G. M., Rejto, P. A., Sherman, C. J., Fogel, D. B., Fogel, L. J., Freer, S.T. (1995) Molecular recognition of the inhibitor AG-1343 by HIV-1 protease: conformationally flexible docking by evolutionary programming. *Chem Biol* **2**(5), 317–324.
12. (b) Gehlhaar, D., Bouzida, D., Rejto, P. A. (1999) Reduced dimensionality in ligand-protein structure prediction: covalent inhibitors of serine proteases and design of site-directed combinatorial libraries, in (Parrill, A. L., Reddy, M. R. eds.) *ACS Symposium Series 719: Rational Drug Design*. ACS Press, New York, pp. 292–311.
13. Marrone, T. J., Luty, B. A., Rose, P. W. (2000) Discovering high-affinity ligands from the computationally predicted structures and affinities of small molecules bound to a target: a virtual screening approach. *Perspect Drug Discov Design* **20**, 209–230.
14. Pipeline Pilot from SciTegic and its molecular fingerprint system: http://www.scitegic.com/

Chapter 17

GLARE: A Tool for Product-Oriented Design of Combinatorial Libraries

Jean-François Truchon

Abstract

Combinatorial chemistry with two or more diversity points often leads to an immense number of theoretical products. It is sensible to select the reagents based on the desired properties of the products in the hope of maximizing the usefulness of the synthesized molecules. The presented tool enables the filtering of reagents such that any further reagent selection will form products matching the desired properties. Virtual combinatorial library leading to thousands of billions of products can be rapidly assessed. The publicly available software (http://glare.sourceforge.net) and key algorithmic elements are discussed.

Key words: Combinatorial library design, computer algorithms, product properties, multi-objective optimization.

1. Introduction

The design of chemical libraries often requires the selection of only a small number of reagents compared to what is available commercially or from proprietary repositories. The combinatorial nature of multi-reagent synthesis can lead to far more products than can be synthesized or screened. It is thus of great interest to apply filtering schemes that are adapted to the goal of the chemical library (1, 2). In order to help the chemist focus on only the reagents that would form a library with desired computed properties, it is attractive to use a computational algorithm to eliminate the unproductive reagents. In practice, it is often difficult to map the desired product properties into reagent-based filtering rules. For example, if the products need to fit within a log

J.Z. Zhou (ed.), *Chemical Library Design*, Methods in Molecular Biology 685,
DOI 10.1007/978-1-60761-931-4_17, © Springer Science+Business Media, LLC 2011

P range, filtering rules applied to the reagents are rather difficult to find. Although one can guess a reagent-based threshold, it would be difficult to account for the fact that some reagent classes are fundamentally greasier than others. This sort of strategy has been found to be misleading (3). The daunting task of generating chemical structures for millions or billions of virtual products and assessing their properties in order to find the best reagents to form the combinatorial matrix is clearly challenging. This is worsened by the multi-objective thresholds needed when more than one property is monitored simultaneously.

A practical solution to this problem, called GLARE (Global Library Assessment of REagents) (4), has been developed and validated in our laboratories and is explained in this chapter. We will focus on a specific chemical combinatorial library to illustrate the workflow and the use of the software.

2. Materials

2.1. Computer Program GLARE

The computer program used in this chapter has been made publicly available under an Open Source Initiative BSD License at http://glare.sourceforge.net. This program is written in C++ and has been successfully compiled on diverse platforms such as Mac OS 10.X, Linux, IBM AIX, and Windows XP. Under its current form, GLARE is mainly a command line application that is invoked identically on any of the platforms. Details of the parameters and options can be found at the aforementioned web site.

2.2. Chemical Databases

There exist a multitude of chemical reagent sources. The Available Chemical Directory^{TM} (ACD) collection, from Symyx Technologies Inc., lists as many as 1,160,000 unique chemicals with chemical structure, pricing, supplier, purity, forms, etc.

2.3. The Oxazolidine Library

The different steps of the protocol to filter chemical reagents based on the product properties will be illustrated with an oxazolidine library (5) for which reagents adding diversity (dimension) are shown in **Fig. 17.1**. Even though only few hundreds of reagents are picked from the ACD in each dimension, the total number of theoretically accessible products is over 59 million.

The objective of GLARE is to filter the reagents such that any further reagent selection would lead to a *good* (*see* **Note 1**) product. **Figure 17.2** shows the property distributions of the products before (black) and after (grey) the use of GLARE that considers a product good if its properties fall within the *goodness* range (dashed vertical lines) for each property.

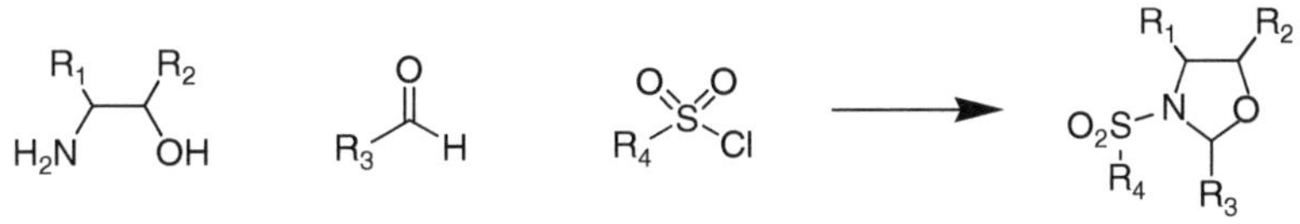

Fig. 17.1. The oxazolidine library used to illustrate how the GLARE tool works. The number of reagents considered and the total number of products they can potentially form are given in parenthesis under each chemical class.

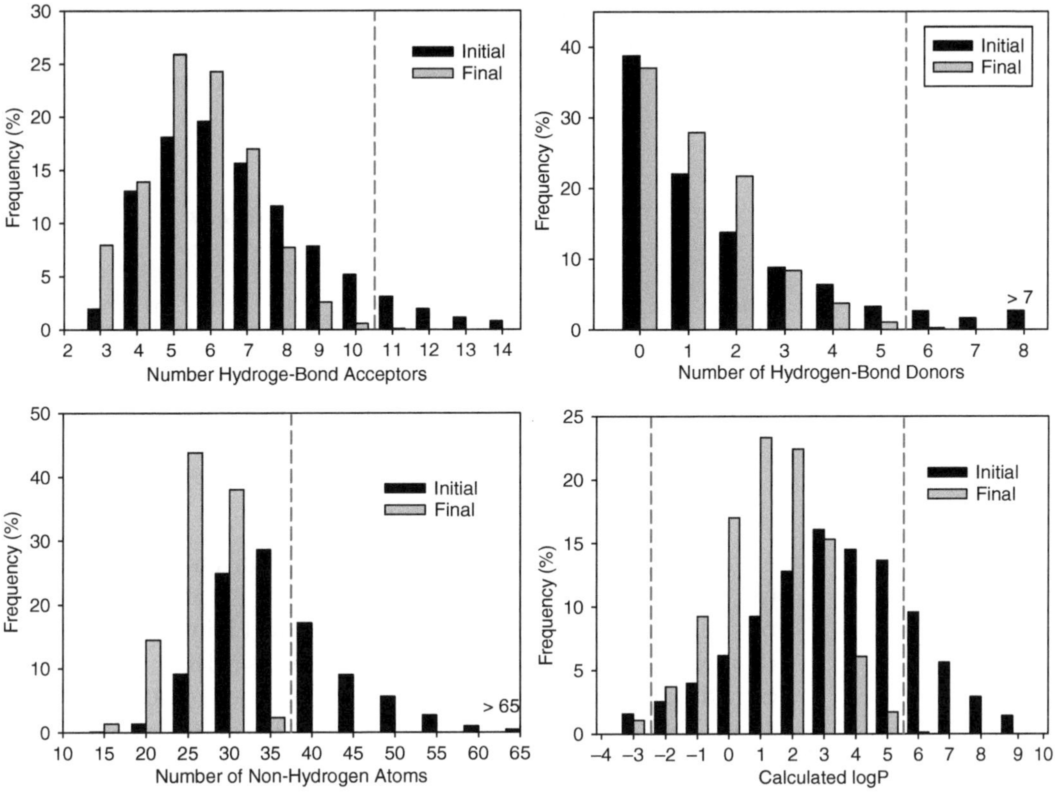

Fig. 17.2. Properties profile of the products in the oxazolidine library formed with all available reagents (*black*) and after filtering the reagents (*grey*) based on the product properties with GLARE. The multi-objective thresholds are illustrated by the *dashed vertical lines*. The initial library is formed by 651 × 637 × 143 products and the filtered library by 144 × 143 × 92 products (aminoalcohols × aldehydes × sulfonyl chlorides).

3. Methods

The different steps, files, and parameters necessary to optimize the oxazolidine library are discussed in this section.

3.1. Selection of the Reagents

It is standard practice to remove chemical functionalities susceptible to interfere with subsequent synthetic steps in the library. We used a Merck & Co proprietary web tool to this end called Virtual Library Toolkit (VLTK) described elsewhere (6).

3.2. Product Properties and Offset Calculations

One of the strategies behind GLARE is to take advantage of the additivity of many of the computed properties in a chemical library. In other words, one can calculate the property of a product by summing the properties of its diversity contributing reagents corrected by an offset kept constant for the entire library. Although this may seem relatively obvious for a property like the number of non-hydrogen atoms, a real-value property such as the calculated logarithm of the octanol/water partition (log *P*) or the polar surface area (PSA) are also well approximated by this scheme (4, 7). We write the property *P* of any product of the chemical library as

$$P\left(product\right) = P_{offset} + \sum_{i}^{N} P\left(reagent_i\right) \qquad [1]$$

where P_{offset} is the constant offset correction for the entire library of property *P*, $P(\text{reagent}_i)$ the property of the *i*th diversity reagent. In practice, the offset is calculated from a single example:

$$P_{offset} = P\left(product\right) - \sum_{i}^{N} P\left(reagent_i\right) \qquad [2]$$

This has been shown to work well for a diverse set of libraries (*see* **Note 2**) (4). In **Fig. 17.3**, the offset is calculated for the oxazolidine library for properties related to Lipinski's rule of five (8): the number of hydrogen bond acceptors (HBA), the number of hydrogen bond donors (HBD), the number of non-hydrogen atoms (NHA), and the calculated log *P* (9).

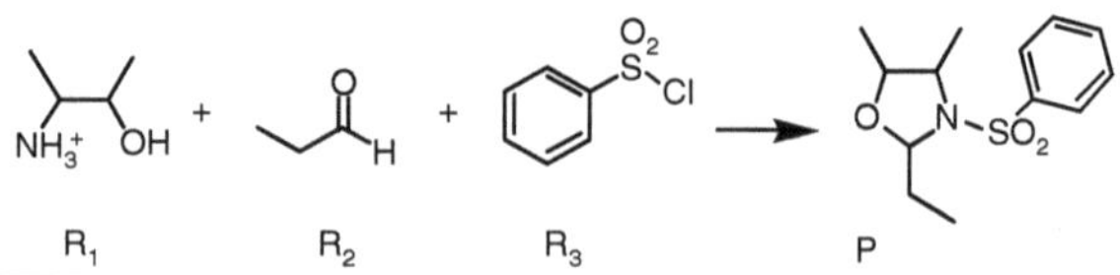

	HBA	HBD	NHA	logP
P	3	0	18	1.3
R1	1	4	6	–0.3
R2	1	0	4	0.3
R3	2	0	10	1.5
Offset	–1	–4	–2	–0.2

Fig. 17.3. Calculation of the oxazolidine offsets from a specific example. From the product (*P*) property, each of the reagent (R_i) property is subtracted. Four properties are considered: the number of hydrogen bond acceptors (HBA), the number of hydrogen bond donors (HBD), the number of non-hydrogen atoms (NHA), and the logarithm of the octanol/water partition constant (log *P*).

The use of an additive scheme has the obvious advantage of avoiding the explicit generation of each product structure, which would be impractical whenever a large number of products are possible. Indeed, the only requirement is the calculation of the properties of each unmodified reagent, avoiding even the complication of forming the synthons. There are many commercial and non-commercial software suitable for this task providing a 2D structure of a reagent. Just to name a few: the OEChem Tk (10), the Molecular Operating Environment (MOE) (11), and JOELib (12). In this specific work, we have used a Merck & Co proprietary cheminformatics platform to calculate the properties of each reagent list, each of which sprang from a substructure query to the ACD.

3.3. Preparation of the Input Files

With GLARE, there are only two file types that need to be prepared. First, the reagent property files contain one reagent per line in a text file starting with a reagent ID followed by a list of numbers corresponding to the reagent properties. The offset information is given in a separate file with the same format and contains only one line.

Second, the virtual library is combined according to the instructions outlined in the library definition file. An example for the oxazolidine library is given in **Fig. 17.4**. The keyword DIMDEF associates a list of reagent property files to one combinatorial dimension identified by a user-defined alias (e.g., *ALDEHYDES*). The listed reagent files are simply appended in the program. The LIBDEF keyword is followed by a user-defined library

Defines the combinatorial dimensions and list the reagent property files that are combined.
DIMDEF *AMINOALCOHOLS* amino_alcohols_acd.gli amino_alcohols_inhouse.gli
DIMDEF *ALDEHYDES* aldehydes_acd.gli
DIMDEF *SULFONYLS* sulfonyl_chlorides_acd.gli
DIMDEF *OFFSET* oxazolidine_offset.gli

Defines a combinatorial library called oxazolidines formed by the matrix of products from the combination of the listed dimensions
LIBDEF *OXAZOLIDINES AMINOALCOHOLS ALDEHYDES SULFONYLS OFFSET*

Defines a property name with the expected minimum and maximum value of the products.
PROPDEF *HBD* 0 5
PROPDEF *HBA* 0 10
PROPDEF *NHA* 0 35
PROPDEF *LOGP* -2.4 5.0

Gives the order of the properties found in the property input files.
INPUTDEF *HBA HBD NHA LOGP*

Fig. 17.4. Example of the library definition file for GLARE. *Bold text* indicates a dedicated keyword, text in *italic* a user-defined alias, lines starting with a *hash mark* are comments, and *normal text* gives the keyword-associated parameters.

name (e.g., *OXAZOLIDINES*) and the list of the dimensions that form the combinatorial matrix (e.g., *AMINOALCOHOLS* × *ALDEHYDES* × *SULFONYLS* × *OFFSET*). When two or more libraries share common intermediates the filtering of the common reagents can be achieved by specifying more than one LIBDEF keyword. This tells GLARE to simultaneously consider all libraries in the filtering. The PROPDEF keyword associates the minimum and maximum values for a "well-behaved" product to a user-defined property name. Finally, the INPUTDEF keyword lists the order of the properties read in the reagent property input files.

3.4. Recommended Optimization Parameters

GLARE uses an iterative filtering that stops when a user-defined fraction of the products formed by the remaining reagents comply with the desired product property ranges. We call this fraction the *goodness*. It is not sufficient to identify a set of reagents leading to *good* products, but one wants to find the largest set of such reagents to provide enough choice to a chemist who also needs to account for other properties. We discuss here the impact of the different optimization parameters on the resulting number of reagents. We measure the efficiency of the filtering with an *effectiveness* metric, which corresponds to the average fraction of reagents left after optimization compared to what was available initially. Quantitatively, the effectiveness E is defined by

$$E = \frac{1}{D}\left[\sum_{i=1}^{D} \frac{N_{i,\,\mathrm{final}}}{N_{i,\,\mathrm{initial}}}\right] \qquad [3]$$

where D corresponds to the number of dimension (three for the oxazolidine library), $N_{i,\,\mathrm{final}}$ to the number of reagents in the dimension i after GLARE has been applied, and $N_{i,\,\mathrm{initial}}$ the number of reagents input before the filtering. When a high compliance to the desired product properties is requested, more reagents are pruned. **Figure 17.5** shows the effectiveness of the oxazolidine library as a function of the *goodness* threshold used. The obvious drawback of a lower goodness threshold is a potential deterioration of the properties of the final library when the reagents are selected for synthesis. We have found, more generally, that whenever a high compliance to the product property rules is needed, a goodness threshold of 95% is the most appropriate as the last 5% unduly reduces the effectiveness.

The scaled pruning strategy is an optional feature that is useful when one of the reagent sets is significantly smaller than the others like the sulfonylchloride reagents of the oxazolidine library. It is often difficult to retain enough diversity in these less populated reagent sets while maintaining high library goodness. The principles of the scaled pruning is to eliminate reagents in the dimensions with more reagents faster. The iterative procedure initially eliminates only reagents from the larger list and progressively starts to prune reagents from the smaller list as the lists become of comparable size. The switching function that turns on the pruning of smaller dimensions depends on a single parameter (α) (4). The final number of reagents in the three dimensions of the oxazolidine library after applying GLARE with different values of α is shown in **Fig. 17.6**. A small α has no effect and as its value increases, proportionally more sulfonylchlorides are kept and less of the other two more populated dimensions. As we found more generally, a value of α between 1 and 10 (a value of 6 is our default) leads to a more evenly distributed diversity across the dimensions.

The third user-defined parameter discussed is related to the partitioning scheme implemented in GLARE. To avoid the combinatorial explosion that makes a product-based filtering algorithm impractical, the reagent sets can be optionally partitioned such that each reagent's ability to form a good product is evaluated in a sub-library formed by combining the individual partitions in a systematic way. This is in contrast with the examination of all combinatorial products. The partitioning approximation systematically leads to libraries matching the desired goodness when verified with all the products (4). However, for a given targeted goodness, the partitioning scheme reduces the effectiveness of the resulting library. **Figure 17.7** shows the effectiveness of the oxazolidine library as a function of the minimum number of reagents in the created partitions. A partition size of 16 (corresponding to a value of 4 on the *x*-axis of **Fig. 17.7**) is generally optimal.

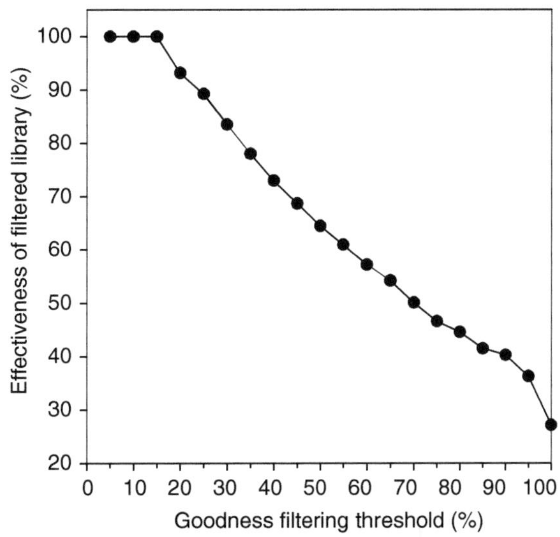

Fig. 17.5. This figure shows how requiring a higher fraction of the products to comply (goodness) with the desired product properties reduces the fraction of retained reagent (effectiveness). The initial goodness of the oxazolidine library used here is 18%.

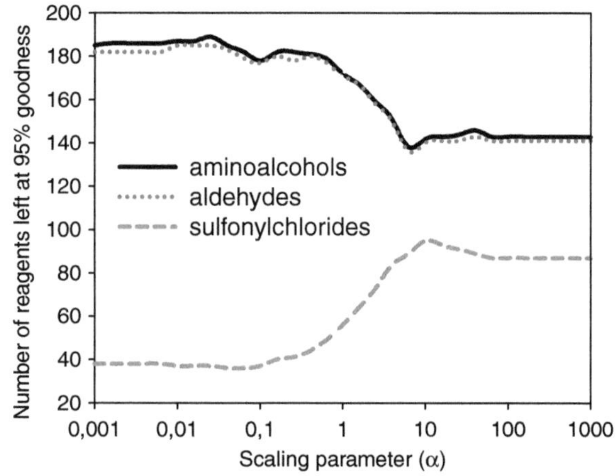

Fig. 17.6. This figure shows the final number of reagents left in each dimension once a 95% goodness threshold is obtained as a function of the scaling parameter (α) displayed on a log scale. The larger is α, the more reagents from the initially less populated dimension are left. We found that $\alpha = 6$ generally leads to useful results.

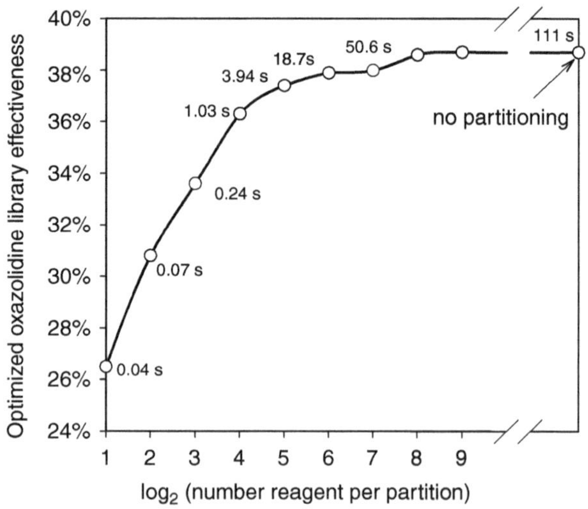

Fig. 17.7. This figure illustrates the advantages and disadvantages of using partitioning. On the one hand, the timings (shown next to the individual points in seconds) are tremendously reduced, on the other hand the effectiveness of the optimized oxazolidine library is sub-optimal with smaller partitions. A partition of 16 reagents seems best overall.

In summary, the two main parameters related to the compliance to the product property rules (goodness) and the number of reagents left for further selection (effectiveness) can be controlled by adjusting the algorithmic goodness threshold, the scaling parameter, and the size of the reagent partitions. Each library being different, it may sometimes be useful to deviate from the proposed defaults. The oxazolidine library is a good surrogate for the relationships normally involved and **Figs. 17.5, 17.6**, and **17.7** can be used to assess sensitivity and expected effects of modifying these parameters.

4. Notes

1. Here the word "good" and "goodness" are strictly related to the binary classification that a product is good only if it fits all the multi-objective criteria. GLARE could easily be adapted to work with a scalar fitness score.
2. Most spectacular exceptions to the property additivity scheme come from nitrogen atoms that can change their basicity, their polar surface area, their number of donors, etc. If this becomes an issue for a library, the reagents can be initially split according to each case and a different offset used.
3. When the partitioning scheme is used, only a small subset of the products is examined and the goodness is then defined as the fraction of the examined products with the desired product properties.

Acknowledgments

The author thanks Dr. Christopher Bayly from Merck Frosst Canada for his initial important contribution to GLARE and for a careful proofreading of this chapter.

References

1. Gillet, V. J. (2008) New directions in library design and analysis. *Curr Opin Chem Biol* **12**, 372–378.
2. Song, C. M., Bernardo, P. H., Chai, C. U., Tong, J. C. (2009) CLEVER: Pipeline for designing in silico chemical libraries. *J Mol Graph Model* **27**, 578–583.
3. Truchon, J. -F., Bayly, C. I. (2006) Is there a single 'Best Pool' of commercial reagents to use in combinatorial library design to conform to a desired product-property profile? *Aust J Chem* **59**, 879–882.
4. Truchon, J. -F., Bayly, C. I. (2006) GLARE: a new approach for filtering large reagent lists in combinatorial library design using product properties. *J Chem Inf Model* **46**, 1536–1548. http://glare.sourceforge.net
5. Conde-Frieboes, K., Schjeltved, R. K., Breinholt, J. (2002) Diastereoselective synthesis of 2-aminoalkyl-3-sulfonyl-1,3-oxazolidines on solid support. *J Org Chem* **67**, 8952–8957.
6. Feuston, B. P., Chakravorty, S. J., Conway, J. F., Culberson, J. C., Forbes, J., Kraker, B., Lennon, P. A., Lindsley, C., McGaughey, G. B., Mosley, R., Sheridan, R. P., Valenciano, M., Kearsley, S. K. (2005) Web enabling technology for the design, enumeration, optimization and tracking of compound libraries. *Curr Top Med Chem* **5**, 773–783.
7. Shi, S. G., Peng, Z. W., Kostrowicki, J., Paderes, G., Kuki, A. (2000) Efficient combinatorial filtering for desired molecular properties of reaction products. *J Mol Graph Model* **18**, 478–496.
8. Lipinski, C. A., Lombardo, F., Dominy, B. W., Feeney, P. J. (1997) Experimental and computational approaches to estimate solubility and permeability in drug discovery and

development settings. *Adv Drug Deliv Rev* **23**, 3–25.
9. Klopman, G., Li, J. Y., Wang, S. M., Dimayuga, M. (1994) Computer automated log P calculations based on an extended group-contribution approach. *J Chem Inf Comput Sci* **34**, 752–781.
10. OpenEye Scientific Software Inc *OEChem Toolkit,* Santa Fe, NM, USA, 2009. www.eyesopen.com
11. *The Molecular Operating Environment (MOE),* Chemical Computing Group Inc., Montreal, QC, Canada, 2008. www.chemcomp.com
12. *JOELib a Java based cheminformatics library,* version 2; University of Tuebingen, Tuebingen, Germany, 2009. http://sourceforge.net/projects/joelib

Chapter 18

CLEVER: A General Design Tool for Combinatorial Libraries

Tze Hau Lam, Paul H. Bernardo, Christina L.L. Chai, and Joo Chuan Tong

Abstract

CLEVER is a computational tool designed to support the creation, manipulation, enumeration, and visualization of combinatorial libraries. The system also provides a summary of the diversity, coverage, and distribution of selected compound collections. When deployed in conjunction with large-scale virtual screening campaigns, CLEVER can offer insights into what chemical compounds to synthesize, and, more importantly, what not to synthesize. In this chapter, we describe how CLEVER is used and offer advice in interpreting the results.

Key words: Virtual combinatorial library, Markush technique, compound analysis, chemoinformatics, chemistry.

1. Introduction

Combinatorial chemistry has become increasingly essential in the modern drug discovery pipeline (1, 2). Through the discovery of new chemical reactions and commercially available reagents, the size of these libraries has amplified exponentially over the past few years (3). Often, such libraries are far too large to be synthesized and screen in their entirety. Moreover, the output frequently produces high level of redundancy in terms of the similarity in the physiochemical properties of the derived compounds. Therefore, a rational approach for combinatorial library design is desirable in order to maximize the outcome of an expensive synthesis and screening campaign (4).

Here we introduce CLEVER (Chemical Library Editing, Visualizing, and Enumerating Resource), a platform-independent

J.Z. Zhou (ed.), *Chemical Library Design*, Methods in Molecular Biology 685,
DOI 10.1007/978-1-60761-931-4_18, © Springer Science+Business Media, LLC 2011

tool that allows not only the enumeration of chemical libraries using customized fragments but also the computation of the physicochemical properties of the generated compounds along with filtering functionalities for evaluating their drug likeness. CLEVER may also be used for visualizing the generated chemical compounds in 3D space, as well as charting various graphs based on the innate properties of the chemical libraries. The system is available at http://datam.i2r.a-star.edu.sg/clever/.

2. Materials

1. Java version 1.6 and above.
2. SmiLib v2.0 (5) for rapid combinatorial library generation in Simplified Molecular Input Line Entry Specification (SMILES) (6).
3. The Chemistry Development Kit (CDK) Application Programming Interface (API) (7), OpenBabel (8), or CORINA (9) for generating 3D coordinates (SDF format) from SMILES strings.
4. Jmol (10) for interactive display of molecular structures in 3D space.
5. JFreeChart (http://www.jfree.org/jfreechart/) for generating histograms and 2D scatter plots for chemical compound analysis.

3. Methods

CLEVER is implemented using the Java 3D API (*see* **Section 4.1**). The main framework is made up of five key modules for chemical library editing, enumeration, conversion, visualization, and analysis. The operations of these functionalities are accomplished by the various applications at the resource layer. For the purpose of illustration, the compound calothrixin B, a secondary metabolite isolated from the Calothrix cyanobacteria (11–13), is used as the scaffold molecule with the variable functional groups [R*n*] attached (**Fig. 18.1**). The calothrixins are redox-active natural products which display potent antimalarial and anticancer properties and thus there is interest in probing the physical as well as biological profiles of their derivatives (14). In this exercise, six functional groups have been selected as the building blocks (**Table 18.1**).

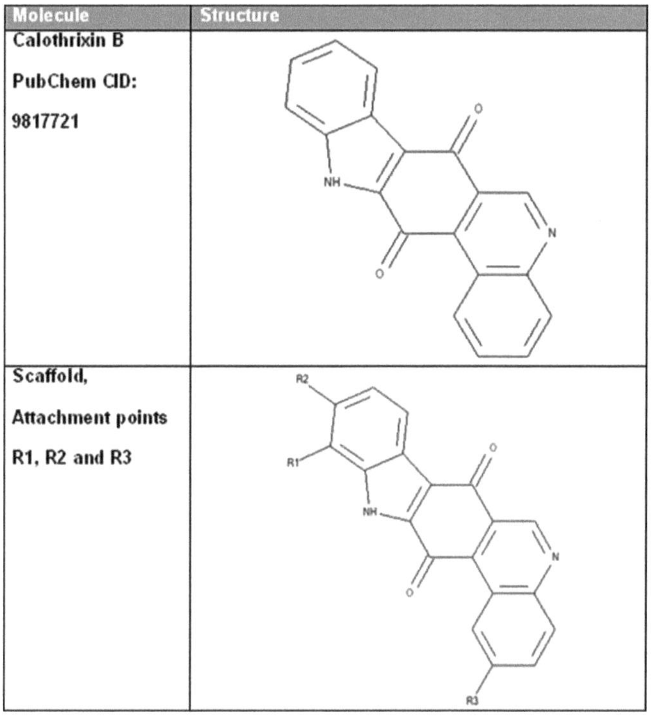

Fig. 18.1. Compound CID: 9817721 and its corresponding scaffold structure for enumerating novel library.

Table 18.1
SMILES string configuration for scaffold and building blocks

Scaffold	SMILES
S1	O=C(C(C(C=C([R3])C=C1)=C1N=C2)=C2C3=O)C4=C3C5=CC=C([R2])C([R1])=C5N4
Attachment blocks	SMILES
B1	C[A]
B2	C(C)(C)([A])
B3	F[A]
B4	CC[A]
B5	C=C[A]
B6	C1=CC=CC=C1[A]

3.1. Data Preparation

1. Use the library editor to create a library file for the compounds under study (**Fig. 18.2**). Library files are essentially plain text files that contain a record on each line, with an entry identifier and a SMILES string for the

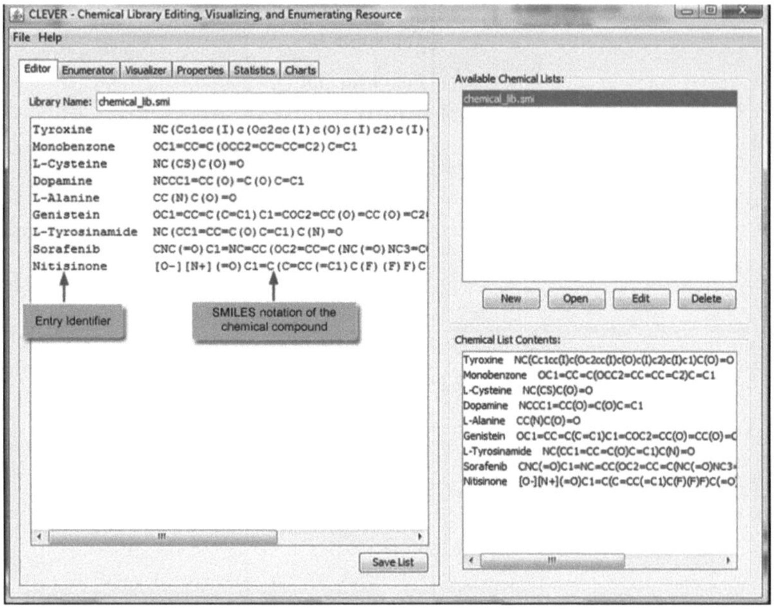

Fig. 18.2. Illustration of the library editor.

scaffold or building blocks (delimited by a tab character) (*see* **Section 4.2**).

2. Define the chemical scaffolds, attachment blocks, linkers, and reaction schemes for the compounds under study. Attachment points on the blocks are represented by '[A]', while functional groups to be permutated on the scaffolds are depicted by '[R*n*]', where *n* is a numerical value unique to each functional group to be varied (**Fig. 18.1**). Linker is the intersection between the scaffolds and the attachment blocks (*see* **Sections 4.3–4.6**).
3. Click on the "Convert SMILES" button to perform the conversion of the linear SMILES strings into 3D coordinates (SDF format). To browse automatically, click the "Start Visualizer" button for the systematic viewing of the 3D molecular structures from the chemical library (**Fig. 18.3**).

3.2. Chemical Library Enumeration

3.2.1. Full Library Enumeration

1. Click on the 'Enumerator' tab to proceed to the library enumeration workspace.
2. Enter the library name.
3. Open both the scaffold and the building blocks text files (**Fig. 18.4a**).
4. Select the appropriate scaffold and building blocks from the scaffold and block lists (**Fig. 18.4b**).
5. Ensure the full combination and the empty linker options are selected.

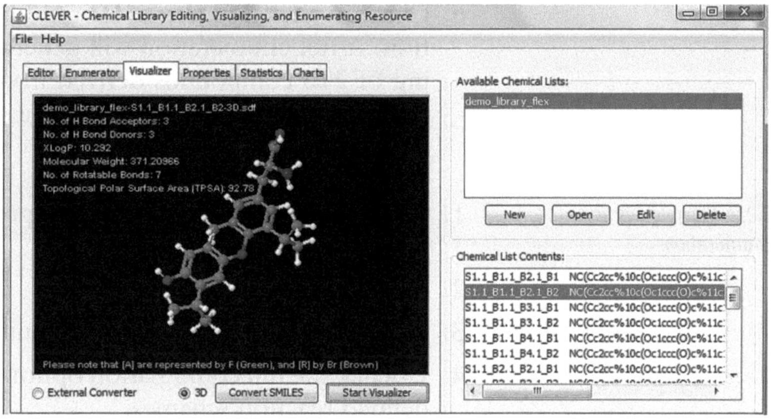

Fig. 18.3. CLEVER SMILES conversion and 3D structure visualization.

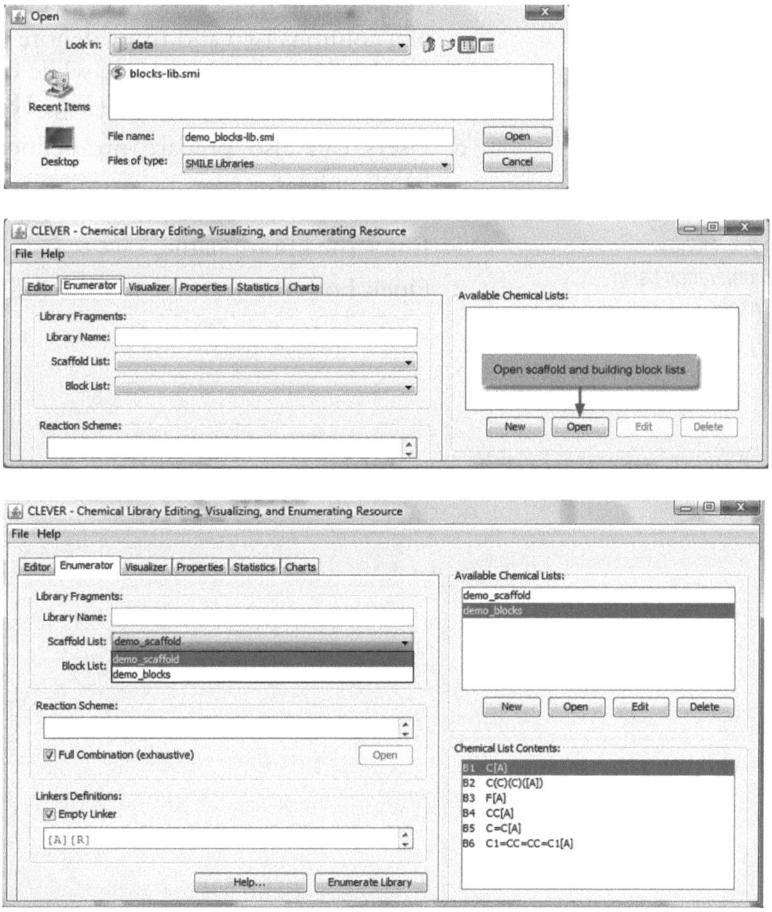

Fig. 18.4. Chemical library enumeration. (**a**) Initiation for the scaffold and block lists. (**b**) Illustration on the usage of the enumerator.

6. Click on the 'Enumerate Library' button to start enumeration. A full enumeration will generate a new library consisting of 216 compounds derived from the systematic permutation of the variable sites with the six attachment blocks on the core scaffold.

3.2.2. Flexible Library Enumeration

1. Click on the 'Enumerator' tab to proceed to the library enumeration workspace.
2. Enter the library name.
3. Open both the scaffold and the building blocks text files.
4. Unclick the full combination option to enable access to user-defined reaction schemes.
5. Within the 'Reaction Scheme' text box, define the scaffold for each reaction scheme in the first column, followed by pairs of linkers and blocks to be used for each attachment site R*n*, where *n* is a numerical value unique to each functional group to be varied (*see* **Sections 4.3–4.6**). For example, columns two and three denote the linker and the blocks for the first attachment site, while columns four and five for the second attachment site (**Fig. 18.5**).
6. Users can also prepare pre-defined reaction schemes for batch upload.

3.2.3. Library Enumeration Using Linkers

1. Enter the library name.
2. Open both the scaffold and the building blocks text files.

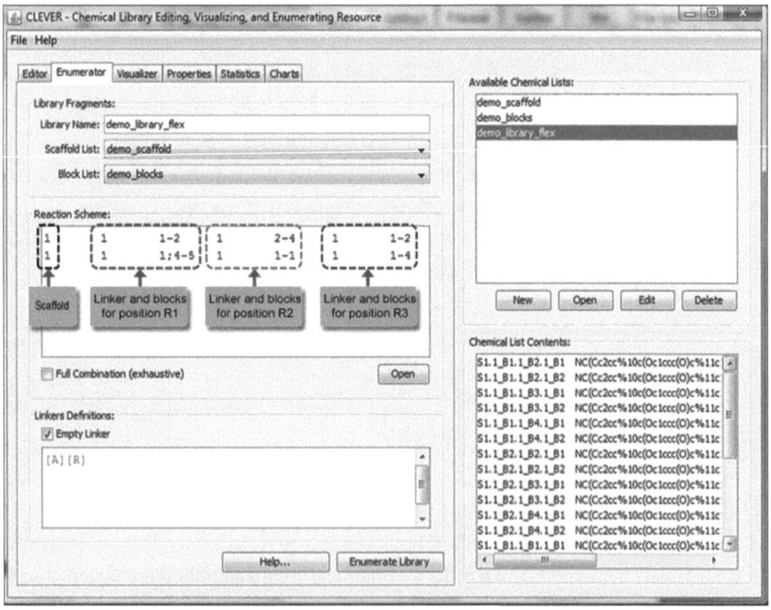

Fig. 18.5. Reaction schemes definition.

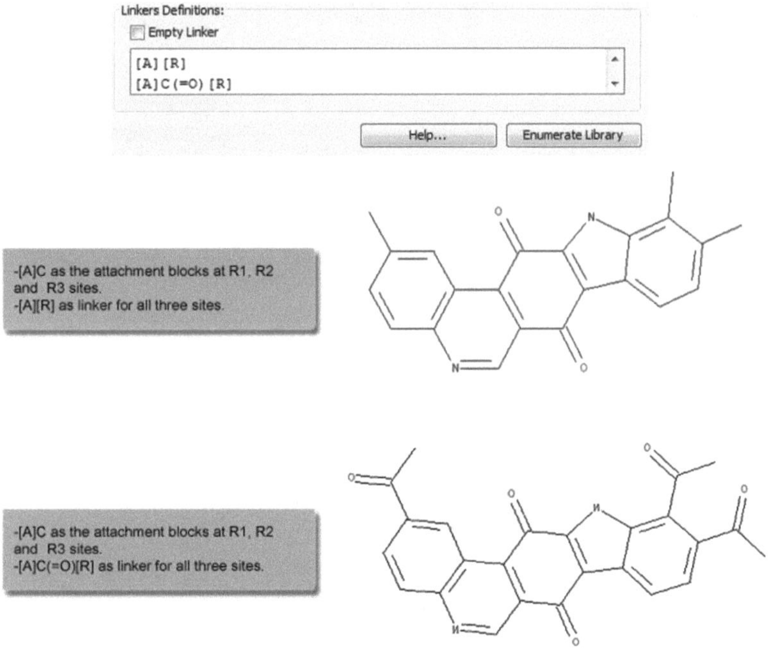

Fig. 18.6. Enumeration using different linkers.

3. Unclick empty linker option to allow addition and modification of the linkers (**Fig. 18.6**). In this exercise, we only demonstrate enumeration using two linkers. More linkers could be included for chemical library construction.

3.3. Chemical Library Analysis

3.3.1. Computation of Physiochemical Properties

1. Click on the 'Properties' tab to proceed to the workspace.
2. Load and select the library for analysis.
3. Click on the 'Compute' button to calculate physiochemical properties including the number of hydrogen bond acceptors and donors, XlogP (partition coefficient) values, molecular weights, number of rotatable bond, and the Topological Polar Surface Area (TPSA) of compounds.
4. User may also save the results for future reference.

3.3.2. Filtering of Chemical Library-Based Predefined Schemes

1. To initiate the filtering function, click on the 'Filter' button, a 'Filter Library' window will appear.
2. User can select one of the six predefined filtering schemes for drug likeness, lead likeness, or fragment likeness from the 'Filter Scheme' dropdown list (**Fig. 18.7**). Users may also define their own criteria for filtering.

3.3.3. Evaluation of Chemical Libraries

To analyse the distribution of chemical compounds of a certain physiochemical property,

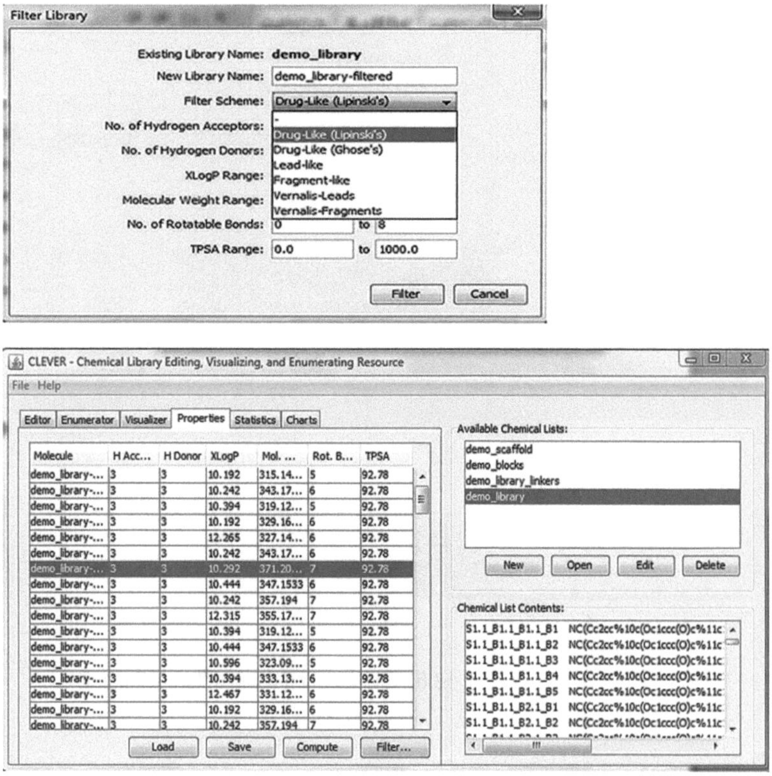

Fig. 18.7. Physiochemical properties computation and the filtration of chemical libraries based on predefined scheme.

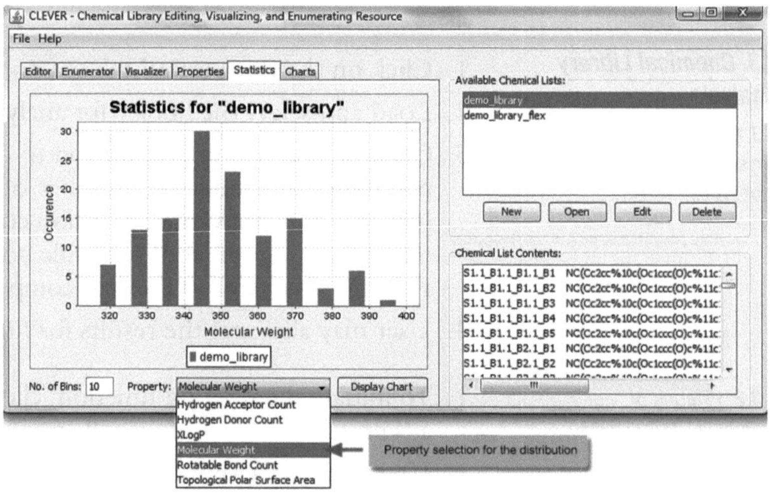

Fig. 18.8. Distribution of compounds of a selection collection(s).

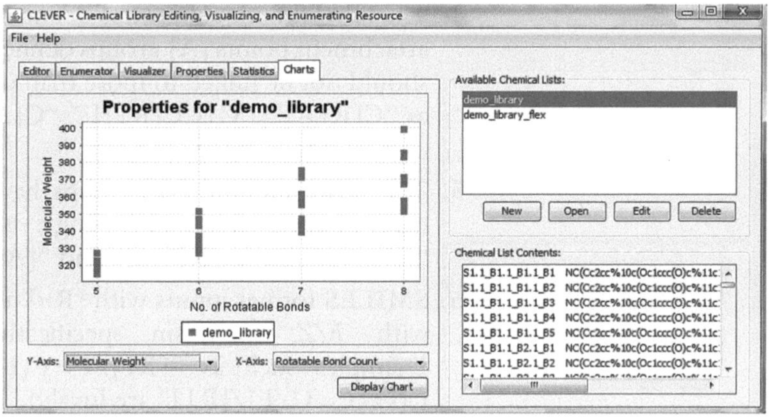

Fig. 18.9. Scatter plot for one or more libraries.

1. Select chemical collection(s) from the 'Available Chemical List' display space.
2. Select the Property combo list to choose a property for the distribution graph.
3. Click on the 'Display Chart' button to display histograms on the distribution of chemical compounds (**Fig. 18.8**).

To analyse the diversity and coverage of the selected chemical library

1. Select chemical collection(s) from the 'Available Chemical List' display space.
2. Select the physicochemical properties for the *X* and *Y* axes.
3. Click on the 'Display Chart' button to show the 2D scatter plot (**Fig. 18.9**).

4. Notes

1. Install a Java Virtual Machine (a runtime version of Java, or JRE 1.6 and above). JVM is compatible to all the major operating systems including Windows, MacOS, and Linux.
2. Ensure the input scaffold and the building block plain text lists are saved in the .smi extension format. Any other extension formats are unrecognizable by the CLEVER enumerator and will generate an error.
3. CLEVER only allows up to a maximum of 90 [R*n*] functional groups to be defined. However, there is no restriction on the number of scaffolds, linkers, and building blocks.

4. The [R*n*] functional groups defined on the scaffolds and the attachment points [A] groups defined on the building blocks should not be linked to more than one atom. Examples such as "C[R1]C", "C1CC[R1]1", "C[A]C", and "C1CC[A]1" are invalid.
5. The [R*n*] and the [A] groups have to be attached to its neighbouring atom by a single bond. Instances such as "[R1]#C", "C(=[R1])C", and "[A]=C" are invalid.
6. SMILES format inputs with [R*n*] groups attached to atoms with *E/Z* isomerism specification are not allowed. Examples such as "[R1]/C=C(F)/I" and "Br/C(Cl)=C(O/C=C/F)/[R1]" are invalid.

References

1. Martin, E. J., Critchlow, R. E. (1999) Beyond mere diversity: tailoring combinatorial libraries for drug discovery. *J Comb Chem* **1**, 32–45.
2. Valler, M. J., Green, D. (2000) Diversity screening versus focussed screening in drug discovery. *Drug Discov Today* **5**, 286–293.
3. Jamois, E. A. (2003) Reagent-based and product-based computational approaches in library design. *Curr Opin Chem Biol* **7**, 326–330.
4. Leach, A. R., Hann, M. M. (2000) The in silico world of virtual libraries. *Drug Discov Today* **5**, 326–336.
5. Schüller, A., Hähnke, V., Schneider, G. (2007) SmiLib v2.0: a Java-based tool for rapid combinatorial library enumeration. *QSAR Comb Sci* **26**, 407–410.
6. Weininger, D. (1988) SMILES, a chemical language and information system. 1. Introduction to methodology and encoding rules. *J Chem Inf Comput Sci* **28**, 31–36.
7. Steinbeck, C., Hoppe, C., Kuhn, S., Floris, M., Guha, R., Willighagen, E. L. (2006) Recent developments of the chemistry development kit (CDK)—an open-source java library for chemo- and bioinformatics. *Curr Pharm Des* **12**, 2111–2120.
8. Guha, R., Howard, M. T., Hutchison, G. R., Murray-Rust, P., Rzepa, H., Steinbeck, C., Wegner, J., Willighagen, E. L. (2006) The Blue Obelisk—interoperability in chemical informatics. *J Chem Inf Model* **46**, 991–998.
9. Sadowski, J. (1997) A hybrid approach for addressing ring flexibility in 3D database searching. *J Comput Aided Mol Des* **11**, 53–60.
10. Angel, H. (2006) Biomolecules in the computer: Jmol to the rescue. *Biochem Educ* **34**, 255–261.
11. Rickards, R. W., Rothschild, J. M., Willis, A. C., de Chazal, N. M., Kirk, J., Kirk, K., Saliba, K. J., Smith, G. D. (1999) Calothrixins A and B, novel pentacyclic metabolites from Calothrix Cyanobacteria with potent activity against malaria parasites and human cancer cells. *Tetrahedron Lett* **55**, 13513–13520.
12. Bernardo, P. H., Chai, C. L. L., Heath, G. A., Mahon, P. J., Smith, G. D., Waring, P., Wilkes, B. A. (2004) Synthesis, electrochemistry, and bioactivity of the Cyanobacterial Calothrixins and related quinones. *J Med Chem* **47**, 4958–4963.
13. Bernardo, P. H., Chai, C. L. L., Le Guen, M.,Geoffrey D., Smith, G. D., Waring, P. (2006) Structure–activity delineation of quinones related to the biologically active Calothrixin B. *Bioorg Med Chem Lett* **17**, 82–85.
14. Khan, Q. A., Lu, J., Hecht, S. M. (2009) Calothrixins, A new class of human DNA topoisomerase I poisons. *J Nat Prod* **72**, 438–442.

Subject Index

J.Z. Zhou (ed.), *Chemical Library Design*, Methods in Molecular Biology 685,
DOI 10.1007/978-1-60761-931-4, © Springer Science+Business Media, LLC 2011

D

E

F

MIX
Papier aus verantwortungsvollen Quellen
Paper from responsible sources
FSC® C105338

If you have any concerns about our products,
you can contact us on
ProductSafety@springernature.com

In case Publisher is established outside the EU,
the EU authorized representative is:
Springer Nature Customer Service Center GmbH
Europaplatz 3, 69115 Heidelberg, Germany

Printed by Libri Plureos GmbH
in Hamburg, Germany